Frontiers in Clinical Drug Research - Dementia

(Volume 2)

Edited by

Dr. José Juan Antonio Ibarra Arias

Centro de Investigación en Ciencias de la Salud (CICSA)
Facultad de Ciencias de la Salud Universidad Anáhuac
México Norte
Mexico

Frontiers in Clinical Drug Research – Dementia

Volume # 2

Editors: Prof. José Juan Antonio Ibarra Arias

ISSN (Online): 2717-5995

ISSN (Print): 2717-5987

ISBN (Online): 978-981-5039-47-4

ISBN (Print): 978-981-5039-48-1

ISBN (Paperback): 978-981-5039-49-8

need for a court order if at any point you breach any terms of this License Agreement. In no event will any delay or failure by Bentham Science Publishers in enforcing your compliance with this License Agreement constitute a waiver of any of its rights.

3. You acknowledge that you have read this License Agreement, and agree to be bound by its terms and conditions. To the extent that any other terms and conditions presented on any website of Bentham Science Publishers conflict with, or are inconsistent with, the terms and conditions set out in this License Agreement, you acknowledge that the terms and conditions set out in this License Agreement shall prevail.

Bentham Science Publishers Pte. Ltd.
80 Robinson Road #02-00
Singapore 068898
Singapore
Email: subscriptions@benthamscience.net

BENTHAM SCIENCE

CONTENTS

PREFACE

Among neurodegenerative diseases, those that lead to a state of dementia are the aim of several investigations. Dementia is a chronic disease whose prevalence is increasing worldwide. The number of dementia patients in the world is approximately 50 million, and it is estimated that the number of patients will reach 131.5 million by 2050. This increase will be accompanied by a significant increase in medical expenditures and other expenses, especially for elderly patients. Therefore, the maintenance cost of dementia in the future is expected to be quite high. For this reason, several investigations aim, firstly, to describe the key mechanisms involved in the origin of dementia and, secondly, to establish preventive and/or therapeutic strategies in order to understand and mitigate this catastrophic pathology. This book aims to discuss the current comorbidities that cause cognitive impairment and the current management alternatives in cases of dementia for a better understanding of the current perspective on the subject.

The book contains five chapters that begin with a clear description of the comorbidities that induce mild cognitive impairment, continues with the description of some mechanisms that contribute to the development of dementia, and then moves on to the discussion of some encouraging therapies.

The editor would like to express his gratitude to the authors of the chapters presented in this book for their invaluable contributions.

Dr. José Juan Antonio Ibarra Arias
Centro de Investigación en Ciencias de la Salud (CICSA)
Facultad de Ciencias de la Salud Universidad Anáhuac México
University of Cambridge
Norte
Mexico

List of Contributors

Alejandra Romo	Centro de Investigación en Ciencias de la Salud (CICSA), Facultad de Ciencias de la Salud Universidad Anáhuac México, Norte, Mexico
Almudena Chávez-Guerra	Centro de Investigación en Ciencias de la Salud (CICSA), Facultad de Ciencias de la Salud Universidad Anáhuac México, Norte, Mexico
Antonio Ibarra	Centro de Investigación en Ciencias de la Salud (CICSA), Facultad de Ciencias de la Salud Universidad Anáhuac México, Norte, Mexico
Anamaria Jurcau	Faculty of Medicine and Pharmacy, Clinical Municipal Hospital, University of Oradea, Romania, Oradea
Adriane Bello Klein	Institute of Basic Health Sciences, Department: Physiology R. Sarmento Leite, Federal University of Rio Grande do Sul (UFRGS), CEP 90035-190- Porto Alegre, RS , Brazil
Alexandre Castro	Institute of Basic Health Sciences, Department: Physiology R. Sarmento Leite, Federal University of Rio Grande do Sul (UFRGS), CEP 90035-190- Porto Alegre, RS , Brazil
Alex Sander da Rosa Araujo	Institute of Basic Health Sciences, Department: Physiology R. Sarmento Leite, Federal University of Rio Grande do Sul (UFRGS), CEP 90035-190- Porto Alegre, RS , Brazil
Aziz Eftekhari	Russian Institute for Advanced Study, Moscow State Pedagogical University, 1/1, Malaya Pirogovskaya St, Moscow, 119991, Russian Federation Pharmacology and Toxicology Department, Maragheh University of Medical Sciences, Maragheh, Iran
Alex France Messias Monteiro	Postgraduate Program in Natural and Bioactive Synthetic Products, Health Sciences Center, Federal University of Paraíba, Castelo Branco Street, João Pessoa, PB – Brazil
Cumali Keskin	Medical Laboratory Techniques, Vocational Higher School of Healthcare Studies, Mardin Artuklu University, Mardin, Turkey
Daiane da Rocha Janner	Carlos Chagas Filho Institute of Biophysics – Neurogenesis Lab Av. Carlos Chagas Filho, Federal University of Rio de Janeiro (UFRJ) , 373–CEP 21941-902, Rio de Janeiro, RJ- Brazil
Elham Ahmadian	Kidney Research Center, Tabriz University of Medical Sciences, Tabriz, Iran
Edeildo Ferreira da Silva-Júnior	Chemistry and Biotechnology Institute, Federal University of Alagoas, 57072-970, Maceió, AL – Brazil
Érika Paiva de Moura	Postgraduate Program in Natural and Bioactive Synthetic Products, Health Sciences Center, Federal University of Paraíba, Castelo Branco Street, João Pessoa, PB – Brazil
Francisco Jaime Bezerra Mendonça Júnior	Laboratory of Synthesis and Drug Delivery, State University of Paraíba, Horácio Trajano de Oliveira Street, João Pessoa, PB – Brazil

Herbert Igor Rodrigues de Medeiros	Postgraduate Program in Natural and Bioactive Synthetic Products, Health Sciences Center, Federal University of Paraíba, Castelo Branco Street, João Pessoa, PB – Brazil
Igor José dos Santos Nascimentos	Chemistry and Biotechnology Institute, Federal University of Alagoas, 57072-970, Maceió, AL – Brazil
Irada Huseynova	Institute of Molecular Biology & Biotechnologies, Azerbaijan National Academy of Sciences, 11 Izzat Nabiyev, Baku AZ 1073, Azerbaijan
Luciana Scotti	Postgraduate Program in Natural and Bioactive Synthetic Products, Health Sciences Center, Federal University of Paraíba, Castelo Branco Street, João Pessoa, PB – Brazil Teaching and Research Management, University Hospital of the Federal University of Paraíba, João Pessoa, PB – Brazil
Macarena Fuentes	Centro de Investigación en Ciencias de la Salud (CICSA), Facultad de Ciencias de la Salud Universidad Anáhuac México, Norte, Mexico
Maria Helena Vianna Metello Jacob	Montfort Hospital – Institut du Savoir 745, Montreal Road, Ottawa, ON K1K 0T1, Canada
Marcus Tullius Scotti	Postgraduate Program in Natural and Bioactive Synthetic Products, Health Sciences Center, Federal University of Paraíba, Castelo Branco Street, João Pessoa, PB – Brazil
Natan Dias Fernandes	Postgraduate Program in Natural and Bioactive Synthetic Products, Health Sciences Center, Federal University of Paraíba, Castelo Branco Street, João Pessoa, PB – Brazil
Roxana Rodríguez-Barrera	Centro de Investigación en Ciencias de la Salud (CICSA), Facultad de Ciencias de la Salud Universidad Anáhuac México, Norte, Mexico
Renad I. Zhdanov	Russian Institute for Advanced Study, Moscow State Pedagogical University, 1/1, Malaya Pirogovskaya St, Moscow, 119991, Russian Federation
Rovshan Khalilov	Russian Institute for Advanced Study, Moscow State Pedagogical University, 1/1, Malaya Pirogovskaya St, Moscow, 119991, Russian Federation Department of Biophysics and Biochemistry, Baku State University, Baku, Azerbaijan Institute of Radiation Problems, Azerbaijan National Academy of Science, Baku, Azerbaijan
Soheila Montazersaheb	Molecular Medicine Research Center, Tabriz University of Medical Sciences, Tabriz, Iran
Solmaz Maleki Dizaj	Dental and Periodontal Research Center, Tabriz University of Medical Sciences, Tabriz, Iran
Soodabeh Davaran	Department of Medicinal Chemistry,School of Pharmacy, Tabriz University of Medical Sciences, Tabriz, Iran
Vharoon Sharma Nunkoo	Faculty of Medicine and Pharmacy, Clinical Municipal Hospital, University of Oradea, Romania, Oradea
Yalda Jahanbani	Department of Medicinal Chemistry,School of Pharmacy, Tabriz University of Medical Sciences, Tabriz, Iran

Yolanda Cruz Centro de Investigación en Ciencias de la Salud (CICSA), Facultad de Ciencias de la Salud Universidad Anáhuac México, Norte, Mexico

CHAPTER 1

Comorbidities Inducing Mild Cognitive Impairment, an Evaluation of the Risk Caused by some Pathological Conditions

Yolanda Cruz[1], Alejandra Romo[1], Roxana Rodríguez-Barrera[1], Almudena Chávez-Guerra[1], Macarena Fuentes[1] and Antonio Ibarra[*, 1]

[1] *Centro de Investigación en Ciencias de la Salud (CICSA), Facultad de Ciencias de la Salud Universidad Anáhuac México Norte. Avenida Universidad Anáhuac 46, Lomas Anáhuac, Huixquilucan, Estado de Mexico, C.P. 52786*

Abstract: Mild cognitive impairment has usually been associated with aging, however, in recent decades with the increase in the prevalence of pathologies such as obesity, diabetes mellitus, cardiovascular diseases, and even spinal cord injury, it has become evident that a significant percentage of people who suffer from one or more of these diseases are at greater risk of suffering from some level of cognitive impairment that can lead to the development of various types of dementia. In this chapter, we review the main characteristics and mechanisms that promote the development of this type of alteration in each of the mentioned pathologies and briefly describe the various ways in which they have been approached.

Keywords: Amnesic Memory, aging, Cognitive Domains, Diabetes Mellitus, Dysbiosis, Hypertension, Long-Term Potentiation, Low-Grade Inflammation, Mild Cognitive Impairment, Neurogenesis, Non-Amnestic Memory, Obesity, Stroke, Spinal Cord Injury.

INTRODUCTION

As life expectancy grows, so does the prevalence of neurodegenerative diseases. Mild cognitive impairment (MCI), with dementia as its most evident prognosis, has a profound impact on public health as well as on patient's life quality. Nowadays, 48 million people worldwide have been diagnosed with dementia, a number that is expected to rise to 131 million by 2050, yet clear diagnosis guidelines and standards of care for patients who suffer this debilitating disease are left wanting [1].

* **Corresponding author: Antonio Ibarra** :Centro de Investigación en Ciencias de la Salud (CICSA), Facultad de Ciencias de la Salud Universidad Anáhuac México Norte;Avenida Universidad Anáhuac 46, Lomas Anáhuac, Huixquilucan, Estado de Mexico ,C.P. 52786, México; Tel: (+55)-87-99610-8311; E-mail: jose.ibarra@anahuac.mx

Cognitive functions are neural processes that help us carry out a task; there are 6 main cognitive domains: learning and memory, social functioning, language, visuospatial function, complex attention and executive functioning [2].

Cognitive impairment refers to a deficit in at least one domain. The term MCI was first used to describe stage 3 of the global deterioration scale (GDS), in which the subject presents subtle deficits in cognition without meeting the criteria for dementia. In the Key Symposium at Sweden (2004), the definition expanded, it now includes the affectation not only in memory but in other cognitive domains, and MCI was sub-classified as: amnestic (aMCI), non-amnestic (naMCI), single and multi-domain. Amnestic subtype refers to the impairment in the ability to recall information; memory being affected. Non-amnestic refers to the impairment in at least one non-memory cognitive domain, whereas memory remains unaffected [2 - 4].

Amnestic MCI is associated with greater risk of developing dementia such as Alzheimer's Disease (AD), whereas naMCI may progress to other syndromes such as frontotemporal dementia, primary progressive aphasia, dementia with Lewis bodies, among others. Multi-domain, as the name says so, refers to the impairment of multiple cognitive domains; therefore, patients manifest subtle problems in daily life activities. It might represent a more advanced stage of the neurodegenerative process [3, 5].

In 2011, the National Institute on Aging-Alzheimer's Association included biochemical and neuroimaging biomarkers in the diagnostic criteria for MCI, as some of these biomarkers are seen in subjects with MCI, and may predict later conversion to AD. These risk factors include: apolipoprotein E (APOE) ε4 allele, lower β amyloid 1-42 (Aβ42), higher phosphorylated tau (P-tau), higher total tau (t-tau), amyloid PET, among others [4].

The prevalence of MCI is mainly reported in people older than 65 years old, and it is estimated to be between 3 to 22%, although currently it is underdiagnosed, as it is not usually recognized by primary care physicians; annually, 5 to 31% will progress to dementia [2, 6].

Cognitive impairment is diagnosed using the criteria established in the Diagnostic and Statistical Manual of Mental Disorders 5th Edition (DSM-V) [2]; it is diagnosed when there is a deterioration of one or more cognitive domains at a higher level than expected at given age and education level, confirmed in an objective manner by a professional, without impairing social nor work abilities [6]. Although, there are no specific tests to diagnose MCI as the differences between normal aging and MCI can be difficult to determine. Furthermore, cognitive impairment is different among patients, with some displaying a single non-memory domain and others involving multiple cognitive domains. Once diagnosed, some people develop further neurodegenerative disorders such as dementia and AD, while others remain stable or even revert to pre-existing cognition levels [7].

The rising numbers of MCI have generated a surge of research from both clinical and investigation perspectives, but while a rising number of older adults suffer from different stages of pre-dementia, most remain undiagnosed [8]. Most doctors diagnose subjects with MCI based on evidence and symptoms provided by the patients themselves while trying to use reliable tools and techniques as to discriminate against those who present normal and pathological signs of aging. Criteria for MCI diagnosis was developed by a workgroup sponsored by the National Institute on Aging and the Alzheimer's Association, who agreed on the following common guidelines [9]:

1. A change in cognition recognized by the affected individual or observers
2. Existence of objective impairment in one or more cognitive domains: memory, planning, following instructions or decision-making processes being hindered
3. Independence in functional activities
4. Absence of dementia

Cognitive impairment is mostly associated to aging, but there are other diseases that can lead to its development, such as obesity, diabetes, cardiovascular diseases such as systemic arterial hypertension (SAH) and ischemia, spinal cord injury (SCI), among others. Each of these diseases has different mechanisms that lead to cognitive impairment, but they also share some of them. This chapter will discuss the mechanisms involved in the development of cognitive impairment in different diseases.

OBESITY AND COGNITIVE IMPAIRMENT

Definition and Epidemiology of Obesity

Obesity has become a social and psychological problem that affects around 650 million adults and 340 million children and adolescents worldwide [10]. It is characterized by being a chronic disease of multifactorial origin, which is defined as the excessive accumulation of adipose tissue in the body linked to a high risk of presenting other diseases. The World Health Organization (WHO) uses body mass index (BMI) as a metric to indicate body fatness, classifying obesity as a BMI $\geq$ 30 kg/m^2 [11].

The worldwide prevalence of obesity increased in around 80% from 1980 to 2015, turning it into a pandemic [11, 12].

In 2016, the WHO indicated that there were more than 1.9 billion adults aged 18 and over who were overweight, of which more than 650 million were obese. In the same year, 39% of adults aged 18 and over (39% of men and 40% of women) were overweight. Therefore, about 13% of the world's adult population (11% of men and 15% of women) were obese [13].

In Mexico, the National Health and Nutrition Survey (ENSANUT, as per its Spanish acronym) 2018th edition, estimated that at national level, the percentage of adults 20 years of age or older who are overweight or obese was 75.2% (39.1% overweight and 36.1% obese), a percentage that has increased by 3.9% since 2012 [14].

According to the Organization for Economic Cooperation and Development (OECD), in 2017, the mean prevalence of obesity in adults was 19.5%. United States and Mexico have the highest prevalence of obesity, with >30% [12].

Etiology

Obesity is mainly linked to energy imbalance, where the energy intake exceeds the energy expenditure, due to adoption of energy and fat-rich diets and physical inactivity. The excess energy is stored in the adipose tissue as triglycerides [11].

Although obesity is linked to a positive energy balance, it is a multifactorial disease that is also associated to genetics, physiological, psychological and social factors, thus classified as an endocrine, nutritional and metabolic disease [12].

Another etiological factor that has been associated with the development of obesity is the gut microbiota composition. Gut microbiota is composed of bacteria, fungi, virus and Archea. Among these, bacteria of the phyla Firmicutes and Bacteroidetes correspond to 90% of the gut bacteria; Firmicutes/Bacteroidetes ratio has been associated to obesity in different experimental studies, but is not completely confirmed in humans. A systemic review carried out by Crovesy L, *et al.,* (2020) showed that individuals with obesity showed higher Firmicutes counts and lower Bacteroidetes counts in the majority of studies [15], this ratio tends to increase with BMI higher than 33 [16]. Proteobacteria have been found to be higher in obesity, whereas some butyric acid producing bacteria such as *Faecalibacterium prausnitzii* are lower, leading to dysbiosis. Gut dysbiosis can cause greater calorie absorption, reduction in anorexigenic hormones such as glucagon like peptide-1 (GLP-1), increase in fat storage and damaged gut barrier, which contributes to lipopolysaccharide translocation and inflammation [15].

Obesity affects nearly all physiological functions, and increases the risk for developing other diseases such as diabetes, cardiovascular disease and cognitive impairment, among others, affecting socioeconomic productivity [11]. Obesity is also associated with decreased life expectancy of around 5 to 20 years lost [12].

Physiopathology of Cognitive Impairment in Obesity

Obesity has now been linked to cognitive decline, as brain imaging has showed neural atrophy in obese individuals [17]. Inflammation is proposed to be the link between obesity and cognitive impairment [18].

The distribution of adipose tissue in different anatomical deposits also has substantial implications for morbidity. In particular, intra-abdominal and subcutaneous abdominal fat are more important than subcutaneous fat present in the lower extremities. The release of fatty acids into the portal circulation has adverse metabolic actions, especially in the liver. It is likely that the adipokines and cytokines secreted from adipocyte deposits are involved in the systemic complications of obesity [19].

The adipose tissue is an endocrine organ that releases hormones, chemokines and cytokines (refered to as adipokines) such as interleukin (IL) -6, tumor necrosis factor alpha (TNF-α), monocyte chemoattractant protein (MCP-1), adiponectin and leptin; these molecules help regulate energy homeostasis, innate immunity and inflammation. Under overnutrition and hyper-anabolic state, adipocytes expand in size (hyperplasia) and number (hypertrophy); adipocyte hypertrophy impairs adipose tissue function and lowers its capacity to store lipids, therefore they reach a threshold, which leads to a stress response in these cells, resulting in hypoxia and death of adipocytes, and initiating an inflammatory response [20 - 22].

Hypertrophic adipocytes release MCP-1, promoting macrophage accumulation [23]. The volume of adipose tissue is correlated with increased levels of TNF-α, IL-6 and IL-1β, as well as C-reactive protein (CRP); it is also associated with a local infiltration of inflammatory cells [22].

Adipocytes also contain immune cells; under normal energy balance, both cells coordinate to regulate the storage and mobilization of energy according to the organism's needs. When overnutrition happens, macrophages greatly increase in number and change to M1 phenotype, secreting pro-inflammatory cytokines such as TNF-α and IL-1β [20]. M1 macrophages induce polarized Th1 responses [22].

As adipocytes increase in size and overpass their capacity to store fatty acids, these get released into the bloodstream as free fatty acids (FFA), which bind to toll like receptors (TLR) such as TLR4 and TLR2; this promotes activation of nuclear factor kappa B (NF-κB), which increases the secretion of pro-inflammatory cytokines, as well as infiltration of macrophages in adipocytes [20].

Obesity is thus associated with low-grade inflammation that spreads from peripheral tissue to the brain. The hippocampus and the cortex are particularly vulnerable to inflammation; these are regions involved in cognitive processing, learning and memory [18].

Obesity, and primarily high BMI and waist circumference, have been linked to cognitive impairment, even in children and adolescents, as it negatively affects brain function and structure. Higher BMI is related to impaired episodic and working memory tasks and verbal learning [17, 24]. As the adipose tissue increases, it releases various adipokines that lead to peripheral low-grade inflammation, as well as systemic insulin resistance; these have been linked to white matter atrophy and disruption of the blood brain barrier (BBB) [17, 25].

Inflammation in the hippocampus inhibits long-term potentiation (LTP) and impairs neurogenesis. It also promotes the production of beta-amyloid, increasing the risk of both cognitive impairment and AD [25].

Obesity also affects executive functions (inhibitions, cognitive flexibility, working memory, decision making, verbal fluency, planning, attention), probably *via* an activation of innate immunity and therefore low-grade inflammation; inhibition and working memory seem to be the most affected functions [26].

Attention is crucial for humans and is considered a core executive function, compromising multiple brain networks, including alerting, orienting and executive control networks; it can be divided into selective and sustained attention. Selective attention refers to processing parts of the sensory input while excluding others; sustained attention refers to maintaining sensitivity to incoming stimuli, which may also be referred to as "concentration" [27, 28].

Obesity has been shown to impair both forms of attention even before being born. Maternal BMI might be linked to attention deficit hyperactivity disorder (ADHD) as a result of increased inflammation, lipotoxicity and oxidative stress in the fetoplacental unit [29]. In a chronic inflammatory state, maintaining attention may require greater cognitive effort [30]. Maternal obesity is also associated with impaired serotonergic (5-HT) and dopaminergic signaling, which may also contribute to the development of ADHD, as 5-HT has a role in neuronal migration, cortical neurogenesis and synaptogenesis in fetal brain development [29].

Obesity also induces changes in brain structures, such as brainstream and diencephalon reduction, lower cortical thickness, decrease gray and white-matter volume and integrity, decline in neuron and myelin viability. All of these changes lead to cognitive impairment [31].

Besides low-grade inflammation, another mechanism linking obesity to cognitive impairment, is the disruption of brain homeostasis caused by endothelial dysfunction. Obesity is linked to changes in nitric oxide (NO), as it disrupts specialized receptors on endothelial cells that facilitate its release. This seems to impair neurovascular coupling, causing neurodegeneration; obesity also seems to deteriorate tight junction proteins Zonula occludens-1 (ZO-1) and claudin-12, breaking down the BBB [31].

On the other hand, obesity will not only affect the person's cognition, but may also affect their offspring's. Both animal and human studies have shown that obesity during pregnancy leads to systemic and placental inflammation, dysregulated metabolic and neuro-endocrine signaling and increase in oxidative stress; this is linked to altered offspring neurogenesis, myelination, and synaptic plasticity in hippocampus and hypothalamus, thus leading to the probable development of cognitive impairment [32].

A cohort study carried out by Monthé-Drèze C, *et al* (2019), showed that pre-pregnant obesity is associated with lower cognitive scores in areas such as fine motor, visual motor and visual spatial function that may be partially mediated by maternal obesity-related inflammation confirmed by higher CRP plasma levels. Although, there is also a socio-demographic factor that should be taken into account, as obese mothers had a lower socioeconomic status [32].

Obesity is also associated with an increase in intestinal permeability, leading to higher lipopolysaccharide (LPS) levels in blood, which may be an important inflammatory trigger [20]. Intestinal permeability may be associated to an alteration in the gut microbiota derives from a poor diet.

Diet and Low-Grade Inflammation

One of the main causes of obesity is high fat and carbohydrate intake, which has been linked to MCI with special emphasis on learning and memory, caused by neurobiological changes in the hippocampus, such as damage to glycoregulation, decreased levels of neurotrophins, neuroinflammation and structural integrity disorders of the BBB [33].

Cohort studies claim that consumption of saturated fatty acids (SFA) is related to MCI, affecting learning and prospective memory capacity [34, 35]. Although the effects of SFA's cannot be measured by BMI, they can be identified by testing for hypertension and/or metabolic syndrome that have a strong correlation with BMI [33].

The intake of simple carbohydrates is also closely related to MCI, since there is evidence that the consumption of foods with high glycemic index (GI) affects postprandial memory, both in patients with Type II diabetes (children and women of normal weight) and in non-diabetics [33, 36]. It has also been suggested that a high GI is a mediating factor for the effects of simple carbohydrates on learning and memory [37].

Dietary factors may also have a role in the development of low-grade inflammation [18]. The immoderate intake of food from the Western diet, aside from the increase of weight and height, generates important neurological and metabolic damage triggered by an inflammatory process. Central inflammation can induce a number of processes such as oxidative stress and neuronal apoptosis [24]. A study conducted in metabolically obese, but normal weight, rodents (MONW) fed a high fat isocaloric diet, with 60% of kcal from fat, found an increase in the expression of mRNA of inflammatory markers (TNF-α) and proapoptotic markers (Casp3) in the hippocampus, even in the absence of body weight gain [38]. Likewise, astrogliosis can be observed, being this a characteristic of damage in the Central Nervous System (CNS) and a sign of MCI [39]. Another indicator of an inflammatory process is the accumulation of mRNA expression of amyloid precursor protein (App) that indicates an Alzheimer-like pathology, increasing the probability of amyloid deposition and pro-inflammatory effects promoting a vicious cycle of neuronal dysfunction [38].

High-fat diets also increase the number of dendritic cells (DC) in adipose tissue; DC induce differentiation of pro-inflammatory Th17 cells and polarization of M1 cells, leading to a pro-inflammatory profile. They are also linked to insulin resistance *via* serine phosphorylation of insulin receptor substrate I (IRS1) by TNF-α [21, 40].

High-fat diets have also been linked to cognitive impairment, as they stimulate LPS receptor and TLR4 on immune cells, initiating an inflammatory cascade *via* NF-κB activation [24]. They also change the gut microbiota composition, leading to dysbiosis. It is now believed that dysbiosis may play a role in cognitive impairment through the "microbiota-gut-brain axis", as it is also associated to systemic inflammation [18].

LPS administration to animals has shown to induce cognitive impairment. LPS induce microglial activation and neuronal cell loss in the hippocampus, as well as an increase in pro-inflammatory cytokines both in serum and in brain, probably *via* activation of cyclooxygenase-2 (COX-2) and NF-κB, which upregulate the expression of pro-inflammatory cytokines. LPS also reduces the expression of anti-inflammatory cytokines such as IL-4 and IL-10 [41].

Treatment

Currently, there are some promising experimental treatments or therapies for cognitive impairment, but none have proved to be completely efficient. Nutrition interventions addressed to glucose control and lowering inflammation show cognitive benefits [17], thus, some possible interventions include dietary approach (for example, Dietary Approaches to Stop Hypertension (DASH) [42] or Mediterranean dietary pattern) [17].

One of the greatest clinical trials that analyzes Mediterranean diet (MedDiet) is the PREDIMED (PREvención con DIeta MEDiterránea) conducted in Spain from 2005 to 2010. In a subsample of this study of 1055 subjects, MedDiet containing either fats from nuts or olive oil, improved global cognition compared to a low-fat diet [43].

MedDiet exerts anti-inflammatory effects; it also reduces gut dysbiosis and improves endothelial function by increasing serum NO and decreasing reactive oxygen species (ROS) production, thus it has been linked to protective effects against cognitive decline [31].

Among MedDiet characteristics associated with neuroprotection is the consumption of mono- and polyunsaturated fats (MUFA and PUFA, respectively), fiber and antioxidants, from fish, extra-virgin olive oil (EVOO), vegetables and fruits. These exert anti-inflammatory and antioxidant effects that seem to be associated with a preservation of both gray and white matter and reduction of cerebrovascular disease [44]. It also reduces vascular risk by improving lipid profile (lower LDL and higher HDL cholesterol levels), and lowering lipid oxidation products, probably due to consumption of rich sources of vitamin E and C [45]. Vitamin C decreases IL-6 and IL-8, thus exerting an anti-inflammatory effect [46].

MedDiet also contributes to cognitive health by lowering the glycemic load and advanced glycation end products (AGEs), related to oxidative stress and inflammation, and also to obesity [47].

Another dietary pattern that has been associated with neuroprotection is the Mediterranean-DASH intervention for neurodegenerative delay (MIND diet), which, as its name says, is a combination of MedDiet and DASH [46]. The MIND diet was actually developed to protect the brain against dementia; the key components of the diet are plant-based foods such are EVOO, berries, green leafy vegetables, beans and nuts, while limiting animal foods (it only includes high amounts of fish) [48].

One of the main components of both MedDiet and MIND diet that promotes cognitive health is EVOO; it contains phenolic compounds, such as oleuropein, that are better antioxidants than vitamin C or E, and can reduce NF-KB nuclear translocation and activation. MUFA in olive oil also has been associated with protection against cognitive impairment, as it prevents inflammasome activation [49].

In obese individuals, weight reduction may have a positive effect in cognition. Bariatric surgery has shown to improve memory and executive functions, and to reduce peripheral inflammation in some individuals [17, 18, 20].

Exercise can also improve cognition or prevent cognitive decline by different mechanisms. Aerobic and resistance training has shown to decrease TNF-α levels in obese subjects and in general can reduce pro-inflammatory mediators such as IL-18, C-reactive protein and IL-1, as well as increase anti-inflammatory markers such as IL-10, and in rodents it has shown to promote a phenotypic conversion of M1 to M2 microglia, promoting an anti-inflammatory state in the hippocampus [31, 50, 51].

Exercise also increases gut microbiome diversity. High intensity training (HIT) and running have shown to improve memory, perhaps, by improving cerebral microcirculation, but also by increasing lactate; lactate seems to be necessary for long-term memory formation, as its accumulation in the hippocampus increases BDNF expression [31, 50].

Acute exercise stimulates synthesis of BDNF, whereas regular exercise has a positive effect on hippocampal volume; aerobic capacity correlates with brain size [50, 52]. Some of these effects might be associated with myokines produced by the contracting muscles and by the electrical stimulation; some of the myokines involved in the muscle-brain cross-talk are *cathepsin B* and *irisin*, which can cross the BBB and induce BDNF expression in the hippocampus *via* the peroxisome proliferator-activated receptor gamma co-activator 1alpha (PGC-1alpha). Irisin has been shown to be reduced in patients with AD [50].

Another pathway in which exercise may improve cognition *via* BDNF is the regulation by PGC-1alpha. Exercise can induce epigenetic modifications which lead to a demethylation of this gene; PGC-1alpha contributes to raising BDNF levels. Contracting muscles also release BDNF, which can cross the BBB [52].

Exercise also improves redox status by enhancing antioxidant capacity in the brain and improving mitochondrial function [50].

Physical activity increases concentration of some neurotransmitters such as dopamine, which is involved in adaptive memory formation [52].

DIABETES MELLITUS AND MILD COGNITIVE IMPAIRMENT

Definition and Epidemiology of Diabetes Mellitus

Diabetes mellitus (DM) includes a set of etiologically and clinically heterogeneous metabolic disorders that share hyperglycemia as a common feature [53]. This disease has rapidly expanded, in the year 2000, the International Diabetes Federation (IDF) estimated around 151 million adults with diabetes worldwide, by 2019 the IDF rose that estimation to 463 million patients, the number tripled in a period shorter than 20 years [54].

The WHO estimates that without preventive measures against diabetes, by the year 2045 there will be 629 million people suffering from diabetes in the world [55], a slightly more optimistic estimate than that of the IDF who calculate 700 million people for the same year [54, 56 - 62].

People with DM are at risk of presenting several complications, especially if they are not properly cared for or if they have some comorbidity such as obesity, hypertension and / or vascular disorders, which can contribute to the development of retinopathy, renal failure, cerebrovascular accidents (stroke), some types of cancer and even cognitive impairment [56 - 65]. These alterations impact directly on the diabetic patient quality of life and their everyday activities as well as their autonomy and are cause of discomfort, pain and depression [66].

The economic impact of diabetes and its complications on families and governments is very high; just the average expenditure of people diagnosed with diabetes in 2017 in the US was $ 16,750, of which $9,600 were directly attributed to diabetes, which implies an expenditure 2.3 times greater than that of people without it [67]. The IDF calculated that in 2019 the total health expenditure for DM was 760 billion dollars worldwide and estimates that this will continue to increase in the next 25 years, with an expenditure as high as 845 billion dollars for the year 2045 [54]. Diabetes is undoubtedly a serious public health and economic problem.

Etiology and Physiology of Diabetes Mellitus

The etiology of diabetes is heterogeneous as it affects populations differently according to age, race, ethnicity, geography, environmental factors, and socioeconomic status [68].

Type I Diabetes Mellitus (TIDM)

Type I diabetes mellitus (TIDM), usually presents during childhood or youth, although a lower percentage occurs in adulthood, it is characterized by the autoimmune destruction of pancreatic β cells causing low or absent insulin production [69]; its development is attributed to genetic alterations, environmental factors and infections [70].

In TIDM, infiltration of macrophages, CD4 + and CD8 + T lymphocytes into the pancreatic islets [69, 71] has been observed, leading to the destruction of β cells. In addition, the production of antibodies that attack self-antigens expressed by these cells is induced, such as insulin (IAA, antibodies against insulin), glutamate decarboxylase (GAD65, antibodies against glutamic acid decarboxylase 65) [72], the transporter of zinc 8 (ZnT8A, zinc transporter 8 autoantibody) [72, 73] and tyrosine phosphatase (IA-2AA, autoantibodies against protein tyrosine phos-phatase antigens) [74]. Predisposition to produce these autoantibodies has been associated with HLA class II alleles, mainly from HLA-DR and HLA-DQ loci [75].

On the other hand, it is also considered that TIDM can be caused by dysregulation of suppressor T cells. In healthy individuals, T reg cells maintain immune tolerance and prevent autoimmunity development [76].

The role of Treg cells in TIDM remains unclear, several studies evaluating the amount of Treg cells with CD4 + CD25 + / FOXP3 markers (immunosuppressive phenotype) in patients with TIDM have observed that they decrease [77], others that they increase in peripheral blood [78, 79] and some did not observe differences in their animal models of diabetes at all [80]; there are also other authors that describe modifications in its function which contribute to the development of the disease [81 - 84], this last theory being the most explored.

Other factors such as vitamin D deficiency [85, 86], infections caused by enteroviruses [87, 88], alterations of the intestinal microbiota either due to the excessive use of antibiotics or a change in diet [89, 90] and the intake of some types of food in certain stages of childhood, have also been linked to the development of TIDM. Although the interlocking of the various factors involved in the development of the disease is not clearly known, various research groups continue looking for answers.

Type II Diabetes Mellitus (TIIDM)

Type II diabetes mellitus (TIIDM) occurs in 90% of all people diagnosed with diabetes [56]. It is considered a heterogeneous, chronic, metabolic disease, characterized by hyperglycemia mainly due to the development of insulin resistance and defective secretion by the β cells of the pancreas [91].

TIIDM is a polygenic disease that has been associated with a large number of genetic variants, only Muhammad *et al*, in 2017 found 50 genes with altered expression that showed interaction with genes associated with the development of TIIDM: ZEB1, USP16, IL6ST, ASPH, Eif4g1, RBL2, MEF2A, vapB and SOS2, these genes that affect the β cells of the pancreas, are involved in the secretion of various cytokines, in pancreatic islet cells and in the peripheral uptake of glucose by the muscles [92].

Insulin is an anabolic hormone that regulates the metabolism of carbohydrates, lipids and proteins; it is also a growth factor that controls cell proliferation and differentiation [93]. It is released by the β cells of the pancreas in response to hyperglycemia, travels through the bloodstream and binds to its receptors which are widely distributed in muscle, adipose tissue, liver and brain [94]. When insulin binds to its receptor, it autophosphorylates and in turn phosphorylates and recruits adapter proteins such as insulin receptor substrate (IRS) 1 and 2 that mediate the activation of signaling pathways: RAS / MAPK and phosphatidylinositol 3-kinase (PI3K) [95].

The RAS / MAPK pathway has been linked to the stimulation of gene expression of proteins associated with cell growth and proliferation [96]; the PI3K / AKT pathway is responsible for metabolic regulation; through the inhibition of GSK-3 it stimulates the synthesis of glycogen; through the activation of different substrates by AKT, it stimulates the translocation of the glucose transporter GLUT-4 from the intracellular compartments to the cell membrane, allowing the incorporation of glucose into the cells and reducing its levels in the blood [97].

Insulin resistance refers to the decrease of the body's biological effects at certain insulin levels in specific tissues and is related to obesity, hepatic steatosis and atherosclerosis [98], this may be due to mutation or loss of insulin receptors, failure or inhibition in some region of insulin signal transduction by the action of cytokines, leptin, adiponectin and others [99].

In obese individuals, hyperglycemia and hyperlipidemia are common, which in the pancreatic cells can cause toxicity (glucotoxicity and lipotoxicity) that in turn induces a greater release of insulin as a compensatory measure [100, 101], toxicity causes the production of inflammatory mediators and free radicals favoring pro-apoptotic signals that lead to the death of β cells [102].

In conditions of obesity, adipose tissue undergoes hypertrophy and hyperplasia [103] which induces changes in its metabolic functions and produces a large amount of inflammatory adipokines such as: IL-6, TNF-α, MCP-1 and resistin causing a local and systemic low-grade inflammatory state [104]; in addition to increasing the release of fatty acids and ceramides that favor lipotoxicity in the pancreas [101] and their accumulation in muscle tissue it alters insulin signaling in the AKT pathway, inhibiting GLUT-4 transport [105] which contributes to hyperglycemia.

Some proinflammatory cytokines have been directly involved in the development of insulin resistance. The TNF-α produced by adipocytes in atrophied state activates the transcriptional factor PIMT that modulates the expression of GLUT-4, so glucose transport is affected in skeletal muscle tissue [106]. IL-6 alters the activation of the PI3K / AKT pathway and glycogen synthesis through the down-regulation of miR-200 in hepatocytes [107]. The low glucose uptake caused by thriving resistance contributes to the atrophy of β cells and adipocytes, generating a vicious cycle that causes TIIDM.

The Role of Insulin in the Brain

Brain tissue has a huge number of insulin receptors, especially in the hypothalamus, hippocampus, olfactory bulb, cortex, cerebellum, and striatum [108, 109].

The insulin receptor has two isoforms, type B, present mainly in peripheral tissues (muscle, liver and adipose tissue) [110] and type A, which has been described mainly in neurons [111]. However, Spencer's team using RT-PCR / FISH (fluorescent *in situ* hybridization) assay techniques, observed that both isoforms are present in astrocytes, microglia and neurons [112]. Insulin, in addition to binding to its receptor with high affinity, can bind to insulin-like growth factor receptors type 1 and 2 (IGFR), although with lower affinity, regulating systemic metabolism [113].

Insulin is synthesized and released in the pancreas, although the type of transporters involved is not yet defined, it is known that it crosses the BBB through a saturable transporter system independent of insulin receptors [114].

Although Kuwabara *et al* (2011), have observed that certain neurons and neural progenitor cells possess the ability to synthesize it / insulin [115].

Cells that possess insulin receptors have the same proteins that participate in signal transduction pathways as in peripheral cells, and many of them have been found to play an important role in brain functions [116]; in addition to intervening in energy homeostasis accompanied by other hormones (anorexins and orexins) in hypothalamic regions that modulate appetite and satiety [117]. For example, Pearson-Lealy *et al.*, (2016), show how GLUT-4 transporters also translocate to the plasmatic membrane after memory training and their selective inhibition affects their acquisition, although prolonged inhibition favors short-term memory and damages the long term memory [118].

Insulin participation in excitatory synaptic transmission and plasticity in the hippocampus has been described by Zhao *et al.*, (2019), they demonstrated that insulin improves the function of NMDA receptors and participates in electrophysiological processes since it influences the induction and expression of LTP and LTD required in memory and learning mechanisms [119]. Soto *et al* (2019), also observed in the same brain structures that the presence of insulin receptors and IGFRs are essential for the formation of the GluA1 subunit of AMPA receptors (α-amino-3-hydroxy-5-methyl-4-isoxazolepropionic acid receptor), which are important for LTP to take place, their lack in the hypothalamus and amygdala cause deficiencies in spatial memory, cognitive and mood disturbances [113]. These data confirm that glutamatergic neurotransmission plays an essential role in synaptic plasticity, also described by Fried *et al*,. (2019), who observed that deficiency of glutamatergic metabolites are associated with poor plasticity [56].

In the same sense, it has also been observed that insulin participates in improving the synaptic currents mediated by GABA (γ-aminobutyric acid) in the brain amygdala, having metabolic and mood implications [120]. This ability of insulin to regulate neurotransmission is also due to the increased availability of receptors at the synapse [119].

On the other hand, insulin is also considered a neurotrophic factor that alongside IGF-1, has a large number of receptors in the hypothalamus and in the amygdala and participates directly in the regulation of neurogenesis [121].

Diabetes and Mild Cognitive Impairment

In recent years, the increase in the prevalence and incidence of TIIDM made evident that patients who acquire this disease are at greater risk of developing cognitive impairment and even dementia [122].

Cognitive alterations in people with DM are determined by various factors, such as; depression [123], low-density lipoprotein cholesterol levels (LDL-C) [124, 125], BMI, lack of physical activity [126], among others. But the two most important are: 1) time to onset of DM; 2) and uncontrolled glucose fluctuations [127]. In a cohort study with 5,653 participants, Tuligenga's team (2014) showed that middle-aged people (54 years) with known diabetes showed during the lapse of ten years a 45% decrease in memory, 29% in reasoning, and 24% decrease in global cognitive function; these data were associated with poor glycemic control [128]. Very similar results were found in the investigation by Xiu *et al* (2019), who mentioned that 58.6% of the patients studied with DM who passed 50 years had MCI [126]. In both cases it is mentioned that MCI is linked to the duration of diabetes.

Several investigations have evaluated the subtypes of MCI presented by patients with DM based on the Petersen classification [129], identifying that this class of patient can develop both amnestic MCI (memory disorders) and non-amnestic (impairment of some type of cognitive domain) [130], which means that patients present memory impairment, reduced attention span (visual and auditory) and information processing, resulting in impaired executive function [131, 132]; this type of cognitive function is essential for the performance of habitual activities since it includes behavioral organization and cognitive flexibility, necessary for solving problems [123, 133].

The diagnosis of MCI in people with DM is usually made by applying cognitive screening tools, such as Mini-mental State Examination (MMSE) and Montreal Cognitive Assessment (MOCA) to evaluate the cognitive function. In many cases, neuroimaging techniques are used to confirm the diagnosis or track the neuroanatomical changes that the brain undergoes over time or to understand how clinical changes are related to structural ones.

With the intention of understanding whether patients with DM have brain structural changes that can induce MCI, Moran *et al* (2019), used magnetic resonance imaging and observed that these patients have a lower baseline cortical thickness which they moderately related to presence of DM [134], showing that DM could contribute to neurodegeneration.

On the other hand, Groeneveld's team (2018) observed that the volume of gray matter is lower in the right temporal lobe and subcortical brain regions in patients with DM. The observed atrophy was present mainly in areas of vascular lesions usually caused by DM. Likewise; they noted that neural connectivity and volume of white matter is also reduced in patients with MCI [135]. Similarly, these structural changes in white matter were observed by Tan *et al* (2016), who, identified the most affected structural tracts – right cingulate, part of the frontal lobe and parietal lobes, and some thalamic regions– structures similar to those observed in people with AD disease [136].

Damage to the integrity of white matter is associated with cognitive alterations in patients with DM. Gao *et al* (2019), observed that lesions of the lower right frontal-occipital tract and the lower right longitudinal tract correlate with disturbances in attention and episodic memory, conditions commonly observed in patients with MCI [137].

The mechanisms that trigger this type of cognitive and structural alterations in the brain of people with DM remain unclear, but it has been observed that insulin resistance, alterations in insulin receptors, high glucose levels, mitochondrial dysfunction and inflammation play an important role in MCI.

The continuous increase in insulin and glucose in the brain causes hyperactivation of insulin signaling, which induces cells to lose sensitivity and modify their responses. It has also been reported that there is a decrease in the amount of insulin receptors [116] that can affect the synaptogenesis in the hypothalamus and amygdala, which directly affects neuroplasticity [113]. Liu *et al,*. (2015), also observed that the synaptic strength of LTP is altered, causing a decrease in synaptic connection forces, essential parameters for memory and learning [138]. At the same time, the incorporation of lipids such as ceramides and inflammatory molecules interfere with proper insulin signaling [139].

In a murine model of insulin resistance, it was observed that a high-fat diet causes the inactivation of IRS1 -adapter proteins that mediate insulin signaling pathways-causing a decrease in the translocation of GLUT-3 and GLUT-4 at cell membranes of neurons and the suppression of the ERK / CREB pathway, essential pathways for the acquisition of memory and learning [138]. Hyperglycemia has also been shown to affect the AMPK / AKT signaling pathway that contributes to insulin resistance and mitochondrial dysfunction *in vitro* [140].

The reduction in PI3K / AKT signaling during insulin resistance causes a greater activation of the enzyme glycogen synthase kinase (GSK-3β) that hyperphosphorylates the tau protein —abundantly found in the CNS— which causes their improper folding, forming fibrils [141]. These formations are directly associated with the development of Alzheimer's, and there is increasing evidence that links metabolic alterations with the development of this dementia [142].

Mitochondrial dysfunction is also caused by alterations in the PI3K / AKT signaling pathway, since it regulates the metabolic function of the   mitochondria, when it's damaged, the mitochondria releases a large amount of free radicals that contribute to the activation of inflammation [143].

The increase in peripheral inflammatory cytokines has also been linked to the presence and development of MCI, mainly IL-6, TNF-α [144] and galactin [145]. TNF-α has been linked to the interruption of glucose transport and to the dysfunction of the BBB facilitating the infiltration of leukocytes [146], mainly granulocytes, in addition to hyperglycemic conditions and a lipid rich diet, it has been observed that there is an increase in the expression of microglia and active astrocytes in the hippocampus [147]. Microglia in the presence of free radicals and TNF-α acquires an inflammatory phenotype by inducing the activation of the enzyme nitric oxide synthase, which produces nitric oxide and generates a neurotoxic environment.

DM also affects neurogenesis through the reduction of BDNF [99] and the transcription factor Neuro D1 present in hippocampal stem cells [148], in both cases due to insulin receptors reduction on hippocampal cells. An investigation carried out by Bonds *et al,.* (2020), demonstrate that in a DM model, neurogenesis is reduced in the subgranular zone of the hippocampus, which implies a smaller number of new neurons that will migrate to the dentate gyrus, which can lead to its dysfunction and cognitive deficits [149].

All these alterations intervene in the gestation and development of MCI, the shape in which all these changes are articulated is not precisely known, their understanding will provide an opportunity to propose therapeutic alternatives that prevent its development.

Therapeutic Alternatives

One of the main factors associated with MCI is hyperglycemia, which is why several investigations propose drugs with a hypoglycemic and neuroprotective nature, we will describe the most important ones.

Metformin: activates AMPK signaling that is affected by insulin resistance and inhibits the activity of NAD (P) H oxidase, one of the enzymes that produces the freest radicals during mitochondrial dysfunction. In this way, the reduction in the amount of ROS reduces inflammation and cell death [150]. It affects TNF-α of different cell types, which decreases the secretion of pro-inflammatory cytokines.

Lithium: there is evidence on the neurotrophic and protective properties of this element, although it has been tested in a very small number of intervention trials. Epidemiological and imaging studies show that it is associated with lower rates of dementia and beneficial brain responses, with higher density of gray matter and greater metabolic integrity of brain tissue [151].

Lycopene: it is a phytochemical that belongs to the carotenoid family, it has antioxidant and anti-inflammatory properties, it has been tested in some models of dementia but there is a lack of evidence to determine its effect more strongly [152]. In a study by Yin *et al,*. (2014), they observed a neuroprotective effect in the treatment of damage caused by insulin resistance [153].

Insulin: since there is a decrease in the synthesis of insulin and in the amount of receptors mainly in the hippocampus, the insulin treatment is the one that has been most evaluated, it has improved the memory and cognition in rats in a model of Parkinson's disease [57]. It is also known that insulin activates the ERK and PI3K signaling cascade, which intervene in multiple functions such as cell growth, proliferative, survival, in protein and lipid synthesis, synaptogenesis and apoptosis. Bayunova *et al,*. (2018), also observed that it has neuroprotective effects on neurons of the cerebral cortex exposed to oxidative stress conditions in an *in vitro* study [154].

Given the great increase in the prevalence of DM and the risk of these people suffering from MCI, it determines the wide importance and urgency that exists to elucidate in greater detail how the mechanisms that generate it are unleashed and with it, to be able to propose effective therapeutic alternatives that allow improving the patients quality of life.

COGNITIVE IMPAIRMENT CAUSED BY VASCULAR ALTERATIONS

Epidemiology of Cardiovascular Diseases

Cardiovascular diseases (CVD) are a set of disorders of the heart and blood vessels [155]; they are the biggest cause of disability and premature deaths worldwide, only in 2017, 17.8 million cases were reported. In recent years, the number of persons affected by CVD has risen, according to the American Heart Association (AHA) net prevalence is 485.6 million, 28.5% more than in 2007 [156].

Among the whole set of CVD (peripheral vascular disease, ischemic heart disease, rheumatic, congenital, myocardiopathy, arterial hypertension and stroke), ischemic heart disease and stroke are the most frequent [157]. In 2016, in the United States, 43.2% of all deaths caused by CVD were due to ischemic heart disease, while 16.9% were caused by stroke and 9.8% by hypertension [158]. Regarding the causes of disability, both ischemic and hemorrhagic stroke account for the larger rate of disability-adjusted life years (DALY), rising from 40 years of age and peaking at 5 DALY in the 74-79 years old group [157].

The AHA 2020 report mentions that, in the United States, direct and indirect costs derived from disability during the 2014-2015 period were $351.2 billion and it is estimated that for 2035 it will reach $749 billion, considering hospital expenses, medication, home care, *etc.*, [156]. Because of the alarmingly high impact on deaths, disability and costs, governments have been working on strategies to ameliorate their population risk of CVD.

Some of the risk factors for developing CVD are: diabetes, hypertension, atherosclerosis, obesity, dyslipidemia, insulin resistance and smoking [158]. These risk factors are associated with well-known lifestyles and habits like diet, education level, physical activity and smoking habits, which can be modified. Which is why the WHO urges governments to promote campaigns that emphasize the benefits of having a healthy weight, increasing physical activity, quit smoking and having blood glucose and lipids checked periodically [159].

ETIOLOGY AND PHYSIOPATHOLOGY OF VASCULAR LESIONS

Atherosclerotic Plaque Formation

Ischemic heart disease and stroke are diseases with their origin in partial or complete disorders of the blood supply either to the brain (ischemic stroke) or heart (ischemic heart disease). The main cause of blood flow obstruction is atherosclerosis formation [160].

Atherosclerosis consists of atheroma plaque formation in medium and large size arteries intima, this plaque is formed mostly by a mixture of different lipids, smooth muscle cells, extracellular matrix, calcium (Ca^{2+}) and immune cells [161]. Formation of this plaques is attributed to hyperlipidemia which is usually present in patients with obesity, insulin resistance and diabetes. High levels of LDL, vLDL (very low density lipids) and triglycerides accompanied with low HDL (high density lipoprotein) levels have also been linked to its development [162].

Plasmatic lipoproteins contain several types of apolipoproteins in its surface, the ones containing ApoB (LDL and vLDL) penetrate the superficial layer of the endotelial intima that covers the inside of the inner surface of the blood vessels, particularly those arranged in arterial branch bifurcations or high shearing degree zones [162].

In normal conditions, several LDL fractions circulating through the blood stream penetrate the endothelium; inside, cell lipase hydrolyze triglycerides and cholesterol is used for synthesizing other molecules; once the cell acquires the cholesterol it needs, it stops synthesizing receptors, thus avoiding cholesterol intake [163].

Baumer *et al* (2017), observed in an *in vitro* model that endothelial cells capture and metabolize LDL and when the inside cholesterol levels are exceeded, they generate crystals; when studied *in vivo,* it was observed how these crystals are stored in the sub-endothelium [164]. LDL molecules attached to the endothelium are then oxidized (ox LDL) and act as Damage Associated Molecular Patterns (DAMP) unleashing an inflammatory process, immune components located in the vascular tunica media participate expanding inflammation, damaged endothelium cells are then activated and express chemokines, cytokines, vascular cell adhesion molecules (VCAM), intercellular adhesion molecules-1 (ICAM-1) and selectin P and E which attract monocytes [165].

Circulating monocytes migrate to the subendothelial space where they differentiate into macrophages which ingest a large amount of oxLDL thus forming foam cells; foam cell accumulation contributes to the atherosclerotic plaque build up [166]. Foam cells release proinflammatory cytokines such as TNF-α, IL-1β and MCP-1 [167]; cholesterol crystals collection induces NLRP3 inflammasome activation, as a consequence, IL-1β and IL-8 expression increases [168]. Cells in the adventitia also contribute to the inflammatory process by producing cytokines, chemokines, ROS and remodeling substances. IL-12 and IL-18 release by the macrophages induce T-lymphocyte differentiation which then arrive to the injury site where they produce TNF-α and INF- ϒ [165].

Macrophages release microparticles with procoagulant tissue factor inhibiting coagulation and fibrinolytic molecules; adhesion molecules, selectin and von Willebrand Factor (vWF) enable platelet adhesion to the endothelium, IL-1β increase their binding to collagen and fibrinogen thus promoting its aggregation [169]. On the other hand, platelets arriving at the developing plaque and promotes atherogenesis and Th-1 for M1 through Socs3 expression increase, as a consequence, IL-6 and other chemokines synthesis are increased thus attracting even more monocytes to the injury site and furthering the plaque development [170].

Endothelial damage signaling induce remodeling mechanisms, this stimulates smooth muscle cells from the vessel's tunica media into losing their contractile capabilities, migrate to the intima, acquire proliferative characteristics [171], further thickening the intima and consequently reducing the vascular lumen and causing endothelial dysfunction.

Endothelial Dysfunction

The anomalies suffered by the vascular endothelium during atherosclerotic plaque formation disrupt its function. The healthy endothelium is a tissue that regulates vascular tone, cell growth, and leukocyte interaction with thrombocytes, and the wall of blood vessels, today it is known that it is capable of synthesizing growth factors and thrombo-regulatory molecules, which respond to physical and chemical signals [172].

Endothelial cells serve several vascular functions, including promoting vascular tone, inhibiting platelet aggregation and adhesion, maintaining anticoagulant, profibrinolytic, and anti-inflammatory effects, and inhibiting smooth muscle cell proliferation and migration. Also, they are responsible for the regulation of a large number of vasodilators, such as prostaglandins, nitric oxide, endothelium-dependent hyperpolarization factors, and endothelial-derived contraction factors, which play an important role in the functioning of endothelial cells [173].

Chronic vascular inflammation is associated with atherogenesis and the subsequent development of diseases of the coronary and cerebrovascular arteries, that is why endothelial dysfunction helps us predict atherosclerosis in adults since elevated levels of inflammatory markers can be seen in atherosclerosis and cardiovascular events [174]. Although the term "endothelial dysfunction" is generally used for the impairment of endothelium-dependent vasodilation; the term also includes abnormalities in the interaction between the endothelium and leukocytes, thrombocytes and regulatory molecules, conditions that result in aberrant endothelial activation [172].

The dysfunctional endothelium is not able to maintain its natural function since it changes towards one of the states described below: (I) reduction in the bioavailability of NO as a consequence of the reduction in production and/or elimination by ROS; (II) weakened endothelium-mediated vasodilation; (III) interrupted fibrinolysis and increased thrombosis; (IV) overproduction of growth factors, chemokines (MCP-1), adhesion molecules such as E-selectin and ICAM-1, pro-inflammatory molecules (IL-6, IL-8, IL-12), ROS and altered angiogenesis [60]. Besides, endothelial dysfunction is usually characterized by an imbalance in the secretion and release of vasoconstrictor and vasodilator agents, which predisposes to prothrombotic and proatherogenic effects [175]. Characteristic mechanisms during the development of atherosclerotic plaque.

Recent studies suggest that macrophage-derived iNOS promotes microvascular endothelial dysfunction by reducing the bioavailability of hydrogen sulfide (H_2S) and plaque instability in atherosclerosis [176]. It has also been observed that endothelial dysfunction may be due to impaired NO synthesis resulting in inadequate perfusion which can contribute to several complications seen in mitochondrial diseases including stroke-like episodes, myopathy, diabetes, and lactic acidosis [177].

Alleviation of endothelial dysfunction might mean better vasodilation and better perfusion of the microvasculature, which could lead to less complications and even improvement of the clinical condition observed in the aforementioned diseases.

Hypertension as a Cause of Cardiovascular Risk

Hypertension is the most common preventable risk factor for CVD, as such, it is considered to be a strong contributor to the total of deaths and disability worldwide [164]. High blood pressure is associated with a higher risk of cardiovascular events; a study by Luo *et al*,. (2020), showed that hypertension in young adults carries with it a higher risk of cardiovascular events in more advanced stages of adult life [179].

There are various mechanisms intervening in adequate arterial pressure maintenance: 1) the Renin-angiotensin-aldosterone system (RAAS), 2) natriuretic peptide and endothelial intervention, 3) the sympathetic nervous system and 4) the immune system; failure of one of this seriously compromises blood pressure homeostasis [178].

Sodium (Na^+) is the most important ion in blood volume regulation, when dietary Na^+ intake is increased there are hemodynamic changes such as renal and peripheral vascular resistance reduction and endothelial NO production increase [180].

Renal Na^+ reabsorption is controlled by aldosterone, which increases Na^+ and water levels in the extravascular space. When blood pressure levels decrease, renin secretion is stimulated in the yuxtaglomerular apparatus in the kidney, renin transforms hepatic angiotensinogen into angiotensin I and II. Once angiotensin II binds to its target cells, Na^+ tubular absorption increases and so does blood pressure [181]. Hence, a high serum Na^+ concentration promotes liquid retention, this increases blood volume and subsequently blood pressure [178].

In this way, long term excessive Na^+ consumption alters RAAS homeostasis and increases the risk for hypertension [182].

On the other hand, under conditions of vascular lesion such as the ones caused by atheroma plaque formation where NO mediated vasodilation mechanisms are absent or altered, there is an increase in blood pressure secondary to Na^+ ion increase, therefore endothelial dysfunction is a risk factor for hypertension [178].

Obesity also contributes greatly to hypertension progression as it is considered to be the cause of 65-75% of primary hypertension [183]. Under this condition, the kidney is compressed beneath the surrounding adipose tissue, aldosterone and angiotensin II levels increase as well as some adipokines like leptin, this causes changes in kidney physiology like increased Na^+ tubular reabsorption that ultimately result in hypertension [184].

Insulin resistance is also closely related to hypertension, clinical trials have shown that 50% of patients with hypertension are also affected by hyperinsulinemia or glucose intolerance. It has been observed that TIIDM patients have increased levels of angiotensin II, over activation of RAAS also stimulates the MAPK pathway that activates insulin, which has been associated with endothelial dysfunction [171]. All of these mechanisms combined promoted hypertension progression.

As arterial blood pressure rises, friction from hematic cells can cause damage to the vascular endothelium thus causing altered vasoactive agents release like NO reduction; mitochondrial oxidative metabolism produces an important amount of ROS which can react with NO producing peroxynitrite anion which induces lipid peroxidation which ends up activating an immune response [176].

Hypertension, besides being a contributing factor for CVD along with dyslipidemia and metabolic syndrome, it also plays an important role in cognitive impairment [186].

Cerebrovascular Accident

Cerebrovascular accidents are the main consequence of the vascular disorders caused by atherosclerotic plaque formation which usually occurs in the presence of concomitant diseases such as diabetes, obesity and / or hypertension [187]. Cerebral arteries usually present some degree of endothelial dysfunction which makes them vulnerable to either occlusion of the lumen or rupture, which would trigger platelet aggregation and consequent thrombi or embolus formation; in both cases, the occluded region of the artery will suffer vascular endothelial damage and as a result will express adhesion molecules, endothelial cell activation and inflammatory response, starting with neutrophil recruitment [188].

Along with blood flow interruption to the cerebral parenchyma in the region of the affected artery, oxygen and glucose are not supplied affecting cell metabolism; the few glucose reserves undergo glycolysis but do not enter aerobic metabolism cycles therefore lactate accumulates [189] causing metabolic acidosis, this generates mitochondrial dysfunction thus producing ROS. Lack of ATP and H^+ increase, render the cell membrane pumps unable to maintain ionic equilibrium and trigger depolarization [190] which drives intracellular Ca^{2+} increase causing sustained excitatory neurotransmitters (*e.g.*, glutamate) release and therefore excitotoxicity. Ca^{2+} activates several proteases, lipases and endonucleases which start degrading the cell components [191].

Cell death mechanisms produce DAMPs and PAMPs and with them activation of the local microglia and perivascular macrophages, which along with granulocyte recruitment in the occluded region, initiate the inflammatory process [192] that, although it tries to limit the damaged area, when not properly modulated will end up causing secondary tissue damage [193]. Inflammation of the tissue surrounding the infarction is characterized by presence of Th1 lymphocytes, M1 phenotype microglia, a set of proinflammatory cytokines such as IL-1β, IL-6, TNF-α and the presence of ROS which exacerbate neural damage [194].

Cerebrovascular events generate two well identified damage areas, the zone in the heart of the infarction which experiences a dramatic drop in blood supply below 20ml/100g/min, as a consequence it causes irreversible lesions due to cellular loss; and the penumbra zone, this is the region of tissue around the heart of the infarction which is hypo perfused so cell viability can be maintained although function is compromised [195]; it is considered a transition region which could potentially be recovered or added to the infarction.

COGNITIVE IMPAIRMENT CAUSED BY VASCULAR ALTERATIONS

Effects Caused by Hypertension

In recent years, hypertension has been associated with the development of cognitive impairment, the leading cause of dementia in the elderly; and although the mechanical bases of this negative effect have yet to be determined, it is well known that the structural and functional alteration of the cerebral blood vessels can contribute to the development of this pathology, and its consequent cognitive deterioration [196]. The main evidence on the effect of hypertension on cognition comes from epidemiological studies, which indicate that, indeed, hypertension results in "pronounced cognitive impairment, poor cognitive performance, as well as MCI and dementia" [197]. Although one of the main secondary effects of hypertension is stroke, the cognitive decline appears to be completely independent.

After blood flow dysfunction due to hypoperfusion, there is a lack of nutrients and oxygen, metabolism becomes anaerobic and glucose decreases; besides, the ion imbalance leads to irreparable cell damage [186]. Because the brain is the main target of hypertension, local vessels must adapt rapidly to protect themselves from impending mechanical stress [196]. Among these adaptive mechanisms is the deposit of collagen and fibronectin, which causes hardening of the arteries, which facilitates later cerebrovascular accidents and cognitive deterioration. Excessive salt intake has been associated with cognitive impairment, an effect mainly attributed to hypertension [186].

The prevalence of cognitive impairment due to hypertension appears to be higher in middle-aged patients, which could explain the reason why there is not much information on the matter since most studies focus on older adults. The areas of cognition that are most impacted by hypertension have been studied; these appear to be executive function, motor speed, and attention, which are the domains related to the subcortical disease. Memory alterations related to hypertension have also been analyzed, and although it can become complicated due to overlapping etiologies, it has been observed that hypertensive patients who also have high levels of the brain β-amyloid protein are more prone to have memory impairments [196].

Recent observational data has shown that antihypertensive drugs are capable of lowering the risk of dementia and cognitive decline, and there are clinical trials to support this claim. More information is still needed to be able to choose the antihypertensive drugs that have a better effect on cognition [198].

Effects Caused by Cerebrovascular Accidents

Ischemic stroke usually brings serious complications, they depend on the length of time the arterial obstruction lasts, the area and volume that has been affected, the number of heart attacks, and the patient's condition, such as age and the presence of morbidities [187].

In addition to the physical manifestations, various studies have reported alterations in different cognitive domains, such as the analysis carried out by the international consortium STROKOG, which included data from 8 countries and evaluated the cognitive alterations of 3146 patients who suffered stroke, of which 44% presented global cognitive alterations and 30-35% showed cognitive alterations in functions such as attention, processing speed and frontal executive function [59]. Alterations very similar to those reported by Sarfo *et al* (2017), They evaluated 147 patients with stroke and found that 53.5% presented some type of cognitive-amnestic alteration with a single domain and multiple domains; non-amnesic with a single domain and multiple domains- characterized by visual-spatial impairment, executive and language dysfunction [199].

The location of the lesions caused by stroke are very heterogeneous and, consequently, there are variations between the affected cognitive domains. Al-Qazzaz team (2014) categorized the affected regions and the possible cognitive damages with the different types of memory. According to the criteria of the National Institute of Neurological Disorders and Stroke, the french Association Internationale pour la Recherché et l'Enseignement en Neurosciences for vascular dementia and, the Diagnostic and Statistical Manual of Mental Disorders; they analyzed several studies evaluating post-stroke brain injuries and cognitive symptoms. In this way, they associated the brain regions with cognitive functioning as follows: 1) in the frontal lobe, the areas of perception and learning of the cognitive domain are found, and short-term memory and working memory are processed; 2) in the medial temporal lobe and hippocampus, long-term episodic and semantic memory are processed –which encode personal information and its visual-personal and language relationship; 3) in the cerebellum and basal ganglia, the procedural domain is processed [200].

The regions typified by Al-Qazzaz are mostly supplied by the middle cerebral artery; it has been observed that the occlusion of this artery causes infarct volumes four times greater than those caused by the anterior and posterior cerebral artery [201]. Although some regions bordering these areas are supplied by the anterior cerebral artery [202]. Meng *et al,*. (2019), observed that even in patients with asymptomatic stenosis of the middle cerebral artery without previous vascular and dementia recordings, they develop evident alterations in executive function, attention, and information processing speed over time [203].

Various investigations have indicated that in addition to the death of neural tissue caused by blood flow occlusion, the loss of tissue in the ipsilateral region continues for months or years after the ischemic event [204]. There is evidence that suggests that the volume of the whole brain in stroke patients is lower, and more obviously the hippocampal region compared to control patients, these data suggest that the brain structure of stroke patients was already exposed before. However if the event occurred, it may also be that there is a progressive loss of neural volume due to tissue vulnerability to microenvironmental changes in the regions near the hippocampus [205].

Neurodegeneration mechanisms may be involved in long-term tissue loss after stroke; Okar *et al,*. (2020), observed that after six months of an ischemic event, patients, in addition to losing volume in the ipsilateral region of the infarction, also lost volume in the contralateral region, showing a monthly loss of 0.8%, which is related to the intensity of the stroke and the degree of inflammation evidenced by the levels of CRP, neutrophils and T lymphocytes [206].

For several decades, the relationship between inflammatory markers (CRP, TNF-α, IL-6, and IL-1β) with changes in the structure and function of the hypothalamus and the presence of cognitive impairment has been analyzed [207, 208]. The effects of inhibitors on some of them have been evaluated at clinical level, such as the case of etanercept, a selective TNF inhibitor that can improve some disorders such as motor impairment, spasticity, sensory impairment, cognition and pain in patients in the chronic phase of stroke, even in those who were more than 10 years old [209].

After an ischemic event, stem cells found in the subventricular and subgranular areas of the dentate gyrus of the hippocampus receive stimuli for their proliferation and consequently for the restoration of the affected tissue [210]. Among the stimuli are various cytokines and growth factors, including BDNF, this neurotrophic factor is capable of increasing neurogenesis in animal models of stroke [211]. Research has shown that this growth factor is reduced in patients with post-ischemic cognitive impairment [212], while its low concentration has been associated with patients with major depressive disorder after stroke [213].

Although attempts have been made to explain the relationship between various molecules, mechanisms, and structural changes in the brain that facilitate the development of cognitive decline after an ischemic event, much research is still needed to limit its development and have alternatives available as therapies that prevent its progression.

Pharmacological and Rehabilitation Alternatives for Cognitive Impairment Caused by Cerebrovascular Accidents

In recent years, various lifestyle factors, including diet and physical activity, have been highlighted as reducing or diminishing vascular disease for the prevention of dementia.

Patients with post-stroke cognitive impairment or dementia may often be less treated with aspirin or warfarin than non-demented individuals. As most of the risk factors are modifiable, the main aim is to influence people's lifestyle. Educational achievement, along with occupational complexity and social commitment, constitutes the paradigm of the "cognitive lifestyle", which has been associated with a lower risk of long-term dementia and is associated with neurotrophic changes in the prefrontal lobe consistent with a compensatory process. Physical activity and a healthy diet, including eating fish, have also been shown to protect against post-stroke dementia [214].

Cognitive rehabilitation (CR) is a non-pharmacological strategy that can compensate or restore some of the sequelae (altered function of visual and auditory perception, attention, memory, language, and executive abilities) observed after stroke [215]. CR consists of a series of programs specifically designed to improve the function of any cognitive domain. Some of these interventions have shown beneficial results; for instance, Swatridge *et al.* in 2017 observed that 20 min of aerobic exercise, improved cortical processes [216]. In 2021 Amorós-Aguilar *et al.* analyzed the effect of a combined therapy (moderate exercise and computer-based cognitive training) on the cognitive function of patients with stroke. They observed a significant improvement in processing speed and attention. In this study, combined therapy presented better results in terms of memory when compared with moderate exercise alone (2017). CR therapies are very promising, but the high variability of patient conditions and the particularities of the interventions limit a more precise analysis of the effects on cognition. This topic should be further addressed.

In the case of pharmacological interventions, The South London Stroke Register (2016) showed that patients with ischemic strokes without a history of atrial fibrillation have a significantly lower risk of cognitive impairment associated with the use of antihypertensive drugs when clinically indicated. Furthermore, there is a tendency to reduce the risk of cognitive impairment associated with the use of a combination of aspirin, dipyridamole and statin. Protective effects against cognitive impairment have also been observed in patients taking a combination of antihypertensives, antithrombotic agents and lipid-lowering drugs (relative risk, 0.55) [187].

To date, the US FDA and the European Medicines Agency have not approved any drug/treatment for cognitive impairment (including vascular dementia). Acetylcholinesterase inhibitors and memantine (an N-methyl-d-aspartate receptor antagonist) have been evaluated in several clinical trials for use in people with vascular dementia, which are established treatments for Alzheimer's disease. Subgroup analyzes indicated that acetylcholinesterase inhibitors have greater benefits in individuals with (multiple) cortical lesions and hippocampal atrophy compared to patients without atrophy and that memantine is more effective in individuals with subcortical vascular dementia than in individuals with other dementias [187].

Other symptomatic therapies evaluated in patients with cognitive impairment are cerebrolysin, actovegin, and nimodipine. Cerebrolysin (a combination of neurotrophic factors that were initially isolated from pig brains) and actovegin (which can promote glucose transport) may have neurotrophic and neuroprotective properties in people with mild to moderate vascular dementia or post-stroke cognitive impairment. However, treatment with cerebrolysin requires a regular intravenous infusion and therefore, the widespread use of this drug for vascular dementia can represent a challenge [187].

The use of immunomodulators such as Cop-1 could have effects on MCI since, it has been observed that it is a neuroprotective agent which stimulates neurogenesis (through the increase of BDNF, NT-3 and IL-10) in a model of cerebral ischemia [211] and learning and memory in young animals with cognitive impairment [218], however, research is still needed.

COGNITIVE IMPAIRMENT AFTER SPINAL CORD INJURY

Epidemiology and Demographics (Spinal Cord Injury)

Spinal Cord Injuries (SCI) are a global public health problem that affects patients physically, psychologically and socially. The majority of reports showed a high male-to-female ratio and an age of peak incidence in people under 30 years old. Traffic accidents were typically the most common cause of SCI, followed by falls in the ageing population [219]. It's incidence has been estimated in 10.5 new cases per 100,000 habitants per year [220]. Singh *et al,* reported in 2014 that the country with the highest prevalence was the United States of America (USA) with 906 per million inhabitants while the Rhone-Alpes region in France and Helsinki, Finland had the lowest prevalence with less than 250 per million inhabitants [219].

Pathophysiology of Spinal Cord Injury

SCI is now considered a major problem for the health care systems of the world due to the morbidity and mortality that this disease entails [221]. This is why, more and more research is being conducted to study its pathophysiology, in order to develop new therapeutic alternatives [222]. It can be appreciated why the SCI represents an extremely complex pathology and with a great impact on the quality of life of the patients. Hence, a great human and economic effort is invested in SCI research, with the hope that unravelling its pathological mechanisms will reveal future molecular therapies [223].

SCI triggers a set of biochemical and cellular events that eventually lead to the death of neurons, oligodendrocytes, astrocytes, and cell precursors. SCI is divided into two phases: primary and secondary, they refer to the stress to which the spinal cord (SC) is subjected from the mechanism of injury, disrupting its axonal connections and cell membranes. These mechanisms are direct cause of cellular death but also, the possible associated hemorrhage will interfere with the anatomical continuity of the SC [224].

The primary phase is caused by the mechanical event, causing damage to the blood-brain barrier (BBB), tissue degeneration and cell death. However, after the initial damage to the SC, the structure and function is lost due to different secondary phase processes that lead to apoptosis and loss of myelin [225].

During the secondary stage there is an ionic imbalance between internal and external concentration of Na^+, Ca^{2+}, K^+ and Mg^{2+} that leads to the depolarization of neuronal membranes and the loss of the ability to regulate the release of excitatory amino acids -such as aspartate and glutamate - to the synaptic space. When there is hyperstimulation of the glutamate N-methyl-D-aspartate (NMDA) receptors, which are permeable to Ca^{2+}, the massive entry of this ion into the intracellular space occurs, activating cell death pathways by apoptosis, enzymes associated with the oxidation of arachidonic acid and generation of free radicals (FR) among others [226, 227].

The main mechanisms of vascular damage within SCI physiopathology will be described below.

Vascular Damage after Spinal Cord Injury

Neurogenic Shock

A disruption of the SC physiology after SCI may be spinal shock with neurogenic shock. It is a temporary loss of function and reflexes below the SC injury level. Involvement from involuntary central centers modulate the activity mediated by alpha-adrenergic receptors, of sympathetic spinal neurons that normally last three to four minutes [228]. Unforeseen disruption of communication between these centers and therefore the sympathetic neurons within the intermediolateral thoracic and lumbar SC results in spinal shock [229].

The neurogenic shock has been described by signs and symptoms like bradycardia, severe hypotension and hypothermia [230]. The sympathetic stimulation is due to massive norepinephrine releasing from the suprarenal glands and disruption of cervical and high thoracic vasoactive neuron and its duration is usually days to weeks with an average of 4 to 6 weeks [231]. Autonomic dysreflexia and coronary heart disease are encountered post injury [232].

Autonomic Dysreflexia

Autonomic dysreflexia is result the sympathetic neurons and altered glutamatergic neurotransmission within the SC, characterized by hypertensive bouts with compensatory bradycardia, after noxious stimuli or bladder or bowel bloating [231].

The brain perceives the hypertensive crisis throughout cervical baroreceptors and IX and X cranial nerves. It generates restrictive impulses that cannot be transmitted beneath the injury. Vasomotor centers from the SC try to lower blood pressure by parasympathetic stimulation of the guts, through the X cranial nerve, generating severe cardiac arrhythmia [232].

Other clinical characteristics comprises severe arterial hypertension, headache and visual impairment due to cerebral vasodilatation, cutaneous pallor below the injury site, secondary to sympathetic activity, profuse sweating and cutaneous vasodilatation above the level of lesion, secondary to parasympathetic activity. Arterial blood pressure can reach up to 300 mmHg, resulting to retinal, intracerebral, or subarachnoid hemorrhage, pulmonary edema, infarction, seizures, confusion and death [232].

Arterial Hypotension

Clinical evidence indicate that the extent and severity of hypotension correlates well with the level and severity of SCI, the interruption of sympatho-excitatory pathways from the brainstem to the spinal sympathetic preganglionic neurons impairs the capability of the blood vessel baroreflex to effectively cause constriction and maintain pressure [233]. This has recently been corroborated in the SCI population, where impaired cerebrovascular and cognitive function has been shown to be associated with low resting blood pressure [234]. Arterial hypotension and deep vein thrombosis are found in both acute and chronic periods of time. Furthermore, arterial hypotension associated with a normal heart rate requires volume loading, with crystalloids and colloids in the first 24-48 hours following SCI [229].

Intraparenchymal Hemorrhage

Initial mechanical trauma tends to damage primarily the central gray matter, with relative preservation of the white matter. This increased propensity to damage gray matter is thought to be due to enhanced vascularization. The immediate effect after mechanical damage to the SC is vasospasm of the superficial vessels, that leads intraparenchymal hemorrhage, causing damage to the gray matter microvasculature [235]. SC microcirculation system is usually broken by SCI, giving rise to intraparenchymal or intramedullary hemorrhage. Reducing perfusion and therefore loss of control of blood flow that causes local infarctions due to hypoxia and ischemia that may occur depending on the severity of the injury, all of which is compounded by neurogenic shock, arterial hypotension, bradycardia, arrhythmia or intraparenchymal hemorrhage of hemorrhagic shock [236].

Vasogenic edema results initially as a consequence of the BBB rupture, secondly it is caused by the loss of ionic regulation, giving way to water accumulation in extracellular spaces. Water accumulation is strongly related to the intensity of the initial trauma and the motor dysfunction possessed by individuals suffering from SCI [237]. Also, systemic and cerebral hemodynamic contribution of SCI cognitive output [238]. Taking into account that the presence of edema in any part of the CNS results in the compression of surrounding tissue. Hence, ischemic events take place promoting the development of other self-destructive mechanisms, such as the release of FR, lipid peroxidation, and inflammation. The presence of hemorrhage within the human SC after SCI has been connected with decreased motor function [239]. In addition, the ensuing inflammatory response disturbs the microenvironment of the neural structure, alters vascular permeability, facilitates the entry of peripheral immune cells, and exposes the adjacent non-injured tissue to doubtless noxious molecule [240].

Inflammatory Response

Immediately after traumatism, which leads to the rupture of the BBB, an inflammatory reaction takes place. This reaction involves the actions of chemical mediators and the participation of inflammatory cells, which originate the activation of resident immunological cells (astrocytes and microglia) and those cells recruited from the periphery: macrophages and lymphocytes [241].

This phase is subdivided into three parts: acute (less than 48h after the injury), sub-acute (between 48 hours and 14 days after the trauma) and chronic (more than 6 months) [242]. The event that will initiate this secondary phase is the inflammation and hemorrhage in the gray matter, which will lead directly to ischemia and cell necrosis. This is in conjunction with the activation of the microglia, T-lymphocytes, astrocytes and pro-inflammatory interleukins that will cause a dysregulation of the BBB permeability [224].

Neutrophils are part of the first cell type to be recruited as they are highly sensitive to chemokines and cytokines from the first 24 hours; These granulocytes adhere and cross the endothelial barriers with the function of releasing FR, secreting metalloproteinases and cytokines with the function of phagocytosis and elimination of cellular debris caused by ischemia, edema and mechanical damage that caused the traumatic event [243]. After 2-3 days post injury, dendritic cells, monocytes and macrophages, mainly from nearby regions such as the choroid plexus, arrive from the cerebrospinal fluid in the M2 phenotype to promote neural repair and type M1 characterized by high expression of reactive oxygen species, which enters through the subarachnoid space and the leptomeninges [244].

Later, during the first and second week (7-14 days post lesion -dpl-) macrophages and mature dendritic cells migrate from the CNS and carry out antigenic presentation in the closest lymph nodes, allowing the arrival of the antigen-specific T-lymphocytes *via* the blood stream and lymphatics to the SC. They are the first cell types to exhibit specificity, diversity, and memory, maintaining high concentrations up to 28 dpl; however, they can present a second and third activation stage after 42 and 180 dpl when it is considered chronic stage [245]. Considering that the proinflammatory peripheral cytokines are capable of interacting with specific brain regions and promote behavioral alterations [246] the perpetuation of a chronic inflammatory phase has been recently elucidated as an organic factor for the development of psychiatric disorders following SCI [247, 248].

Furthermore, studies showed that depressive-like behavior after SCI was marked by microglial hyperactivity with M1-polarised phenotype in the hippocampus and cerebral cortex [249] where the severity of SCI positively correlated with the rodent's brain inflammation. In this way it can be seen how all the mechanisms involved in this phase lead to a cellular loss that conditions the characteristic demyelination of the SCI, thus explaining the neurodegeneration to which these patients are exposed [224]. The discovery of these molecular events has served for the innovation of approach techniques in patients, such as, neuronal assessment in the hippocampus, as a parameter of cytotoxicity and prognosis in patients [250].

Nowadays, research continues on more mechanisms of damage involved in the development of this pathology, with the purpose of continuing to provide even more diagnostic and therapeutic tools to injured patients [251]. Recently, SCI has also been associated with cognitive impairment. Most important cognitive complications are described below:

Cognitive Impairment After Spinal Cord Injury

Damage to the SC can be a devastating event, pioneer work in this area on cognitive feature health when SCI showed that up to 64% of injured people were cognitively impaired [252]. Several later studies have confirmed significant impairment in varied cognitive feature domains further because the presence of depression and anxiety [253 - 256], dementia [255, 257] and learning and memory impairment [258 - 260].

Depression and Anxiety

Depression and anxiety disorders and/or symptoms are frequently reported after SCI [254]. in accord with the incidence of depression, suicide risk and suicidal ideation following SCI is calculable to be three or additional times larger than within the general population [261]. The fatigue and lack of motivation associated with depression considerably cut back adherence to rehabilitation protocols and depression itself is related with negative long-term complications in recovery [255, 261]. In more than 40 percent, multiple studies have shown a predisposition of women with adverse effects in recovery processes such as loss of appetite, psychomotor anxiety, anhedonia and/or depressed mood and hopelessness may be part of the depressive and anxious neurological rehabilitation process scenario [262, 263]. In addition, in a study with female rats it was shown that the lesion promotes neuropsychiatric disorders such as depression and anxiety related to the imbalance in the production of pro and anti-inflammatory cytokines after SCI.

The gut brain axis is now known as a strong physiological regulator of mood and mental wellbeing. When the gut is actively colonized with distinct types of bacteria, it is possible to activate signaling networks in the brain that induce anxiety-like behaviors, whereas anxiety-like behaviors in germ-free mice are decreased compared to specific pathogen-free control mice [264]. Fecal transplantation in rats reduces gut dysbiosis and anxiety-like behavior following SCI [265]. A first study of adults with symptoms of depression or anxiety will be recruited from the Kingston area. They assessed metabolic changes, neurotransmitter levels, inflammatory markers, and the level of engraftment of the fecal samples that alleviate psychiatric symptoms. Finally, evaluate its safety and tolerability [266]. Further studies should be designed in order to provide more evidence about the efficacy of fecal transplantation.

Dementia

Dementia due to many causes is becoming common in the aging. Particularly, abnormal protein deposition has been shown to co-exist in dementia with injured neuro vasculature at different stages of the disease. The association between vascular dysfunction and dementia has been characterized several decades ago. Pathologists also recognized small blood vessels in the brain since the nineteenth century and reflected their association with hypoperfusion resulting in brain injury [257]. A cohort study in Taiwan establish that SCI patients have a significantly higher risk for developing dementia, providing the epidemiologic evidence that SCI will contribute to cognitive impairment [267]. Numerous epidemiological studies suggest the high prevalence of depression and dementia are increasingly recognized as serious secondary complications that may cut off recovery of patients [255, 257, 268]. The disturbances in adult neurogenesis that can be seen in dementia and learning and memory impairment.

Inflammation modulation, posttraumatic therapy is a potential mechanism for the high risk of dementia in patients with SCI [255]. Neuroinflammation and the neurodegeneration associated with it as the primary cellular feature of the microglia, an innate immune system in the CNS, plays a vital role in responding to CNS trauma. Microglia may generate neuroprotective factors, clear cellular debris and orchestrate response to injury with neurorestorative mechanisms which are useful for the neurology recovery. However, high levels of pro-inflammatory and cytotoxic mediators that impede CNS recovery may also be released by dysregulated microglia [269]. Rehabilitation studies have shown that the sensorimotor biomechanical effect (orientation-dependent regulation of reaction times) for mental rotation of foot images (absent in pre-physiotherapy). This illustrates that the representation of the body is adaptable to contingent situations, under which dependence on sensorimotor or visuospatial techniques can be altered and restored as a physiotherapy feature, at least partially [270]. Systemic administration after SCI of the selective cyclin-dependent kinase inhibitor CR8 significantly decreased the expression of cell cycle genes and proteins, microglial activation and brain neurodegeneration, cognitive impairment and depression. These studies show that a chronic brain neurodegenerative response can be triggered by SCI, possibly linked to delayed, prolonged induction of M1-type microglia and related activation of the cell cycle, resulting in cognitive deficits and physiological depression [251]. Other potent microglial activator, the cysteine-cysteine chemokine ligand 21 (CCL21), was elevated in the brain sites after SCI in association with increased microglial activation. These findings indicate that in major brain regions associated with cognitive impairment and physiological depression, SCI induces chronic neuroinflammation that leads to neuronal loss, impaired hippocampal neurogenesis and increased neuronal endoplasmic reticulum tension. CCL21 accumulation in the brain can play a pathophysiological role [254]. Therefore, more studies are needed to know about its microglial activation in SCI.

Learning and Memory Impairment

Learning refers to the processes used to encode new environmental stimuli and relations, whereas memory is seen as a mechanism that preserves the consequences of learning over time [272]. Hippocampal neurogenesis is related to hippocampal-dependent learning under physiological conditions, while deficiencies in adult hippocampal neurogenesis have been shown to associate with disturbances in spatial learning and memory. Initial demonstrations of habituation and sensitization in the SC provided fundamental evidence that the SC may learn from repetitive activity, and revealed a form of spinal memory that manifested behaviorally [260]. After SCI, neurogenesis around the injury site is reduced in chronic phase [273, 274], but this damage is not local level. In a study, it has been shown to reduce neurogenesis and induces glial reactivity in the hippocampus after SCI [258]. Additionally, a study revealed that SCI resulted in long-term decrease in the number of newly developed immature neurons in the hippocampal dentate gyrus, followed by evidence of higher stress in the neuronal endoplasmic reticulum. Also, by stereological analysis found that moderate and severe SCI in the cerebral cortex and hippocampus decreased neuronal survival and increased the amount of active microglia chronically [271]. Likewise, reduction of neurogenesis can be due to inactivation of neural stem cells and inhibition of amplifying neural progenitors proliferation at later points in time. Only in the chronic process did the number of granular cells and CA1 pyramidal neurons decrease. At the chronic stage, the release of pro-inflammatory cytokines can involve reducing neurogenesis and neurodegeneration of hippocampal neurons. SCI has also contributed to improvements in the hippocampus that may be involved in cognitive deficits found in rodents and humans [275]. Another work showed substantially poorer test output compared to healthy controls in people with paraplegia for new learning and memory testing. The group with tetraplegia, on the other hand showed substantially impaired performance on a processing speed task compared with healthy controls, and both the tetraplegia and paraplegia groups were equally impaired on a verbal fluency scale [276]. Authors suggest further investigations in learning and memory impairment.

CONCLUSION

The patients with MCI present one or more comorbidities mainly related to lifestyle; the mechanisms in almost all the pathologies described here, develop for long periods until they become evident. when cognitive changes are diagnosed early and there is an intervention on habits that can be modifiable, they are usually reversible. on the contrary, when the injuries are irreparable and the damage is more than notorious, few therapeutic proposals provide improvement, it is therefore of great importance that investigations continue so that the mechanisms leading to these damages are understood in greater depth and thus, provide other alternatives that improve the quality of life of this type of patients.

CONSENT FOR PUBLICATION

Not applicable.

CONFLICT OF INTEREST

The authors declare no conflict of interest, financial or otherwise.

ACKNOWLEDGEMENTS

None declared.

REFERENCES

[1]   World Alzheimer R. Improving healthcare for people living with dementia. 2016.

[2]   Sanford AM. Mild cognitive impairment. Clin geriatr med 2017; 33(3): 325-37.
       [http://dx.doi.org/10.1016/j.cger.2017.02.005] [PMID: 28689566]

[3]   Glynn K, O'Callaghan M, Hannigan O, Bruce I, Gibb M, Coen R, *et al.* Clinical utility of mild
       cognitive impairment subtypes and number of impaired cognitive domains at predicting progression to
       dementia: A 20-year retrospective study. Int J Geriatr Psychiatry 2021; 36(1): 31-7.
       [PMID: 32748438]

[4]   Cheng Y-W, Chen T-F, Chiu M-J. From mild cognitive impairment to subjective cognitive decline:
       conceptual and methodological evolution. Neuropsychiatr Dis Treat 2017; 13: 491-8.
       [http://dx.doi.org/10.2147/NDT.S123428] [PMID: 28243102]

[5]   Tangalos EG, Petersen RC. Mild cognitive impairment in geriatrics. Clin Geriatr Med 2018; 34(4):
       563-89.
       [http://dx.doi.org/10.1016/j.cger.2018.06.005] [PMID: 30336988]

[6]   Pinto TCC, Machado L, Bulgacov TM, *et al.* Is the Montreal Cognitive Assessment (MoCA) screening
       superior to the Mini-Mental State Examination (MMSE) in the detection of mild cognitive impairment
       (MCI) and Alzheimer's Disease (AD) in the elderly? Int Psychogeriatr 2019; 31(4): 491-504.
       [http://dx.doi.org/10.1017/S1041610218001370] [PMID: 30426911]

[7]   Morris JC. Revised criteria for mild cognitive impairment may compromise the diagnosis of
       Alzheimer disease dementia. Arch Neurol 2012; 69(6): 700-8.
       [http://dx.doi.org/10.1001/archneurol.2011.3152] [PMID: 22312163]

[8]   Custodio N, Herrera E, Lira D, Montesinos R, Linares J, Bendezú L, Eds. Deterioro cognitivo leve:¿
       dónde termina el envejecimiento normal y empieza la demencia?. 2012. UNMSM. Facultad de
       Medicina.

[9]   Winblad B, Palmer K, Kivipelto M, *et al.* Mild cognitive impairment--beyond controversies, towards a
       consensus: report of the International Working Group on Mild Cognitive Impairment. J Intern Med
       2004; 256(3): 240-6.
       [http://dx.doi.org/10.1111/j.1365-2796.2004.01380.x] [PMID: 15324367]

[10]   Niccolai E, Boem F, Russo E, Amedei A. The gut brain axis in the neuropsychological disease model
       of obesity: A classical movie revised by the emerging director "Microbiome". Nutrients 2019; 11(1):
       156.
       [http://dx.doi.org/10.3390/nu11010156] [PMID: 30642052]

[11]   Chooi YC, Ding C, Magkos F. The epidemiology of obesity. Metabolism 2019; 92: 6-10.
       [http://dx.doi.org/10.1016/j.metabol.2018.09.005] [PMID: 30253139]

[12]   Blüher M. Obesity: global epidemiology and pathogenesis. Nat Rev Endocrinol 2019; 15(5): 288-98.
       [http://dx.doi.org/10.1038/s41574-019-0176-8] [PMID: 30814686]

[13]   Salud OMdl. Obesidad y sobrepeso. Datos sobre el sobrepeso yla obesidad 2020. Available from
       .https://www.who.int/es/news-room/fact-sheets/detail/obesity-and-overweight.

[14]   Encuesta Nacional de Salud y N. Presentación de resultados. 2018.

[15]   Crovesy L, Masterson D, Rosado EL. Profile of the gut microbiota of adults with obesity: a systematic
       review. Eur J Clin Nutr 2020; 74(9): 1251-62.
       [http://dx.doi.org/10.1038/s41430-020-0607-6] [PMID: 32231226]

[16]   Castaner O, Goday A, Park Y-M, Lee S-H, Magkos F, Shiow S-ATE, *et al.* The gut microbiome
       profile in obesity: a systematic review Int J Endocrinol 2018; 2018; 4095789.
       [http://dx.doi.org/10.1155/2018/4095789]

[17]   Dye L, Boyle NB, Champ C, Lawton C. The relationship between obesity and cognitive health and
       decline. Proc Nutr Soc 2017; 76(4): 443-54.

[http://dx.doi.org/10.1017/S0029665117002014] [PMID: 28889822]

[18] Solas M, Milagro FI, Ramírez MJ, Martínez JA. Inflammation and gut-brain axis link obesity to cognitive dysfunction: plausible pharmacological interventions. Curr Opin Pharmacol 2017; 37: 87-92.
[http://dx.doi.org/10.1016/j.coph.2017.10.005] [PMID: 29107872]

[19] Flier JS, Maratos-Flier E. Harrison. Principios de Medicina Interna.Harrison Principios de Medicina Interna. 20th ed., New York: McGraw-Hill Education 2018.

[20] Reilly SM, Saltiel AR. Adapting to obesity with adipose tissue inflammation. Nat Rev Endocrinol 2017; 13(11): 633-43.
[http://dx.doi.org/10.1038/nrendo.2017.90] [PMID: 28799554]

[21] Asghar A, Sheikh N. Role of immune cells in obesity induced low grade inflammation and insulin resistance. Cell Immunol 2017; 315: 18-26.
[http://dx.doi.org/10.1016/j.cellimm.2017.03.001] [PMID: 28285710]

[22] Hersoug LG, Møller P, Loft S. Role of microbiota-derived lipopolysaccharide in adipose tissue inflammation, adipocyte size and pyroptosis during obesity. Nutr Res Rev 2018; 31(2): 153-63.
[http://dx.doi.org/10.1017/S0954422417000269] [PMID: 29362018]

[23] Engin AB. Adipocyte-macrophage cross-talk in obesity. Adv Exp Med Biol 2017; 960: 327-43.
[http://dx.doi.org/10.1007/978-3-319-48382-5_14] [PMID: 28585206]

[24] Miller AA, Spencer SJ. Obesity and neuroinflammation: a pathway to cognitive impairment. Brain Behav Immun 2014; 42: 10-21.
[http://dx.doi.org/10.1016/j.bbi.2014.04.001] [PMID: 24727365]

[25] Liu Y, Yu J, Shi Y-C, Zhang Y, Lin S. The role of inflammation and endoplasmic reticulum stress in obesity-related cognitive impairment. Life Sci 2019; 233: 116707.
[http://dx.doi.org/10.1016/j.lfs.2019.116707] [PMID: 31374234]

[26] Yang Y, Shields GS, Guo C, Liu Y. Executive function performance in obesity and overweight individuals: A meta-analysis and review. Neurosci Biobehav Rev 2018; 84: 225-44.
[http://dx.doi.org/10.1016/j.neubiorev.2017.11.020] [PMID: 29203421]

[27] Fisher AV. Selective sustained attention: a developmental foundation for cognition. Curr Opin Psychol 2019; 29: 248-53.
[http://dx.doi.org/10.1016/j.copsyc.2019.06.002] [PMID: 31284233]

[28] Hanć T, Cortese S. Attention deficit/hyperactivity-disorder and obesity: A review and model of current hypotheses explaining their comorbidity. Neurosci Biobehav Rev 2018; 92: 16-28.
[http://dx.doi.org/10.1016/j.neubiorev.2018.05.017] [PMID: 29772309]

[29] Edlow AG. Maternal obesity and neurodevelopmental and psychiatric disorders in offspring. Prenat Diagn 2017; 37(1): 95-110.
[http://dx.doi.org/10.1002/pd.4932] [PMID: 27684946]

[30] Balter LJT, Bosch JA, Aldred S, Drayson MT. 2019 Available from .https://www.proquest .com/docview /2307698908?accountid=41021&pq-origsite=primo

[31] Buie JJ, Watson LS, Smith CJ, Sims-Robinson C. Obesity-related cognitive impairment: The role of endothelial dysfunction. Neurobiol Dis 2019; 132: 104580.
[http://dx.doi.org/10.1016/j.nbd.2019.104580] [PMID: 31454547]

[32] Monthé-Drèze C, Rifas-Shiman SL, Gold DR, Oken E, Sen S. Maternal obesity and offspring cognition: the role of inflammation. Pediatr Res 2019; 85(6): 799-806.
[http://dx.doi.org/10.1038/s41390-018-0229-z] [PMID: 30420706]

[33] Kanoski SE, Davidson TL. Western diet consumption and cognitive impairment: links to hippocampal dysfunction and obesity. Physiol Behav 2011; 103(1): 59-68.
[http://dx.doi.org/10.1016/j.physbeh.2010.12.003] [PMID: 21167850]

[34] Morris MC, Evans DA, Tangney CC, *et al.* Dietary copper and high saturated and trans fat intakes

associated with cognitive decline. Arch Neurol 2006; 63(8): 1085-8.
[http://dx.doi.org/10.1001/archneur.63.8.1085] [PMID: 16908733]

[35]   Eskelinen MH, Ngandu T, Helkala EL, *et al.* Fat intake at midlife and cognitive impairment later in life: a population-based CAIDE study. Int J Geriatr Psychiatry 2008; 23(7): 741-7.
[http://dx.doi.org/10.1002/gps.1969]

[36]   Papanikolaou Y, Palmer H, Binns MA, Jenkins DJA, Greenwood CE. Better cognitive performance following a low-glycaemic-index compared with a high-glycaemic-index carbohydrate meal in adults with type 2 diabetes. Diabetologia 2006; 49(5): 855-62.
[http://dx.doi.org/10.1007/s00125-006-0183-x] [PMID: 16508776]

[37]   Kanoski SE, Meisel RL, Mullins AJ, Davidson TL. The effects of energy-rich diets on discrimination reversal learning and on BDNF in the hippocampus and prefrontal cortex of the rat. Behav Brain Res 2007; 182(1): 57-66.
[http://dx.doi.org/10.1016/j.bbr.2007.05.004] [PMID: 17590450]

[38]   Cifre M, Palou A, Oliver P. Cognitive impairment in metabolically-obese, normal-weight rats: identification of early biomarkers in peripheral blood mononuclear cells. Mol Neurodegener 2018; 13(1): 14.
[http://dx.doi.org/10.1186/s13024-018-0246-8] [PMID: 29566703]

[39]   Wan Y, Xu J, Meng F, *et al.* Cognitive decline following major surgery is associated with gliosis, β-amyloid accumulation, and τ phosphorylation in old mice. Crit Care Med 2010; 38(11): 2190-8.
[http://dx.doi.org/10.1097/CCM.0b013e3181f17bcb] [PMID: 20711073]

[40]   Singh M, Benencia F. Inflammatory processes in obesity: focus on endothelial dysfunction and the role of adipokines as inflammatory mediators. Int Rev Immunol 2019; 38(4): 157-71.
[http://dx.doi.org/10.1080/08830185.2019.1638921] [PMID: 31286783]

[41]   Zhao J, Bi W, Xiao S, *et al.* Neuroinflammation induced by lipopolysaccharide causes cognitive impairment in mice. Sci Rep 2019; 9(1): 5790.
[http://dx.doi.org/10.1038/s41598-019-42286-8] [PMID: 30962497]

[42]   Lozano R, Fullman N, Mumford JE, *et al.* Measuring universal health coverage based on an index of effective coverage of health services in 204 countries and territories, 1990–2019: a systematic analysis for the Global Burden of Disease Study 2019. The Lancet.

[43]   Martínez-Lapiscina EH, Clavero P, Toledo E, *et al.* Mediterranean diet improves cognition: the PREDIMED-NAVARRA randomised trial. J Neurol Neurosurg Psychiatry 2013; 84(12): 1318-25.
[http://dx.doi.org/10.1136/jnnp-2012-304792] [PMID: 23670794]

[44]   Loughrey DG, Lavecchia S, Brennan S, Lawlor BA, Kelly ME. The impact of the mediterranean diet on the cognitive functioning of healthy older adults: A systematic review and meta-analysis. Adv Nutr 2017; 8(4): 571-86.
[http://dx.doi.org/10.3945/an.117.015495] [PMID: 28710144]

[45]   Petersson SD, Philippou E. Mediterranean diet, cognitive function, and dementia: A systematic review of the evidence. Adv Nutr 2016; 7(5): 889-904 .
[http://dx.doi.org/10.3945/an.116.012138] [PMID: 27633105]

[46]   Angeloni C, Businaro R, Vauzour D. The role of diet in preventing and reducing cognitive decline. Curr Opin Psychiatry 2020; 33(4): 432-8.
[http://dx.doi.org/10.1097/YCO.0000000000000605] [PMID: 32149739]

[47]   Rodríguez JM, Leiva Balich L, Concha MJ, *et al.* Reduction of serum advanced glycation end-products with a low calorie Mediterranean diet. Nutr Hosp 2015; 31(6): 2511-7.
[http://dx.doi.org/10.3305/nh.2015.31.6.8936] [PMID: 26040359]

[48]   van den Brink AC, Brouwer-Brolsma EM, Berendsen AAM, van de Rest O. The Mediterranean, Dietary Approaches to Stop Hypertension (DASH), and Mediterranean-DASH Intervention for Neurodegenerative Delay (MIND) Diets Are Associated with Less Cognitive Decline and a Lower

Risk of Alzheimer's Disease-A Review. Adv Nutr 2019; 10(6): 1040-65.
[http://dx.doi.org/10.1093/advances/nmz054] [PMID: 31209456]

[49] Omar SH. Mediterranean and MIND diets containing olive biophenols reduces the prevalence of alzheimer's aisease. Int J Mol Sci 2019; 20(11): 2797.
[http://dx.doi.org/10.3390/ijms20112797] [PMID: 31181669]

[50] Valenzuela PL, Castillo-García A, Morales JS, *et al.* Exercise benefits on Alzheimer's disease: State-of-the-science. Ageing Res Rev 2020; 62: 101108.
[http://dx.doi.org/10.1016/j.arr.2020.101108] [PMID: 32561386]

[51] Scheffer DDL, Latini A. Exercise-induced immune system response: Anti-inflammatory status on peripheral and central organs. Biochim Biophys Acta Mol Basis Dis 2020 ; 1866(10): 165823.
[http://dx.doi.org/10.1016/j.bbadis.2020.165823] [PMID: 32360589]

[52] Di Liegro CM, Schiera G, Proia P, Di Liegro I. Physical activity and brain health. Genes (Basel) 2019; 10(9): 720.
[http://dx.doi.org/10.3390/genes10090720] [PMID: 31533339]

[53] Petersmann A, Müller-Wieland D, Müller UA, Landgraf R, Nauck M, Freckmann G, *et al.* Definition, classification and diagnosis of diabetes mellitus. Exp Clin Endocrinol Diabetes 2019; 127(S 01): S1-7.

[54] Federation ID. IDF Diabetes Atlas. International Diabetes Federation. 2019; Belgium: 2019; 1-176.Available from:https://www.diabetesatlas.org

[55] Organization WH. Classification of diabetes mellitus. World Health Organization 2019; p. 36.

[56] Fried PJ, Pascual-Leone A, Bolo NR. Diabetes and the link between neuroplasticity and glutamate in the aging human motor cortex. Clin Neurophysiol 2019; 130(9): 1502-10.
[http://dx.doi.org/10.1016/j.clinph.2019.04.721] [PMID: 31295719]

[57] Green H, Tsitsi P, Markaki I, Aarsland D, Svenningsson P. Novel treatment opportunities against cognitive impairment in parkinson's disease with an emphasis on diabetes-related pathways. CNS Drugs 2019; 33(2): 143-60.
[http://dx.doi.org/10.1007/s40263-018-0601-x] [PMID: 30687888]

[58] Federation ID. IDF Diabetes Atlas 2019.Available from:https://www.diabetesatlas.org

[59] Lo JW, Crawford JD, Desmond DW, *et al.* Profile of and risk factors for poststroke cognitive impairment in diverse ethnoregional groups. Neurology 2019; 93(24): e2257-71.
[http://dx.doi.org/10.1212/WNL.0000000000008612] [PMID: 31712368]

[60] Maamoun H, Abdelsalam SS, Zeidan A, Korashy HM, Agouni A. Endoplasmic reticulum stress: A critical molecular driver of endothelial dysfunction and cardiovascular disturbances associated with diabetes. Int J Mol Sci 2019; 20(7): 1658.
[http://dx.doi.org/10.3390/ijms20071658] [PMID: 30987118]

[61] Tomiyama H, Ishizu T, Kohro T, *et al.* Longitudinal association among endothelial function, arterial stiffness and subclinical organ damage in hypertension. Int J Cardiol 2018; 253: 161-6.
[http://dx.doi.org/10.1016/j.ijcard.2017.11.022] [PMID: 29174285]

[62] Zhang T, Wang D, Li X, *et al.* Excess salt intake promotes M1 microglia polarization *via* a p38/MAPK/AR-dependent pathway after cerebral ischemia in mice. Int Immunopharmacol 2020; 81: 106176.
[http://dx.doi.org/10.1016/j.intimp.2019.106176] [PMID: 32044667]

[63] Oliva J, Fernández-Bolaños A Fau-Hidalgo A, Hidalgo A. Health-related quality of life in diabetic people with different vascular risk. MC public health 2012; 12,812(1471-2458 (Electronic).)

[64] Vaidya V, Gangan N Fau - Sheehan J, Sheehan J. Impact of cardiovascular complications among patients with Type 2 diabetes mellitus: a systematic review Expert Rev Pharmacoecon Outcomes Res 2015; 15(3): 487-97.

[65] Pal KA-O, Mukadam N, Petersen I, Cooper C. Mild cognitive impairment and progression to dementia

in people with diabetes, prediabetes and metabolic syndrome: a systematic review and meta-analysis (1433-9285 (Electronic)).

[66]   Oliva J, Fernández-Bolaños A Fau-Hidalgo A, Hidalgo A. Health-related quality of life in diabetic people with different vascular risk. MC public health 2012; 12,812(1471-2458 (Electronic).)

[67]   Association AD. Economic Costs of Diabetes in the U.S. in 2017. Diabetes care 2018; 41(1935-5548 (Electronic).)

[68]   Skyler JS, Bakris GL, Bonifacio E, *et al.* Differentiation of diabetes by pathophysiology, natural history, and prognosis. Diabetes 2017; 66(2): 241-55.
        [http://dx.doi.org/10.2337/db16-0806] [PMID: 27980006]

[69]   Rodriguez-Calvo T, Suwandi JS, Amirian N, *et al.* Heterogeneity and lobularity of pancreatic pathology in type 1 diabetes during the prediabetic phase. J Histochem Cytochem 2015; 63(8): 626-36.

[70]   Saberzadeh-Ardestani B, Karamzadeh R, Basiri M, *et al.* Type 1 diabetes mellitus: cellular and molecular pathophysiology at a glance. Cell J 2018; 20(3): 294-301.
        [PMID: 29845781]

[71]   Knoop J, Gavrisan A, Kuehn D, *et al.* GM-CSF producing autoreactive CD4$^+$ T cells in type 1 diabetes. Clin Immunol 2018; 188: 23-30.
        [http://dx.doi.org/10.1016/j.clim.2017.12.002] [PMID: 29229565]

[72]   Zhao LP, Papadopoulos GK, Kwok WW, *et al.* Motifs of Three HLA-DQ Amino acid residues (α44, β57, β135) capture full association with the risk of type 1 diabetes in DQ2 and DQ8 children. Diabetes 2020; 69(7): 1573-87.
        [http://dx.doi.org/10.2337/db20-0075] [PMID: 32245799]

[73]   Bhatty A, Baig S, Fawwad A, Rubab ZE, Shahid MA, Waris N. Association of Zinc Transporter-8 Autoantibody (ZnT8A) with Type 1 diabetes mellitus. Cureus 2020; 12(3): e7263-e.

[74]   Barker JM, Barriga Kj Fau - Yu L, *et al.* Prediction of autoantibody positivity and progression to type 1 diabetes: Diabetes Autoimmunity Study in the Young (DAISY). J Clin Endocrinol Metab 2004; 89(8): 3896-902.

[75]   Singh G, Singh U, Singh SK, Singh S. Immunogenetic study of diabetes mellitus in relation to HLA DQ and DR. Indian J Endocrinol Metab 2020; 24(4): 325-32.
        [http://dx.doi.org/10.4103/ijem.IJEM_564_19] [PMID: 33088755]

[76]   Gao P, Uzun Y, He B, *et al.* Risk variants disrupting enhancers of T_H1 and T_{REG} cells in type 1 diabetes. Proc Natl Acad Sci USA 2019; 116(15): 7581-90.
        [http://dx.doi.org/10.1073/pnas.1815336116] [PMID: 30910956]

[77]   Zhao Y, Alard P, Kosiewicz MM. High thymic output of effector CD4$^+$ cells may lead to a Treg : T effector imbalance in the periphery in NOD Mice. J Immunol Res 2019; 2019: 8785263.
        [http://dx.doi.org/10.1155/2019/8785263] [PMID: 31281853]

[78]   Viisanen T, Gazali AM, Ihantola E-L, *et al.* FOXP3+ regulatory T cell compartment is altered in children with newly diagnosed type 1 diabetes but not in autoantibody-positive at-risk children. Front Immunol 2019; 10: 19.
        [http://dx.doi.org/10.3389/fimmu.2019.00019] [PMID: 30723474]

[79]   Kaur N, Minz RW, Bhadada SK, Dayal D, Singh J, Anand S. Deranged regulatory T-cells and transforming growth factor-β1 levels in type 1 diabetes patients with associated autoimmune diseases. J Postgrad Med 2017; 63(3): 176-81.
        [http://dx.doi.org/10.4103/jpgm.JPGM_608_16] [PMID: 28695870]

[80]   Holohan DR, Van Gool F, Bluestone JA. Thymically-derived Foxp3+ regulatory T cells are the primary regulators of type 1 diabetes in the non-obese diabetic mouse model. PLoS One 2019; 14(10): e0217728.
        [http://dx.doi.org/10.1371/journal.pone.0217728] [PMID: 31647813]

[81]   Hamari S, Kirveskoski T, Glumoff V, *et al.* Analyses of regulatory CD4+ CD25+ FOXP3+ T cells and observations from peripheral T cell subpopulation markers during the development of type 1 diabetes in children. Scandinavian J immunology 2016; 83: 279-87.

[82]   Marwaha AK, Panagiotopoulos C, Biggs CM, *et al.* Pre-diagnostic genotyping identifies T1D subjects with impaired Treg IL-2 signaling and an elevated proportion of FOXP3⁺IL-17⁺ cells. Genes Immun 2017; 18(1): 15-21.
[http://dx.doi.org/10.1038/gene.2016.44] [PMID: 28053319]

[83]   Sebastiani G, Ventriglia G, Stabilini A, *et al.* Regulatory T-cells from pancreatic lymphnodes of patients with type-1 diabetes express increased levels of microRNA miR-125a-5p that limits CCR2 expression. Sci Rep 2017; 7(1): 6897.
[http://dx.doi.org/10.1038/s41598-017-07172-1] [PMID: 28761107]

[84]   Dwyer CJ, Bayer AL, Fotino C, *et al.* Altered homeostasis and development of regulatory T cell subsets represent an IL-2R-dependent risk for diabetes in NOD mice. Sci Signal 2017; 10(510): eaam9563.
[http://dx.doi.org/10.1126/scisignal.aam9563] [PMID: 29259102]

[85]   Rasoul MA, Al-Mahdi M, Al-Kandari H, Dhaunsi GS, Haider MZ. Low serum vitamin-D status is associated with high prevalence and early onset of type-1 diabetes mellitus in Kuwaiti children. BMC Pediatr 2016; 16: 95.
[http://dx.doi.org/10.1186/s12887-016-0629-3] [PMID: 27422640]

[86]   Ismail MM, Abdel Hamid TA, Ibrahim AA, Marzouk H. Serum adipokines and vitamin D levels in patients with type 1 diabetes mellitus. Arch Med Sci 2017; 13(4): 738-44.
[http://dx.doi.org/10.5114/aoms.2016.60680] [PMID: 28721140]

[87]   Lietzen N, An LTT, Jaakkola MK, *et al.* Enterovirus-associated changes in blood transcriptomic profiles of children with genetic susceptibility to type 1 diabetes. Diabetologia 2018; 61(2): 381-8.
[http://dx.doi.org/10.1007/s00125-017-4460-7] [PMID: 29119244]

[88]   Ifie E, Russell MA, Dhayal S, *et al.* Unexpected subcellular distribution of a specific isoform of the Coxsackie and adenovirus receptor, CAR-SIV, in human pancreatic beta cells. Diabetologia 2018; 61(11): 2344-55.
[http://dx.doi.org/10.1007/s00125-018-4704-1] [PMID: 30074059]

[89]   Brown K, Godovannyi A, Ma C, *et al.* Prolonged antibiotic treatment induces a diabetogenic intestinal microbiome that accelerates diabetes in NOD mice. ISME J 2016; 10(2): 321-32.
[http://dx.doi.org/10.1038/ismej.2015.114] [PMID: 26274050]

[90]   Vatanen T, Franzosa EA, Schwager R, *et al.* The human gut microbiome in early-onset type 1 diabetes from the TEDDY study. Nature 2018; 562(7728): 589-94.
[http://dx.doi.org/10.1038/s41586-018-0620-2] [PMID: 30356183]

[91]   Roden M, Shulman GI. The integrative biology of type 2 diabetes. Nature 2019; 576(7785): 51-60.
[http://dx.doi.org/10.1038/s41586-019-1797-8] [PMID: 31802013]

[92]   Muhammad SA, Raza W, Nguyen T, Bai B, Wu X, Chen J. Cellular signaling pathways in insulin resistance-systems biology analyses of microarray dataset reveals new drug target gene signatures of type 2 diabetes mellitus. Front Physiol 2017; 8: 13.
[http://dx.doi.org/10.3389/fphys.2017.00013] [PMID: 28179884]

[93]   Smith RM, Harada S, Jarett L. Insulin internalization and other signaling pathways in the pleiotropic effects of insulin. Int Rev Cytol 1997; 173: 243-80.
[http://dx.doi.org/10.1016/S0074-7696(08)62479-1] [PMID: 9127955]

[94]   Spencer B, Rank L, Metcalf J, Desplats P. Identification of insulin receptor splice variant B in neurons by *in situ* detection in human brain samples. Sci Rep 2018; 8(1): 4070.
[http://dx.doi.org/10.1038/s41598-018-22434-2] [PMID: 29511314]

[95]   Hölscher C. Brain insulin resistance: role in neurodegenerative disease and potential for targeting.

Expert Opin Investig Drugs 2020; 29(4): 333-48.
[http://dx.doi.org/10.1080/13543784.2020.1738383] [PMID: 32175781]

[96]   Choi E, Kikuchi S, Gao H, *et al.* Mitotic regulators and the SHP2-MAPK pathway promote IR endocytosis and feedback regulation of insulin signaling. Nat Commun 2019; 10(1): 1473.
[http://dx.doi.org/10.1038/s41467-019-09318-3] [PMID: 30931927]

[97]   Haeusler RA, McGraw TE, Accili D. Biochemical and cellular properties of insulin receptor signalling. Nat Rev Mol Cell Biol 2018; 19(1): 31-44.
[http://dx.doi.org/10.1038/nrm.2017.89] [PMID: 28974775]

[98]   Samuel VT, Shulman GI. The pathogenesis of insulin resistance: integrating signaling pathways and substrate flux. J Clin Invest 2016; 126(1): 12-22.
[http://dx.doi.org/10.1172/JCI77812] [PMID: 26727229]

[99]   Ma Q, Li Y, Wang M, *et al.* Progress in metabonomics of type 2 diabetes mellitus. Molecules 2018; 23(7): 1834.
[http://dx.doi.org/10.3390/molecules23071834] [PMID: 30041493]

[100] Cerf ME. High fat programming of beta cell compensation, exhaustion, death and dysfunction. Pediatr Diabetes 2015; 16(2): 71-8.
[http://dx.doi.org/10.1111/pedi.12137] [PMID: 25682938]

[101] Paschen M, Moede T, Valladolid-Acebes I, *et al.* Diet-induced β-cell insulin resistance results in reversible loss of functional β-cell mass. FASEB J 2019; 33(1): 204-18.
[http://dx.doi.org/10.1096/fj.201800826R] [PMID: 29957055]

[102] Tomita T. Apoptosis in pancreatic β-islet cells in Type 2 diabetes. Bosn J Basic Med Sci 2016; 16(3): 162-79.
[http://dx.doi.org/10.17305/bjbms.2016.919] [PMID: 27209071]

[103] McLaughlin T, Craig C, Liu LF, *et al.* Adipose cell size and regional fat deposition as predictors of metabolic response to overfeeding in insulin-resistant and insulin-sensitive humans. Diabetes 2016; 65(5): 1245-54.
[http://dx.doi.org/10.2337/db15-1213] [PMID: 26884438]

[104] Makki K, Froguel P, Wolowczuk I. Adipose tissue in obesity-related inflammation and insulin resistance: cells, cytokines, and chemokines. ISRN Inflamm 2013; 2013: 139239.
[http://dx.doi.org/10.1155/2013/139239] [PMID: 24455420]

[105] Petersen MC, Shulman GI. Mechanisms of insulin action and insulin resistance. Physiol Rev 2018; 98(4): 2133-223.
[http://dx.doi.org/10.1152/physrev.00063.2017] [PMID: 30067154]

[106] Kain V, Kapadia B, Viswakarma N, *et al.* Co-activator binding protein PIMT mediates TNF-α induced insulin resistance in skeletal muscle *via* the transcriptional down-regulation of MEF2A and GLUT4. Sci Rep 2015; 5: 15197.
[http://dx.doi.org/10.1038/srep15197] [PMID: 26468734]

[107] Dou L, Zhao T, Wang L, *et al.* miR-200s contribute to interleukin-6 (IL-6)-induced insulin resistance in hepatocytes. J Biol Chem 2013; 288(31): 22596-606.
[http://dx.doi.org/10.1074/jbc.M112.423145] [PMID: 23798681]

[108] Zahniser NR, Goens MB, P J Hanaway PJ, Vinych JV. Characterization and regulation of insulin receptors in rat brain. J Neurochem 1984; 42: 1354-62.

[109] Salameh TS, Bullock KM, Hujoel IA, *et al.* Central nervous system delivery of intranasal insulin: mechanisms of uptake and effects on cognition. J Alzheimers Dis 2015; 47(3): 715-28.
[http://dx.doi.org/10.3233/JAD-150307] [PMID: 26401706]

[110] Moller DE, Yokota A, Caro JF, Flier JS. Tissue-specific expression of two alternatively spliced insulin receptor mRNAs in man. Mol Endocrinol 1989; 3(8): 1263-9.
[http://dx.doi.org/10.1210/mend-3-8-1263] [PMID: 2779582]

[111] Garwood CJ, Ratcliffe LE, Morgan SV, *et al.* Insulin and IGF1 signalling pathways in human astrocytes *in vitro* and *in vivo;* characterisation, subcellular localisation and modulation of the receptors. Mol Brain 2015; 8: 51.
[http://dx.doi.org/10.1186/s13041-015-0138-6] [PMID: 26297026]

[112] Spencer B, Rank L, Metcalf J, Desplats P. Identification of insulin receptor splice variant B in neurons by *in situ* detection in human brain samples. Scientific Reports 2018; (2045-2322 (Electronic)):

[113] Soto M, Cai W, Konishi M, Kahn CR. Insulin signaling in the hippocampus and amygdala regulates metabolism and neurobehavior. Proc Natl Acad Sci USA 2019; 116(13): 6379-84.
[http://dx.doi.org/10.1073/pnas.1817391116] [PMID: 30765523]

[114] Rhea EM, Rask-Madsen C, Banks WA. Insulin transport across the blood-brain barrier can occur independently of the insulin receptor. J Physiol 2018; 596(19): 4753-65.
[http://dx.doi.org/10.1113/JP276149] [PMID: 30044494]

[115] Kuwabara T, Kagalwala MN, Onuma Y, *et al.* Insulin biosynthesis in neuronal progenitors derived from adult hippocampus and the olfactory bulb. EMBO Mol Med 2011; 3: 742-54.

[116] Spinelli M, Fusco S, Grassi C. Brain insulin resistance and hippocampal plasticity: mechanisms and biomarkers of cognitive decline. Front Neurosci 2019; 13: 788.
[http://dx.doi.org/10.3389/fnins.2019.00788] [PMID: 31417349]

[117] Marino JS, Xu Y, Hill JW. Central insulin and leptin-mediated autonomic control of glucose homeostasis. Trends Endocrinol Metab 2011; 22(7): 275-85.
[http://dx.doi.org/10.1016/j.tem.2011.03.001] [PMID: 21489811]

[118] Pearson-Leary J, McNay EC. Novel roles for the insulin-regulated glucose transporter-4 in hippocampally dependent memory. J Neurosci 2016; 36(47): 11851-64.
[http://dx.doi.org/10.1523/JNEUROSCI.1700-16.2016] [PMID: 27881773]

[119] Zhao F, Siu JJ, Huang W, Askwith C, Cao L. Insulin modulates excitatory synaptic transmission and synaptic plasticity in the mouse hippocampus. Neuroscience 2019; 411: 237-54.
[http://dx.doi.org/10.1016/j.neuroscience.2019.05.033] [PMID: 31146008]

[120] Korol SV, Tafreshiha A, Bhandage AK, Birnir B, Jin Z. Insulin enhances GABA(A) receptor-mediated inhibitory currents in rat central amygdala neurons. Neuroscience Letters 2018; 671(1872-7972): 76-81.

[121] McNay EC, Recknagel AK. Brain insulin signaling: a key component of cognitive processes and a potential basis for cognitive impairment in type 2 diabetes. Neurobiol Learn Mem 2011; 96(3): 432-42.
[http://dx.doi.org/10.1016/j.nlm.2011.08.005] [PMID: 21907815]

[122] Zhang J, Chen C, Hua S, *et al.* An updated meta-analysis of cohort studies: Diabetes and risk of Alzheimer's disease. Diabetes Res Clin Pract 2017; 124: 41-7.
[http://dx.doi.org/10.1016/j.diabres.2016.10.024] [PMID: 28088029]

[123] Li W, Sun L, Li G, Xiao S. Prevalence, influence factors and cognitive characteristics of mild cognitive impairment in type 2 diabetes mellitus. Front Aging Neurosci 2019; 11: 180.
[http://dx.doi.org/10.3389/fnagi.2019.00180] [PMID: 31417393]

[124] Albai O, Frandes M, Timar R, Roman D, Timar B. Risk factors for developing dementia in type 2 diabetes mellitus patients with mild cognitive impairment. Neuropsychiatr Dis Treat 2019; 15: 167-75.
[http://dx.doi.org/10.2147/NDT.S189905] [PMID: 30655669]

[125] Xia S-S, Xia W-L, Huang J-J, Zou H-J, Tao J, Yang Y. The factors contributing to cognitive dysfunction in type 2 diabetic patients. Ann Transl Med 2020; 8(4): 104.
[http://dx.doi.org/10.21037/atm.2019.12.113] [PMID: 32175397]

[126] Xiu S, Liao Q, Sun L, Chan P. Risk factors for cognitive impairment in older people with diabetes: a community-based study. Ther Adv Endocrinol Metab 2019; 10: 2042018819836640.

[http://dx.doi.org/10.1177/2042018819836640] [PMID: 31156800]

[127]   Matsubara M, Makino H, Washida K, *et al.* A prospective longitudinal study on the relationship between glucose fluctuation and cognitive function in type 2 diabetes: PROPOSAL study protocol. Diabetes Ther 2020; 11(11): 2729-37.
[http://dx.doi.org/10.1007/s13300-020-00916-9] [PMID: 32889699]

[128]   Tuligenga RH, Dugravot A, Tabák AG, *et al.* Midlife type 2 diabetes and poor glycaemic control as risk factors for cognitive decline in early old age: a post-hoc analysis of the Whitehall II cohort study. Lancet Diabetes Endocrinol 2014; 2(3): 228-35.
[http://dx.doi.org/10.1016/S2213-8587(13)70192-X] [PMID: 24622753]

[129]   Petersen RC. Mild cognitive impairment as a diagnostic entity. J Intern Med 2004; 256(3): 183-94.
[http://dx.doi.org/10.1111/j.1365-2796.2004.01388.x] [PMID: 15324362]

[130]   Roberts RO, Knopman DS, Geda YE, *et al.* Association of diabetes with amnestic and nonamnestic mild cognitive impairment. Alzheimers Dement 2014; 10(1): 18-26.
[http://dx.doi.org/10.1016/j.jalz.2013.01.001] [PMID: 23562428]

[131]   Liu Z, Liu J, Yuan H, *et al.* Identification of cognitive dysfunction in patients with T2DM using whole brain functional connectivity. Genomics Proteomics Bioinformatics 2019; 17(4): 441-52.
[http://dx.doi.org/10.1016/j.gpb.2019.09.002] [PMID: 31786312]

[132]   Valenza S, Paciaroni L, Paolini S, *et al.* Mild cognitive impairment subtypes and type 2 diabetes in elderly subjects. J Clin Med 2020; 9(7): 2055.
[http://dx.doi.org/10.3390/jcm9072055] [PMID: 32629878]

[133]   Zhao Q, Zhang Y, Liao X, Wang W. Executive function and diabetes: A clinical neuropsychology perspective. Front Psychol 2020; 11: 2112.
[http://dx.doi.org/10.3389/fpsyg.2020.02112] [PMID: 32973635]

[134]   Moran C, Beare R, Wang W, Callisaya M, Srikanth V. Type 2 diabetes mellitus, brain atrophy, and cognitive decline. Neurology 2019; 92(8): e823-30.
[http://dx.doi.org/10.1212/WNL.0000000000006955] [PMID: 30674592]

[135]   Groeneveld O, Reijmer Y, Heinen R, *et al.* Brain imaging correlates of mild cognitive impairment and early dementia in patients with type 2 diabetes mellitus. Nutr Metab Cardiovasc Dis 2018; 28(12): 1253-60.
[http://dx.doi.org/10.1016/j.numecd.2018.07.008] [PMID: 30355471]

[136]   Tan X, Fang P, An J, *et al.* Micro-structural white matter abnormalities in type 2 diabetic patients: a DTI study using TBSS analysis. Neuroradiology 2016; 58(12): 1209-16.
[http://dx.doi.org/10.1007/s00234-016-1752-4] [PMID: 27783100]

[137]   Gao S, Chen Y, Sang F, *et al.* White matter microstructural change contributes to worse cognitive function in patients with type 2 diabetes. Diabetes 2019; 68(11): 2085-94.
[http://dx.doi.org/10.2337/db19-0233] [PMID: 31439643]

[138]   Liu Z, Patil IY, Jiang T, *et al.* High-fat diet induces hepatic insulin resistance and impairment of synaptic plasticity. PloS one 2015; 10(5): e0128274-e.
[http://dx.doi.org/10.1371/journal.pone.0128274]

[139]   Lyn-Cook LE Jr, Lawton M, Tong M, *et al.* Hepatic ceramide may mediate brain insulin resistance and neurodegeneration in type 2 diabetes and non-alcoholic steatohepatitis. J Alzheimers Dis 2009; 16(4): 715-29.
[http://dx.doi.org/10.3233/JAD-2009-0984] [PMID: 19387108]

[140]   Peng Y, Liu J, Shi L, *et al.* Mitochondrial dysfunction precedes depression of AMPK/AKT signaling in insulin resistance induced by high glucose in primary cortical neurons. J Neurochem 2016; 137(5): 701-13.
[http://dx.doi.org/10.1111/jnc.13563] [PMID: 26926143]

[141]   de la Monte SM. Brain insulin resistance and deficiency as therapeutic targets in Alzheimer's disease.

Curr Alzheimer Res 2012; 9(1): 35-66.
[http://dx.doi.org/10.2174/156720512799015037] [PMID: 22329651]

[142] Berlanga-Acosta J, Guillén-Nieto G, Rodríguez-Rodríguez N, *et al.* Insulin resistance at the crossroad of alzheimer disease pathology: A review. Front Endocrinol (Lausanne) 2020; 11: 560375.
[http://dx.doi.org/10.3389/fendo.2020.560375] [PMID: 33224105]

[143] Muriach M, Flores-Bellver M, Romero FJ, Barcia JM. Diabetes and the brain: oxidative stress, inflammation, and autophagy. Oxid Med Cell Longev 2014; 2014: 102158..
[http://dx.doi.org/10.1155/2014/102158] [PMID: 25215171]

[144] Zheng M, Chang B, Tian L, *et al.* Relationship between inflammatory markers and mild cognitive impairment in Chinese patients with type 2 diabetes: a case-control study. BMC Endocr Disord 2019; 19(1): 73.
[http://dx.doi.org/10.1186/s12902-019-0402-3] [PMID: 31296192]

[145] Ma S, Li S, Lv R, Hou X, Nie S, Yin Q. Prevalence of mild cognitive impairment in type 2 diabetes mellitus is associated with serum galectin-3 level. J Diabetes Investig 2020; 11(5): 1295-302.
[http://dx.doi.org/10.1111/jdi.13256] [PMID: 32196999]

[146] Takechi R, Lam V, Brook E, *et al.* Blood-brain barrier dysfunction precedes cognitive decline and neurodegeneration in diabetic insulin resistant mouse model: An implication for causal link. Front Aging Neurosci 2017; 9: 399.
[http://dx.doi.org/10.3389/fnagi.2017.00399] [PMID: 29249964]

[147] Wanrooy BJ, Kumar KP, Wen SW, Qin CX, Ritchie RH, Wong CHY. Distinct contributions of hyperglycemia and high-fat feeding in metabolic syndrome-induced neuroinflammation. J Neuroinflammation 2018; 15(1): 293.
[http://dx.doi.org/10.1186/s12974-018-1329-8] [PMID: 30348168]

[148] Fujimaki S, Kuwabara T. Diabetes-induced dysfunction of mitochondria and stem cells in skeletal muscle and the nervous system. Int J Mol Sci 2017; 18(10): 2147.
[http://dx.doi.org/10.3390/ijms18102147] [PMID: 29036909]

[149] Bonds JA, Shetti A, Stephen TKL, Bonini MG, Minshall RD, Lazarov O. Deficits in hippocampal neurogenesis in obesity-dependent and -independent type-2 diabetes mellitus mouse models. Sci Rep 2020; 10(1): 16368.
[http://dx.doi.org/10.1038/s41598-020-73401-9] [PMID: 33004912]

[150] Díaz CEM, Salazar JV. M.D., Anchundia AA, M.D. MA, M.D., Altamirano ZM, M.D., Molina JJ, M.D. Metformina: más allá de la diabetes mellitus. Arch Venez Farmacol Ter 2020; 39(4): 272-6.

[151] Forlenza OVRM, Radanovic M, Talib LL, Gattaz WF. Clinical and biological effects of long-term lithium treatment in older adults with amnestic mild cognitive impairment: randomised clinical trial. Br J Psychiatry 2019; 215(5): 668-74.
[http://dx.doi.org/10.1192/bjp.2019.76] [PMID: 30947755]

[152] Crowe-White KA-O, Phillips TA, Ellis AC. Lycopene and cognitive function. J nutritional sci 2019; 8: 2048-6790. (Print)
[http://dx.doi.org/10.1017/jns.2019.16]

[153] Yin Q, Ma Y, Hong Y, *et al.* Lycopene attenuates insulin signaling deficits, oxidative stress, neuroinflammation, and cognitive impairment in fructose-drinking insulin resistant rats. Neuropharmacology 2014; 86: 389-96.
[http://dx.doi.org/10.1016/j.neuropharm.2014.07.020] [PMID: 25110828]

[154] Bayunova LV, Zorina II, Zakharova IO, Avrova NF. Insulin increases viability of neurons in rat cerebral cortex and normalizes Bax/Bcl-2 ratio under conditions of oxidative stress. Bull Exp Biol Med 2018; 165(1): 14-7.
[http://dx.doi.org/10.1007/s10517-018-4088-8] [PMID: 29797135]

[155] WHO. Cardiovascular Diseases 2020. Available from:https://www.who.int/health-topics/cardio-

vascular-diseases/#tab=tab_1.

[156] Virani SS, Alonso A, Benjamin EJ, *et al.* Heart Disease and Stroke Statistics-2020 Update: A Report From the American Heart Association. Circulation 141: e139-e1596.

[157] Roth GA, Johnson C, Abajobir A, *et al.* Global, regional, and national burden of cardiovascular diseases for 10 causes, 1990 to 2015. J Am Coll Cardiol 2017; 70(1): 1-25.
[http://dx.doi.org/10.1016/j.jacc.2017.04.052] [PMID: 28527533]

[158] Benjamin EJMP, Muntner P, Alonso A, *et al.* Heart disease and stroke statistics - 2019 update: a report from the American Heart Association. Circulation 2019; 139(10): e56-e528.
[http://dx.doi.org/10.1161/CIR.0000000000000659] [PMID: 30700139]

[159] Organization WH. 2020. Available from:https://www.who.int/cardiovascular_diseases/priorities/es/

[160] Frostegård J. Immunity, atherosclerosis and cardiovascular disease. BMC Med 2013; 11: 117.
[http://dx.doi.org/10.1186/1741-7015-11-117] [PMID: 23635324]

[161] Shah PK, Lecis D. Inflammation in atherosclerotic cardiovascular disease. F1000Res 2019; 8 F1000 Faculty Rev-1402.

[162] Sandesara PB, Virani SS, Fazio S, Shapiro MD. The forgotten lipids: triglycerides, remnant cholesterol, and atherosclerotic cardiovascular disease risk. Endocr Rev 2019; 40(2): 537-57.
[http://dx.doi.org/10.1210/er.2018-00184] [PMID: 30312399]

[163] Pentikäinen MO, Oörni K Fau - Ala-Korpela M, Ala-Korpela M Fau - Kovanen PT, Kovanen PT. Modified LDL - trigger of atherosclerosis and inflammation in the arterial intima. J Intern Med 2000 Mar; 247(3): 359-70.

[164] Baumer Y, McCurdy S, Weatherby TM, *et al.* Hyperlipidemia-induced cholesterol crystal production by endothelial cells promotes atherogenesis. Nat Commun 2017; 8(1): 1129.
[http://dx.doi.org/10.1038/s41467-017-01186-z] [PMID: 29066718]

[165] Marchio P, Guerra-Ojeda S, Vila JM, Aldasoro M, Victor VM, Mauricio MD. Targeting early atherosclerosis: A focus on oxidative stress and inflammation. Oxid Med Cell Longev 2019; 2019: 8563845.
[http://dx.doi.org/10.1155/2019/8563845] [PMID: 31354915]

[166] Bobryshev YV, Ivanova EA, Chistiakov DA, Nikiforov NG, Orekhov AN. Macrophages and their role in atherosclerosis: pathophysiology and transcriptome analysis. BioMed Res Int 2016; 2016: 9582430.
[http://dx.doi.org/10.1155/2016/9582430] [PMID: 27493969]

[167] De Paoli F, Staels B, Chinetti-Gbaguidi G. Macrophage phenotypes and their modulation in atherosclerosis. Circ J 2014; 78(8): 1775-81.
[http://dx.doi.org/10.1253/circj.CJ-14-0621] [PMID: 24998279]

[168] Bäck M, Yurdagul A Jr, Tabas I, Öörni K, Kovanen PT. Inflammation and its resolution in atherosclerosis: mediators and therapeutic opportunities. Nat Rev Cardiol 2019; 16(7): 389-406.
[http://dx.doi.org/10.1038/s41569-019-0169-2] [PMID: 30846875]

[169] Koupenova M, Clancy L, Corkrey HA, Freedman JE. Circulating platelets as mediators of immunity, inflammation, and thrombosis. Circ Res 2018; 122(2): 337-51.
[http://dx.doi.org/10.1161/CIRCRESAHA.117.310795] [PMID: 29348254]

[170] Barrett TJ, Schlegel M, Zhou F, *et al.* Platelet regulation of myeloid suppressor of cytokine signaling 3 accelerates atherosclerosis. Sci Transl Med 2019; 11(517): eaax0481.
[http://dx.doi.org/10.1126/scitranslmed.aax0481] [PMID: 31694925]

[171] Finney AC, Stokes KY, Pattillo CB, Orr AW. Integrin signaling in atherosclerosis. Cell Mol Life Sci 2017; 74(12): 2263-82.
[http://dx.doi.org/10.1007/s00018-017-2490-4] [PMID: 28246700]

[172] Konukoglu D, Uzun H. Endothelial dysfunction and hypertension. Adv Exp Med Biol 2017; 956: 511-40.

[173] Yamagata K, Yamori Y. Inhibition of endothelial dysfunction by dietary flavonoids and preventive effects against cardiovascular disease. J Cardiovasc Pharmacol 2020; 75(1): 1-9.

[174] Chrysohoou C, Kollia N, Tousoulis D. The link between depression and atherosclerosis through the pathways of inflammation and endothelium dysfunction. Maturitas 2018; 109: 1-5.

[175] Kwaifa IK, Bahari H, Yong YK, Noor SM. Endothelial dysfunction in obesity-induced inflammation: molecular mechanisms and clinical implications. Biomolecules 2020; 10(2): 291.
[http://dx.doi.org/10.3390/biom10020291] [PMID: 32069832]

[176] Sun H-J, Wu Z-Y, Nie X-W, Bian J-S. Role of endothelial dysfunction in cardiovascular diseases: The link between inflammation and hydrogen sulfide. Front Pharmacol 2020; 10: 1568.
[http://dx.doi.org/10.3389/fphar.2019.01568] [PMID: 32038245]

[177] Al Jasmi F, Al Zaabi N, Al-Thihli K, Al Teneiji AM, Hertecant J, El-Hattab AW. Endothelial dysfunction and the effect of arginine and citrulline Supplementation in Children and Adolescents With Mitochondrial Diseases. J Cent Nerv Syst Dis 2020; 12: 1179573520909377.
[http://dx.doi.org/10.1177/1179573520909377] [PMID: 32165851]

[178] Oparil S, Acelajado MC, Bakris GL, *et al.* Hypertension. Nat Rev Dis Primers 2018; 4: 18014.
[http://dx.doi.org/10.1038/nrdp.2018.14] [PMID: 29565029]

[179] Luo D, Cheng Y, Zhang H, *et al.* Association between high blood pressure and long term cardiovascular events in young adults: systematic review and meta-analysis. BMJ. 2020; 370: p. m3222.

[180] Grillo A, Salvi L, Coruzzi P, Salvi P, Parati G. Sodium Intake and Hypertension. Nutrients 2019; 11(9): 1970.
[http://dx.doi.org/10.3390/nu11091970] [PMID: 31438636]

[181] Rossier BA-O, Bochud M, Devuyst O. The hypertension pandemic: An evolutionary perspective. Physiology (Bethesda) 2017; 32(2): 112-25.

[182] Tanaka M, Itoh H. Hypertension as a metabolic disorder and the novel role of the gut. Curr Hypertens Rep 2019; 21(8): 63.
[http://dx.doi.org/10.1007/s11906-019-0964-5] [PMID: 31236708]

[183] Jiang P, Ma D, Wang X, *et al.* Astragaloside IV prevents obesity-associated hypertension by improving pro-inflammatory reaction and leptin resistance. Mol Cells 2018; 41(3): 244-55.
[PMID: 29562733]

[184] Hall JE, do Carmo JM, da Silva AA, Wang Z, Hall ME. Obesity, kidney dysfunction and hypertension: mechanistic links. Nat Rev Nephrol 2019; 15(6): 367-85.
[http://dx.doi.org/10.1038/s41581-019-0145-4] [PMID: 31015582]

[185] Ormazabal V, Nair S, Elfeky O, Aguayo C, Salomon C, Zuñiga FA. Association between insulin resistance and the development of cardiovascular disease. Cardiovasc Diabetol 2018; 17(1): 122.
[http://dx.doi.org/10.1186/s12933-018-0762-4] [PMID: 30170598]

[186] Santisteban MM, Iadecola C. Hypertension, dietary salt and cognitive impairment. J Cereb Blood Flow Metab 2018; 38(12): 2112-28.
[http://dx.doi.org/10.1177/0271678X18803374] [PMID: 30295560]

[187] Kalaria RN, Akinyemi R, Ihara M. Stroke injury, cognitive impairment and vascular dementia. Biochim Biophys Acta 2016; 1862(5): 915-25.
[http://dx.doi.org/10.1016/j.bbadis.2016.01.015] [PMID: 26806700]

[188] Anrather J, Iadecola C. Inflammation and stroke: An overview. Neurotherapeutics 2016; 13(4): 661-70.
[http://dx.doi.org/10.1007/s13311-016-0483-x] [PMID: 27730544]

[189] Henriksen O, Gideon P, Sperling B, Olsen TS, Jørgensen HS, Arlien-Søborg P. Cerebral lactate production and blood flow in acute stroke. J Magn Reson Imaging 1992; 2(5): 511-7.

[http://dx.doi.org/10.1002/jmri.1880020508] [PMID: 1392243]

[190]   Hofmeijer J, van Putten MJ. Ischemic cerebral damage: an appraisal of synaptic failure. Stroke . 2012; 43: pp. (2)07-15.
[http://dx.doi.org/10.1161/STROKEAHA.111.632943]

[191]   Cross JL, Meloni BP, Bakker AJ, Lee S, Knuckey NW. Modes of neuronal calcium entry and homeostasis following cerebral ischemia. Stroke Res Treat 2010; 2010: 316862.
[http://dx.doi.org/10.4061/2010/316862] [PMID: 21052549]

[192]   Kim E, Cho S. Microglia and monocyte-derived macrophages in stroke. Neurotherapeutics 2016; 13(4): 702-18.
[http://dx.doi.org/10.1007/s13311-016-0463-1] [PMID: 27485238]

[193]   Kanazawa M, Ninomiya I, Hatakeyama M, Takahashi T, Shimohata T. Microglia and monocytes/macrophages polarization reveal novel therapeutic mechanism against stroke. Int J Mol Sci 2017; 18(10): E2135.
[http://dx.doi.org/10.3390/ijms18102135] [PMID: 29027964]

[194]   Vidale S, Consoli A, Arnaboldi M, Consoli D. Postischemic inflammation in acute stroke. J Clin Neurol 2017; 13(1): 1-9.
[http://dx.doi.org/10.3988/jcn.2017.13.1.1] [PMID: 28079313]

[195]   Castillo J, Alvarez-Sabin J, Dávalos A, *et al.* Consensus review. Pharmacological neuroprotection in cerebral ischemia: is it still a therapeutic option?. Neurologia18 Spain . 2003; pp. 368-84.

[196]   Iadecola C, Yaffe K, Biller J, *et al.* Impact of hypertension on cognitive function: A scientific statement from the american heart association. Hypertension (Dallas, Tex : 1979) 2016; 68(6): e67-94.
[http://dx.doi.org/10.1161/HYP.0000000000000053]

[197]   Tadic M, Cuspidi C, Hering D. Hypertension and cognitive dysfunction in elderly: blood pressure management for this global burden. BMC Cardiovasc Disord 2016; 16(1): 208.
[http://dx.doi.org/10.1186/s12872-016-0386-0] [PMID: 27809779]

[198]   Aronow WS. Hypertension and cognitive impairment. Ann Transl Med 2017; 5(12): 259.
[http://dx.doi.org/10.21037/atm.2017.03.99] [PMID: 28706927]

[199]   Sarfo FS, Akassi J, Adamu S, Obese V, Ovbiagele B. Burden and predictors of poststroke cognitive impairment in a sample of ghanaian stroke survivors. Journal of stroke and cerebrovascular diseases : the official journal of National Stroke Association 2017; 26(11): 2553-62.
[http://dx.doi.org/10.1016/j.jstrokecerebrovasdis.2017.05.041]

[200]   Al-Qazzaz NK, Ali SH, Ahmad SA, Islam S, Mohamad K. Cognitive impairment and memory dysfunction after a stroke diagnosis: a post-stroke memory assessment. Neuropsychiatr Dis Treat 2014; 10: 1677-91.
[http://dx.doi.org/10.2147/NDT.S67184] [PMID: 25228808]

[201]   Kim D-E, Park J-H, Schellingerhout D, *et al.* Mapping the Supratentorial Cerebral Arterial Territories Using 1160 Large Artery Infarcts. JAMA Neurol 2019; 76(1): 72-80.
[http://dx.doi.org/10.1001/jamaneurol.2018.2808] [PMID: 30264158]

[202]   Dale Purves GJA, David Fitzpatrick, Lawrence C Katz, Anthony-Samuel LaMantia, James O McNamara, S Mark Williams. Neuroscience 2nd edition. 2001.

[203]   Meng Y, Yu K, Zhang L, Liu Y. Cognitive Decline in Asymptomatic Middle Cerebral Artery Stenosis Patients with Moderate and Poor Collaterals: A 2-Year Follow-Up Study. Med Sci Monit 2019; 25: 4051-8.
[http://dx.doi.org/10.12659/MSM.913797] [PMID: 31148547]

[204]   Werden E, Cumming T, Bird L, Khlif M, Brodtmann A. Brain volume loss precedes cognitive decline in the first year after ischaemic stroke. Journal of Neurology, Neurosurgery & Psychiatry 2017; 88(5): e1..

[205] Werden E, Cumming T, Li Q, *et al.* Structural MRI markers of brain aging early after ischemic stroke. Neurology 2017; 89(2): 116-24.
[http://dx.doi.org/10.1212/WNL.0000000000004086] [PMID: 28600458]

[206] Okar SV, Topcuoglu MA, Yemisci M, Cakir Aktas C, Oguz KK, Arsava EM. Post-stroke inflammatory response is linked to volume loss in the contralateral hemisphere. J Neuroimmunol 2020; 344: 577247..
[http://dx.doi.org/10.1016/j.jneuroim.2020.577247] [PMID: 32388192]

[207] Kliper E, Bashat Db Fau - Bornstein NM,, Bornstein Nm Fau - Shenhar-Tsarfaty S, Shenhar-Tsarfaty S Fau - Hallevi H, Hallevi H Fau - Auriel E, Auriel E Fau - Shopin L, *et al.* Cognitive decline after stroke: relation to inflammatory biomarkers and hippocampal volume. Stroke 2013; 44(1524-4628 (Electronic)): 1433-5.

[208] Kulesh Aa Fau - Drobakha VE, Drobakha Ve Fau - Nekrasova IV, Nekrasova Iv Fau - Kuklina EM, Kuklina Em Fau - Shestakov VV, Shestakov VV. [Neuroinflammatory, Neurodegenerative and Structural Brain Biomarkers of the Main Types of Post-Stroke Cognitive Impairment in Acute Period of Ischemic Stroke]. Vestn Ross Akad Med Nauk 2016; 71(4): 304-12.
[PMID: 29297648]

[209] Tobinick E, Kim NM, Reyzin G, Rodriguez-Romanacce H, DePuy V. Selective TNF inhibition for chronic stroke and traumatic brain injury: an observational study involving 629 consecutive patients treated with perispinal etanercept. CNS Drugs 2012; 26(12): 1051-70.
[http://dx.doi.org/10.1007/s40263-012-0013-2] [PMID: 23100196]

[210] Marques BL, Carvalho GA, Freitas EMM, Chiareli RA, Barbosa TG, Di Araújo AGP, *et al.* The role of neurogenesis in neurorepair after ischemic stroke. Semin Cell Dev Biol. 2019; 95: pp. 98-110.
[http://dx.doi.org/10.1016/j.semcdb.2018.12.003]

[211] Cruz Y, Lorea J, Mestre H, *et al.* Copolymer-1 promotes neurogenesis and improves functional recovery after acute ischemic stroke in rats. PLoS One 2015; 10(3): e0121854..
[http://dx.doi.org/10.1371/journal.pone.0121854] [PMID: 25821957]

[212] Hassan TM, Yarube IU. Peripheral brain-derived neurotrophic factor is reduced in stroke survivors with cognitive impairment. Pathophysiology. 2018; 25: pp. (4)405-10.
[http://dx.doi.org/10.1016/j.pathophys.2018.08.003]

[213] Chang WH, Shin MA, Lee A, Kim H, Kim YH. Relationship between Serum BDNF Levels and Depressive Mood in Subacute Stroke Patients: A Preliminary Study. Int J Mol Sci 2018; 19(10): E3131.
[http://dx.doi.org/10.3390/ijms19103131] [PMID: 30322026]

[214] van der Flier WM, Skoog I, Schneider JA, Pantoni L, Mok V, Chen CLH, *et al.* Vascular cognitive impairment. Nat Rev Dis Primers. 4 England. 2018; p. 18003.

[215] Cognitive rehabilitation 2021.https://www.uhcprovider.com/content/dam/ provider/docs/ public/ policies/comm-medical-drug/cognitive-rehabilitation.pdf

[216] Swatridge K, Regan K, Staines WR, Roy E, Middleton LE. The Acute Effects of Aerobic Exercise on Cognitive Control among People with Chronic Stroke. J Stroke Cerebrovasc Dis 2017; 26(12): 2742-8.
[http://dx.doi.org/10.1016/j.jstrokecerebrovasdis.2017.06.050] [PMID: 28774794]

[217] Amorós-Aguilar L, Rodríguez-Quiroga E, Sánchez-Santolaya S, Coll-Andreu M. Effects of Combined Interventions with Aerobic Physical Exercise and Cognitive Training on Cognitive Function in Stroke Patients: A Systematic Review. Brain Sci 2021; 11(4): 473.
[http://dx.doi.org/10.3390/brainsci11040473] [PMID: 33917909]

[218] Nieto-Vera R, Kahuam-López N, Meneses A, *et al.* Copolymer-1 enhances cognitive performance in young adult rats. PLoS One 2018; 13(3): e0192885..
[http://dx.doi.org/10.1371/journal.pone.0192885] [PMID: 29494605]

[219] Singh A, Tetreault L, Kalsi-Ryan S, Nouri A, Fehlings MG. Global prevalence and incidence of

traumatic spinal cord injury. Clin Epidemiol 2014; 6: 309-31.
[PMID: 25278785]

[220] Kumar R, Lim J, Mekary RA, *et al.* Traumatic spinal injury: global epidemiology and worldwide volume. world neurosurg. 113. United States: © 2018 Elsevier Inc;. 2018; pp. e345-63.

[221] Farzaneh M, Anbiyaiee A, Khoshnam SE. Human Pluripotent Stem Cells for Spinal Cord Injury. Curr Stem Cell Res Ther 2020; 15(2): 135-43.
[http://dx.doi.org/10.2174/1574362414666191018121658] [PMID: 31656156]

[222] Furlan JC, Liu Y, Dietrich WD III, Norenberg MD, Fehlings MG. Age as a determinant of inflammatory response and survival of glia and axons after human traumatic spinal cord injury. Exp Neurol 2020; 332: 113401.
[http://dx.doi.org/10.1016/j.expneurol.2020.113401] [PMID: 32673621]

[223] Wang S, Smith GM, Selzer ME, Li S. Emerging molecular therapeutic targets for spinal cord injury. Expert Opin Ther Targets 2019; 23(9): 787-803.
[http://dx.doi.org/10.1080/14728222.2019.1661381] [PMID: 31460807]

[224] Venkatesh K, Ghosh SK, Mullick M, Manivasagam G, Sen D. Spinal cord injury: pathophysiology, treatment strategies, associated challenges, and future implications. Cell Tissue Res 2019; 377(2): 125-51.
[http://dx.doi.org/10.1007/s00441-019-03039-1] [PMID: 31065801]

[225] Alizadeh A, Dyck SM, Karimi-Abdolrezaee S. Traumatic Spinal Cord Injury: An Overview of Pathophysiology, Models and Acute Injury Mechanisms. Front Neurol 2019; 10: 282.
[http://dx.doi.org/10.3389/fneur.2019.00282] [PMID: 30967837]

[226] Oyinbo CA. Secondary injury mechanisms in traumatic spinal cord injury: a nugget of this multiply cascade. Acta Neurobiol Exp (Warsz) 2011; 71(2): 281-99.
[PMID: 21731081]

[227] Ahuja CS, Wilson JR, Nori S, *et al.* Traumatic spinal cord injury. Nat Rev Dis Primers 2017; 3: 17018.
[http://dx.doi.org/10.1038/nrdp.2017.18] [PMID: 28447605]

[228] Gondim FA, Lopes AC Jr, Oliveira GR, *et al.* Cardiovascular control after spinal cord injury. Curr Vasc Pharmacol 2004; 2(1): 71-9.
[http://dx.doi.org/10.2174/1570161043476474] [PMID: 15320835]

[229] Popa C, Popa F, Grigorean VT, *et al.* Vascular dysfunctions following spinal cord injury. J Med Life 2010; 3(3): 275-85.
[PMID: 20945818]

[230] Krassioukov A, Claydon VE. The clinical problems in cardiovascular control following spinal cord injury: an overview. Prog Brain Res 152. Netherlands 2006; pp. 223-9.
[http://dx.doi.org/10.1016/S0079-6123(05)52014-4]

[231] Krassioukov AV, Karlsson AK, Wecht JM, Wuermser LA, Mathias CJ, Marino RJ. Assessment of autonomic dysfunction following spinal cord injury: rationale for additions to International Standards for Neurological Assessment. J Rehabil Res Dev 2007; 44(1): 103-12.
[http://dx.doi.org/10.1682/JRRD.2005.10.0159] [PMID: 17551864]

[232] Grigorean VT, Sandu AM, Popescu M, *et al.* Cardiac dysfunctions following spinal cord injury. J Med Life 2009; 2(2): 133-45.
[PMID: 20108532]

[233] Saadeh YS, Smith BW, Joseph JR, *et al.* The impact of blood pressure management after spinal cord injury: a systematic review of the literature. Neurosurg Focus 2017; 43(5): E20.
[http://dx.doi.org/10.3171/2017.8.FOCUS17428] [PMID: 29088944]

[234] Phillips AA, Krassioukov AV. Contemporary cardiovascular concerns after spinal cord injury: mechanisms, maladaptations, and management. J Neurotrauma 2015; 32(24): 1927-42.
[http://dx.doi.org/10.1089/neu.2015.3903] [PMID: 25962761]

[235]   Mautes AE, Weinzierl MR, Donovan F, Noble LJ. Vascular events after spinal cord injury: contribution to secondary pathogenesis. Phys Ther 2000; 80(7): 673-87.
[http://dx.doi.org/10.1093/ptj/80.7.673] [PMID: 10869130]

[236]   Sinescu C, Popa F, Grigorean VT, *et al.* Molecular basis of vascular events following spinal cord injury. J Med Life 2010; 3(3): 254-61.
[PMID: 20945816]

[237]   Wagner I, Volbers B, Hilz MJ, Schwab S, Doerfler A, Staykov D. Radiopacity of intracerebral hemorrhage correlates with perihemorrhagic edema. Eur J Neurol 2012; 19(3): 525-8.
[http://dx.doi.org/10.1111/j.1468-1331.2011.03526.x] [PMID: 21951394]

[238]   Wecht JM, Weir JP, Katzelnick CG, *et al.* Systemic and cerebral hemodynamic contribution to cognitive performance in spinal cord injury. J Neurotrauma 2018; 35(24): 2957-64.
[http://dx.doi.org/10.1089/neu.2018.5760] [PMID: 30113243]

[239]   Flanders AE, Spettell CM, Tartaglino LM, Friedman DP, Herbison GJ. Forecasting motor recovery after cervical spinal cord injury: value of MR imaging. Radiology 1996; 201(3): 649-55.
[http://dx.doi.org/10.1148/radiology.201.3.8939210] [PMID: 8939210]

[240]   Losey P, Young C, Krimholtz E, Bordet R, Anthony DC. The role of hemorrhage following spinal-cord injury. Brain Res . 2014; 1569: pp. 9-18.
[http://dx.doi.org/10.1016/j.brainres.2014.04.033]

[241]   Gensel JC, Zhang B. Macrophage activation and its role in repair and pathology after spinal cord injury. Brain Res 2015; 1619: 1-11.
[http://dx.doi.org/10.1016/j.brainres.2014.12.045] [PMID: 25578260]

[242]   Shao A, Tu S, Lu J, Zhang J. Crosstalk between stem cell and spinal cord injury: pathophysiology and treatment strategies. Stem Cell Res Ther 2019; 10(1): 238.
[http://dx.doi.org/10.1186/s13287-019-1357-z] [PMID: 31387621]

[243]   Neirinckx V, Coste C, Franzen R, Gothot A, Rogister B, Wislet S. Neutrophil contribution to spinal cord injury and repair. J Neuroinflammation 2014; 11: 150.
[http://dx.doi.org/10.1186/s12974-014-0150-2] [PMID: 25163400]

[244]   Shechter R, Miller O, Yovel G, *et al.* Recruitment of beneficial M2 macrophages to injured spinal cord is orchestrated by remote brain choroid plexus. Immunity 2013; 38(3): 555-69.
[http://dx.doi.org/10.1016/j.immuni.2013.02.012] [PMID: 23477737]

[245]   Popovich PG, Wei P, Stokes BT. Cellular inflammatory response after spinal cord injury in Sprague-Dawley and Lewis rats. J Comp Neurol 377. 443-64. United States1997

[246]   Miller AH, Raison CL. The role of inflammation in depression: from evolutionary imperative to modern treatment target. Nat Rev Immunol 2016; 16(1): 22-34.
[http://dx.doi.org/10.1038/nri.2015.5] [PMID: 26711676]

[247]   Allison DJ, Ditor DS. Targeting inflammation to influence mood following spinal cord injury: a randomized clinical trial. J Neuroinflammation 2015; 12: 204.
[http://dx.doi.org/10.1186/s12974-015-0425-2] [PMID: 26545369]

[248]   Maldonado-Bouchard S, Peters K, Woller SA, *et al.* Inflammation is increased with anxiety- and depression-like signs in a rat model of spinal cord injury. Brain Behav Immun 2016; 51: 176-95.
[http://dx.doi.org/10.1016/j.bbi.2015.08.009] [PMID: 26296565]

[249]   Wu J, Zhao Z, Sabirzhanov B, *et al.* Spinal cord injury causes brain inflammation associated with cognitive and affective changes: role of cell cycle pathways. J Neurosci 2014; 34(33): 10989-1006.
[http://dx.doi.org/10.1523/JNEUROSCI.5110-13.2014] [PMID: 25122899]

[250]   Chen SY, Liu S, Zhang LL, *et al.* Construction of injectable silk fibroin/polydopamine hydrogel for treatment of spinal cord injury. Chem Eng J 2020; 399.

[251]   Invernizzi M, de Sire A, Carda S, *et al.* Bone muscle crosstalk in spinal cord injuries: pathophysiology

and implications for patients' quality of life. Curr Osteoporos Rep 2020; 18(4): 422-31.
[http://dx.doi.org/10.1007/s11914-020-00601-7] [PMID: 32519284]

[252]   Sachdeva R, Gao F, Chan CCH, Krassioukov AV. Cognitive function after spinal cord injury: A systematic review. Neurology 2018; 91(13): 611-21.
[http://dx.doi.org/10.1212/WNL.0000000000006244] [PMID: 30158159]

[253]   North NT. The psychological effects of spinal cord injury: a review. Spinal Cord 1999; 37(10): 671-9.
[http://dx.doi.org/10.1038/sj.sc.3100913] [PMID: 10557122]

[254]   Sakakibara BM, Miller WC, Orenczuk SG, Wolfe DL. A systematic review of depression and anxiety measures used with individuals with spinal cord injury. Spinal Cord 2009; 47(12): 841-51.
[http://dx.doi.org/10.1038/sc.2009.93] [PMID: 19621021]

[255]   Li Y, Cao T, Ritzel RM, He J, Faden AI, Wu J. Dementia, depression, and associated brain inflammatory mechanisms after spinal cord injury. Cells 2020; 9(6): E1420.
[http://dx.doi.org/10.3390/cells9061420] [PMID: 32521597]

[256]   do Espírito Santo CC, da Silva Fiorin F, Ilha J, Duarte M, Duarte T, Santos ARS. Spinal cord injury by clip- compression induces anxiety and depression-like behaviours in female rats: The role of the inflammatory response. Brain Behav Immun 78 Netherlands: © 2019 Elsevier Inc. 2019; pp. 91-104.

[257]   Raz L, Knoefel J, Bhaskar K. The neuropathology and cerebrovascular mechanisms of dementia. J Cereb Blood Flow Metab 2016; 36(1): 172-86.
[http://dx.doi.org/10.1038/jcbfm.2015.164] [PMID: 26174330]

[258]   Grau JW. Learning from the spinal cord: how the study of spinal cord plasticity informs our view of learning. Neurobiol Learn Mem 2014; 108: 155-71.
[http://dx.doi.org/10.1016/j.nlm.2013.08.003] [PMID: 23973905]

[259]   Wecht JM, Bauman WA. Decentralized cardiovascular autonomic control and cognitive deficits in persons with spinal cord injury. J Spinal Cord Med 2013; 36(2): 74-81.
[http://dx.doi.org/10.1179/2045772312Y.0000000056] [PMID: 23809520]

[260]   Ferguson AR, Huie JR, Crown ED, et al. Maladaptive spinal plasticity opposes spinal learning and recovery in spinal cord injury. Front Physiol 2012; 3: 399.
[http://dx.doi.org/10.3389/fphys.2012.00399] [PMID: 23087647]

[261]   Brakel K, Aceves AR, Aceves M, Hierholzer A, Nguyen QN, Hook MA. Depression-like behavior corresponds with cardiac changes in a rodent model of spinal cord injury Exp Neurol 320. United States: Published by Elsevier Inc. 2019; p. 112969.

[262]   Fuhrer MJ, Rintala DH, Hart KA, Clearman R, Young ME. Depressive symptomatology in persons with spinal cord injury who reside in the community. Arch Phys Med Rehabil. 1993; 74: pp. (2)255-60.

[263]   Khazaeipour Z, Taheri-Otaghsara SM, Naghdi M. Depression following spinal cord injury: Its relationship to demographic and socioeconomic indicators. Top Spinal Cord Inj Rehabil 2015; 21(2): 149-55.
[http://dx.doi.org/10.1310/sci2102-149] [PMID: 26364284]

[264]   Kigerl KA, Zane K, Adams K, Sullivan MB, Popovich PG. The spinal cord-gut-immune axis as a master regulator of health and neurological function after spinal cord injury. Exp Neurol 2020; 323: 113085..
[http://dx.doi.org/10.1016/j.expneurol.2019.113085] [PMID: 31654639]

[265]   Schmidt EKA, Torres-Espin A, Raposo PJF, et al. Fecal transplant prevents gut dysbiosis and anxiety-like behaviour after spinal cord injury in rats. PLoS One 2020; 15(1): e0226128..
[http://dx.doi.org/10.1371/journal.pone.0226128] [PMID: 31940312]

[266]   Chinna Meyyappan A, Forth E, Wallace CJK, Milev R. Effect of fecal microbiota transplant on symptoms of psychiatric disorders: a systematic review. BMC Psychiatry 2020; 20(1): 299.
[http://dx.doi.org/10.1186/s12888-020-02654-5] [PMID: 32539741]

[267] Huang SW, Wang WT, Chou LC, Liou TH, Lin HW. Risk of dementia in patients with spinal cord injury: A nationwide population-based cohort study. J Neurotrauma 2017; 34(3): 615-22.
[http://dx.doi.org/10.1089/neu.2016.4525] [PMID: 27539630]

[268] Craig A, Guest R, Tran Y, Middleton J. Cognitive impairment and mood states after spinal cord injury. J Neurotrauma 2017; 34(6): 1156-63.
[http://dx.doi.org/10.1089/neu.2016.4632] [PMID: 27717295]

[269] Plemel JR, Wee Yong V, Stirling DP. Immune modulatory therapies for spinal cord injury--past, present and future Exp Neurol. 2014; 258: pp. 91-104.

[270] Scandola M, Dodoni L, Lazzeri G, *et al.* Neurocognitive benefits of physiotherapy for spinal cord injury. J Neurotrauma 2019; 36(12): 2028-35.
[http://dx.doi.org/10.1089/neu.2018.6123] [PMID: 30526335]

[271] Wu J, Zhao Z, Kumar A, *et al.* Endoplasmic reticulum stress and disrupted neurogenesis in the brain are associated with cognitive impairment and depressive-like behavior after spinal cord injury. J Neurotrauma 2016; 33(21): 1919-35.
[http://dx.doi.org/10.1089/neu.2015.4348] [PMID: 27050417]

[272] Grau JW, Baine RE, Bean PA, *et al.* Learning to promote recovery after spinal cord injury. Exp Neurol 2020; 330: 113334.
[http://dx.doi.org/10.1016/j.expneurol.2020.113334] [PMID: 32353465]

[273] Rodríguez-Barrera R, Flores-Romero A, García E, *et al.* Immunization with neural-derived peptides increases neurogenesis in rats with chronic spinal cord injury. CNS Neurosci Ther 2020; 26(6): 650-8.
[http://dx.doi.org/10.1111/cns.13368] [PMID: 32352656]

[274] Rodríguez-Barrera R, Flores-Romero A, Buzoianu-Anguiano V, *et al.* Use of a Combination Strategy to Improve Morphological and Functional Recovery in Rats With Chronic Spinal Cord Injury. Front Neurol 2020; 11: 189.
[http://dx.doi.org/10.3389/fneur.2020.00189] [PMID: 32300328]

[275] Jure I, Pietranera L, De Nicola AF, Labombarda F. Spinal cord injury impairs neurogenesis and induces glial reactivity in the hippocampus. Neurochem Res. 2017; 42: pp. (8)2178-90.
[http://dx.doi.org/10.1007/s11064-017-2225-9]

[276] Chiaravalloti ND, Weber E, Wylie G, Dyson-Hudson T, Wecht JM. The impact of level of injury on patterns of cognitive dysfunction in individuals with spinal cord injury. J Spinal Cord Med 2020; 43(5): 633-41.
[http://dx.doi.org/10.1080/10790268.2019.1696076] [PMID: 31859606]

CHAPTER 2

Tau-Targeted Therapy in Alzheimer's Disease - History and Current State

Anamaria Jurcau[1,*] and **Vharoon Sharma Nunkoo**[1]

[1] University of Oradea, Faculty of Medicine and Pharmacy, Clinical Municipal Hospital, Dr. G. Curteanu", Oradea, Romania

Abstract: The two main histopathological hallmarks still required for the diagnosis of Alzheimer's disease are the presence of amyloid plaques and intraneuronal neurofibrillary tangles formed mainly of tau protein. Normally, tau protein regulates intracellular trafficking and provides microtubule stability. However, in AD as well as in other tauopathies, there is a disruption in the normal function of tau, leading to the development of neurofibrillary tangles with disease-dependent ultrastructure of the tau filaments.

After several failures of trials with drugs trying to prevent the accumulation of amyloid, tau protein became another target of molecules designed to modify the course of AD.

Each stage in the development of tau pathology, from the expression of tau protein to its post-translational modifications, with the protein's aggregation and impaired clearance, presents opportunities for therapeutic intervention: reducing tau expression with antisense oligonucleotides, reducing tau phosphorylation with kinase inhibitors, inhibiting tau acetylation, tau deglycosylation, tau aggregation, modulating tau degradation, stabilizing the microtubules, as well as active or passive anti-tau immunotherapies (with various monoclonal antibodies), have been attempted or are still in trials, with rather inconclusive results so far. It appears that an efficient disease-modifying therapy is not yet available. Given the complex pathophysiology of Alzheimer's disease, most likely, a multi-targeted approach would be more effective.

Keywords: Alzheimer's disease, Anti-tau therapy, Microtubules, Mitochondrial dysfunction, Tauopathies, Tau protein.

* **Corresponding author Anamaria Jurcau:** The University of Oradea, Faculty of Medicine and Pharmacy, Oradea, Romania; E-mail:anamaria.jurcau@gmail.com

Dr. José Juan Antonio Ibarra Arias (Eds.)

HISTORICAL BACKGROUND

Alois Alzheimer, a German physician and a pioneer of linking disease symptoms to microscopic brain changes, first met and examined Auguste D. in 1901 in Frankfurt. Although 1 year later he took a position in Munich, he was haunted by this case. Thus, in April 1906, when the patient died, he examined her brain and found impressive shrinkage and abnormal depositions in and around nerve cells. Today, more than 100 years after the presentation of Alzheimer's findings at the Conference held in Tübingen in November 1906 [1, 2], amyloid plaques and neurofibrillary tangles are still required for the pathological diagnosis of Alzheimer's disease [3].

Using the newly discovered electron microscopy technique, Terry and Kidd described the intraneuronal deposits in 1963 as being paired helical filaments [4, 5]. Further, in 1975, Weingarten and coworkers characterized these filaments as being a protein named tau, which is crucial for the assembly of tubulin into microtubules [6]. Interestingly, 1975 was the same year in which tau the lepton was also discovered by Perl *et al.* [7]. Soon thereafter, Cleveland and coworkers provided a biochemical characterization of tau [8, 9].

However, because monogenic mutations in the amyloid precursor protein (APP) or the presenilins involved in its processing can lead to phenotypes similar to AD, until recent years, research has focused mainly on these molecules. The discovery of tau mutations able to cause neurodegenerative diseases on their own [10] as well as of intracellular tau aggregates in several neurodegenerative diseases like progressive supranuclear palsy, frontotemporal lobar degeneration, corticobasal degeneration, or Pick's disease (collectively referred to as tauopathies) [11], boosted tau research and led to exploring several therapeutic strategies in neurodegenerative diseases [7, 12, 13].

NORMAL TAU PROTEIN STRUCTURE AND FUNCTION

The Tau Gene and Tau Isoforms

Human tau is encoded by the MAPT (microtubule-associated protein tau) gene located on chromosome 17q21 [14]. Alternative splicing generates mainly 6 isoforms of 37-46 kDa in the central nervous system (CNS), while a "big tau" isoform is found mainly in the peripheral nervous system [15, 16]. The six isoforms found in the CNS (Fig. 1) differ by the presence of zero, one, or two N-terminal inserts and either 3 (3R) or 4 (4R) microtubule binding C-terminal inserts

[17]. Due to the developmentally regulated expression of tau, all 6 isoforms can be found in the adult human brain, while the fetal brain expresses only the shortest isoform (0N3R) [18].

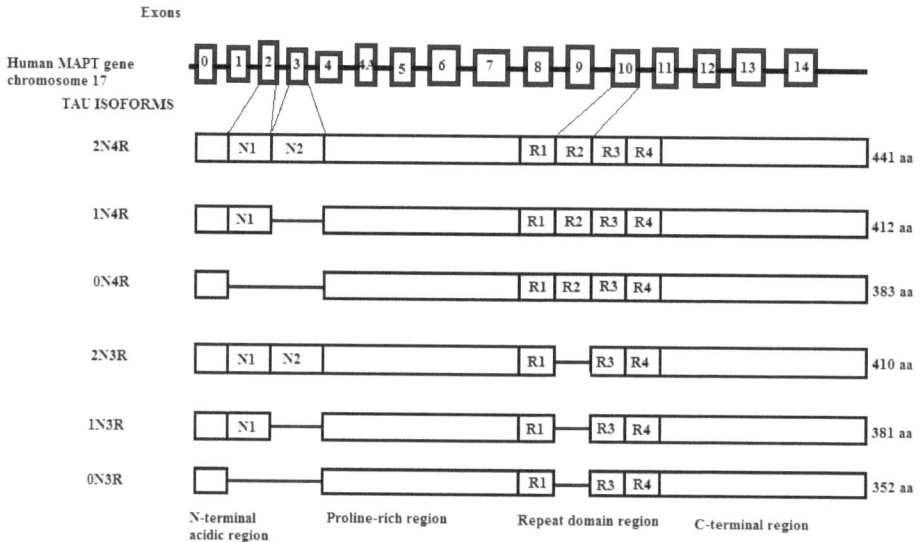

Fig. (1). Genomic structure of the human tau gene; of the 14 exons, exons 2, 3, and 10 are alternatively spliced, generating six tau isoforms in the adult brain. Exons 9, 11, and 12 each encode for a microtubule-binding repeat generating 3R tau isoforms. The presence of E10 adds an extra MT-binding repeat generating 4R tau isoforms. 3R and 4R tau isoforms further differ depending on the presence of exon 2 (1N) or exons 2 and 3 together (2N), while the absence of both exons generates 0N3R and 0N4R isoforms of tau. The number of aminoacids in each isoform is shown on the right.

Tau Protein Structure

The amino acid composition of tau is unusually hydrophilic [19], and the protein has an overall basic character. The charges have an asymmetrical distribution, with the amino-terminal being acidic and the carboxy-terminal being neutral. The middle region contains numerous prolines which harbor many epitopes of antibodies that are hyperphosphorylated in Alzheimer's disease [12].

Due to its hydrophilic character, the polypeptide chain of the protein is flexible and mobile. Tau is normally unfolded as opposed to most cytosolic proteins, which have a compact folded structure [20, 21]. To date, the "paperclip" conformation of tau is widely accepted, in which the C-terminus is folded over the microtubule-binding domain, and the N-terminus folds over the C-terminus [22], as shown in Fig. (2). This conformation is disrupted by tau phosphorylation [9].

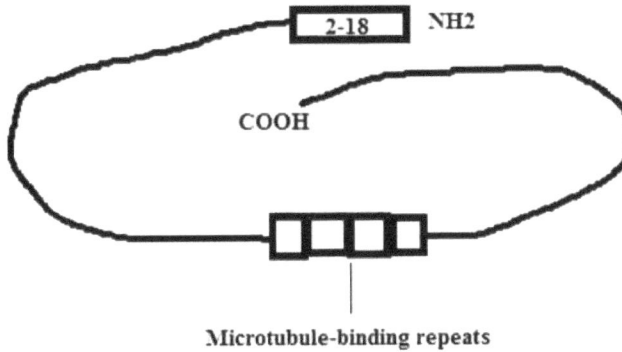

Fig. (2). In the free tau folded into the paperclip conformation, the C-terminal domain lies close to the N-terminus and protects the domain comprising aminoacids 2-18, known as the phosphatase activating domain to interact with protein phosphatase 1 (PP1). When the C-terminal domain moves away from the N-terminal domain, the phosphatase activating domain activates PP1, which activates further glycogen synthase kinase 3β (GSK3β) and initiates the phosphorylation cascade.

Tau Localization

Tau Localization in Neurons

Tau is located mainly in axons [23] but can be found in lower concentrations in the plasma membrane, nucleus, mitochondria, as well as in dendrites and at the level of synapses [24].

Multiple mechanisms that contribute to the predominantly axonal localization of tau are as follows:

- cytosolic tau diffuses freely between the different compartments but is retained in the axon through its binding to microtubules favored by the low phosphorylation level and the retrograde barrier formed by the initial segment of the axon, which "traps" tau in the axon [25].
- tau is transported actively by motor proteins such as the kinesin family members [26].

- the MAPT gene has a region specifically targeted to the axonal compartment [27], following the transport of MAPT mRNA into the axon and rapamycin-p70S6 kinase (m-TOR-p70S6 K) pathway upregulates the translation of tau [28].

Tau Localization in Glial Cells

Although predominantly produced by neuronal in AD, tau is expressed at low levels in oligodendrocytes, astrocytes and is found in microglial cells as well across tauopathies [29]. In oligodendrocytes, tau forms fine, coiled branching bodies and argyrophilic threads, which line myelinating processes [30], while in astrocytes, tau can take on various appearances in the different tauopathies [30 - 32]:

- in the tufted astrocytes, characteristic of progressive supranuclear palsy, granular phosphorylated tau is clustered around a nucleus of dense tangles;
- in corticobasal degeneration, tau is localized in the distal astrocytic processes, forming annular astrocytic plaques;
- more common are the thorn-shaped astrocytes, observed in PSP, AD, Pick's disease, argyrophilic grain disease, and in the brain parenchyma of cognitively normal elderly individuals, featuring perinuclear tau deposits.

Ultrastructurally, these glial tau inclusions are straight and/or twisted filaments that mimic the neuronal tau inclusions [33]. However, the functional implications of each phenotype are still unknown. Different distributions of glial fibrillary acidic protein (GFAP), a marker that is upregulated in activated astrocytes, have been found in each astrocytic phenotype [34]. As for microglia, although it does not express tau, microglial tau inclusions have also been reported [35].

Functions of Tau Protein

Since the work of Weingarten *et al.* [6], it became clear that tau protein is essential in the self-assembly of tubulin into microtubules. Microtubules are cytoskeletal protein polymers that stabilise cell shape, serve as tracks for intracellular transport mediated by motor proteins [36], and have an important role in mitosis as well [7]. In neuroblasts, the microtubules have very dynamic structures because the probability of assembly is equal to that of depolymerisation in all directions. This results in the spheric morphology of the proliferating cell. During the differentiation of neuroblasts into neurons, the microtubules become stabilized in certain directions leading to the formation of cytoplasmic extensions which will differentiate into dendrites or axons [37]. After the microtubules have formed, tau and other MAPs (microtubule-associated proteins) bind to tubulin (Fig. **3**), stabilize the microtubule structure [7], and protect the microtubule ends against length fluctuations (dynamic instability) [38], an effect used in cancer chemotherapy with taxol [39].

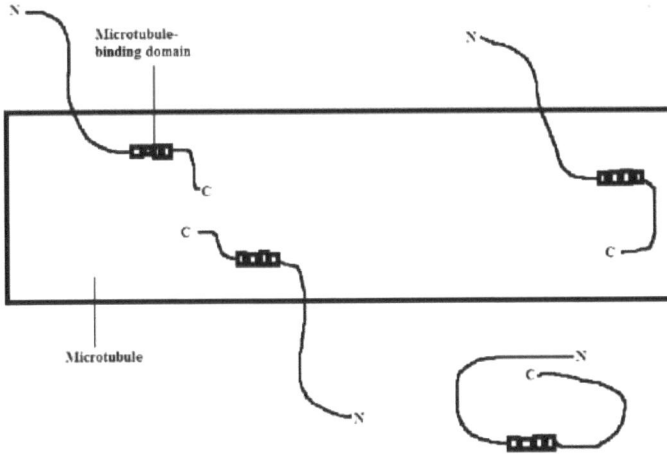

Fig. (3). Tau binds to microtubules mainly through the microtubule-binding domain. Upon binding to microtubules, the "paperclip" conformation is altered and the N terminus of tau projects away from the microtubule surface.

MAPs also exhibit a spacer function establishing a clear zone around microtubules in the cells, a function which is more pronounced for large MAPs, like MAP2 [40].

The role of tau in dendrites is currently not well understood, but it seems to be involved in the regulation of synaptic plasticity in hippocampal neurons under the influence of brain-derived neurotrophic factor [41] and the formation of dendritic spines and post-synaptic densities [17].

In addition, many domains of tau can bind to membranes. This may have a role in neurite development by binding the microtubules to the membrane in the growth cones [42]. Interaction of tau with cell surface proteins, like the GluR2/3 subunits of the AMPA receptors, can modulate synaptic signaling [43], a function highly dependent on the phosphorylation level of tau because the conformational changes induced by phosphorylation abolish the interaction with cell membranes [44, 45].

In the nucleus, most of the tau is non-phosphorylated [46], which enables the protein to bind and protect DNA against free radical-induced damage [47] and also possibly to participate in DNA repair mechanisms [48]. In addition, tau may modulate gene expression by unwinding DNA [49], thereby enhancing the activity of many protein-DNA complexes and enhancing or inhibiting gene transcription. Following tau depletion, the transcription of at least 14 genes was increased [50].

Furthermore, research has pointed to the role of tau in the regulation of neuronal activity, synaptic plasticity, and long-term depression especially in hippocampal neurons [51] aside from its role in neuronal migration [52], survival of neurons, the response of neurons to external stimuli [53], and maturation of neuroblasts [36].

Oligodendrocytes are also rich in microtubules. In these cells, tau is expressed mainly in the cellular processes where it establishes early axonal contact and stabilizes the microtubules during the formation of oligodendroglial processes and myelination [54, 55].

In astrocytes, tau is not a major cytoskeletal component and is normally found only at trace levels; the functional implications of tau accumulations in the various tauopathies being still unknown. Microglia do contain microtubule networks and it is yet unclear if tau is normally present in these cells [29, 56].

Post-Translational Modifications of Tau

After its biosynthesis, tau protein can undergo a large number of post-translational modifications which impact the function of the protein.

Tau Phosphorylation

Tau contains 85 phosphorylation sites [57], which make phosphorylation the most common post-translational modification of tau.

The degree of tau phosphorylation is influenced by the balance between protein kinases and phosphatases.

Protein kinases which phosphorylate tau can be classified into:

- Proline-directed serine/threonine-protein kinases, like glycogen synthase kinase (GSK)3α/β, mitogen-activated protein kinases (MAPKs), *etc.*
- Non-proline-directed serine/threonine-protein kinases such as cAMP-dependent protein kinase A (PKA), protein kinase C, calcium/calmodulin-dependent

protein kinase II (CAMKII), *etc.*
• Protein kinases specific for tyrosine residues, like Fyn, Src, Syk [17].

Dephosphorylation is controlled by protein phosphatase 2A (PP2A) and protein phosphatase 5 (PP5), the activity of both being significantly decreased in Alzheimer's disease [58,59]. There is a complex interplay between phosphorylation/dephosphorylation because PP2A can dephosphorylate GSK3β at Ser9, while activation of GSK3β inhibits PP2A [60, 61]. In addition, Akt inhibits GSK3β. In Alzheimer's disease, attenuation of the PI3 K/Akt signaling pathway increases GSK3β activity resulting in tau hyperphosphorylation and tangle formation [17].

Tau in a phosphorylated state has a low affinity for microtubules and can detach from these, resulting in destabilisation of the cytoskeleton particularly in neurons [62]. Further, the detached tau self-aggregates forming oligomers and tau aggregates [63], compromises axonal microtubule integrity, and induces synaptic dysfunction [64, 65].

Tau Acetylation

Tau acetylation is mainly mediated by cAMP-response element-binding protein (CREB-binding protein = CBP) but tau protein also has an intrinsic acetyl-trans-ferases activity which can catalyse its auto-acetylation [66, 67]. Deacetylation depends on sirtuin 1 (SIRT1) and histone deacetylase 6 (HDAC6) activity [68].

Depending on the lysine residue involved, acetylation protects tau from increased phosphorylation and suppresses tau aggregation or, conversely, inhibits tau degradation and causes accumulation of hyperphosphorylated tau [66, 68, 69]. Particularly acetylation of tau at the lysine residue 280 delays tau turnover and induces tau toxicity [70].

In addition, tau acetylation may impact synaptic function through disruptions in AMPA (α-amino-3-hydroxy-5-methyl-4-isoxazole propionic acid) receptor membrane insertion [71], and destabilize the cytoskeleton of the initial axonal segment by reducing ankrin and β-spectrin [72].

Thus, tau acetylation can be both beneficial or detrimental to tau function depending on the target lysine residue, and correcting tau acetylation could be a promising therapeutic strategy in tauopathies [17].

Tau Glycosylation

Tau glycosylation occurs mainly at the N-terminal through a yet unknown mechanism and facilitates tau phosphorylation, thereby affecting its conformation [73]. In contrast, the addition of O-linked N-acetylglucosamine (O-GlcNAc) to threonine or serine residues, a reaction mediated by O-GlcNAc transferase, protects tau from phosphorylation and can suppress tau aggregation [74, 75]. The enzyme has been found significantly reduced in Alzheimer's disease brains [76] leading to the hypothesis that O-GlcNAcylation might be a target for therapeutic approaches in tauopathies [17].

Tau Ubiquitination

This reaction occurs also at lysine residues (mainly Lys48) and is mediated by CHIP (a ubiquitin E3 ligase) and tumor necrosis factor receptor-associated factor 6 (TRAF6) leading to increased proteasomal degradation of tau [77]. Due to the fact that ubiquitination and acetylation occur at similar sites, possibly in competition, tau acetylation could influence ubiquitination-dependent tau degradation [50].

Tau Sumoylation

Sumoylation of tau by small ubiquitin-like modifier protein 1 (SUMO-1) correlates in cultured cells with increased tau phosphorylation because the reaction counteracts the effects of ubiquitination [78]. Since in AD, SUMO-1 colocalizes with phosphorylated tau, it can be hypothesized that sumoylation increases tau phosphorylation and thereby inhibits ubiquitin-mediated tau degradation [78].

Tau Methylation

Methylation of tau occurs on lysine and arginine residues but the functional implications of this reaction are yet unclear [79]. It is likely that methylation of the lysine residues within the microtubule-binding domain reduces the propensity of tau to bind to and stabilise the microtubules.

Other Post-translational Modifications of Tau

Tau protein can undergo a number of other post-translational modifications, including:

- Glycation
- Deamidation
- Isomerisation
- Abnormal nitration of tyrosine residues

All of these modifications have been detected in tau extracted from AD brains but not in tau stemming from normal brains [80].

To summarize, there are many post-translational modifications of tau in both physiological and pathological states, making it difficult to select the most important pathways which modify tau in pathological states and target them therapeutically. However, lysine residues appear to have a strategic role, since at least four potentially competing modifications (glycation, acetylation, methylation, and ubiquitination) occur on them [17].

Tau Clearance

Tau is degraded both by the ubiquitin-proteasome system (UPS) and the autophagic-lysosomal system [17].

The proteasome is a key structure in clearing soluble cytosolic proteins [81]. The regulatory cap with chaperone proteins unfolds the target protein by removing the ubiquitin tag in an ATP-dependent process after which the target protein is fed into the catalytic core and is degraded by the proteasomal enzymes. However, the 20S proteasome can clear unfolded proteins through an ATP- and ubiquitin-independent mechanism. Tau is a substrate for both proteasomal forms [82, 83].

Autophagy is the main mechanism for clearing larger protein aggregates and organelles such as mitochondria or peroxisomes [84], as well as bacterial pathogens [85]. In the process, a double membrane autophagophore is formed which expands and engulfs a region of cytoplasm containing the substrates which need to be degraded, then encloses and forms the autophagosome which further fuses with a lysosome and forms an autophagic vacuole [81].

As mentioned above, tau as a monomer, natively unfolded, is likely a proteasomal substrate, however after many modifications it undergoes during AD pathogenesis and it is degraded mainly by autophagy. Ubiquitin targets substrates to both pathways by using chaperone proteins, although there is a significant overlap between chaperones in the two pathways [86, 87]. Two modifications of tau, phosphorylation, and truncation, seem critical in selecting the clearing mechanism. Caspase-3 cleaved tau is preferentially degraded by autophagy [88], while phosphorylation drives tau mainly to the proteasomal pathway of clearing

[89]. The two degradative systems are influencing each other: blocking the proteasome causes an increase in the autophagic flux [90], while impairment of autophagy further inhibits the proteasomal system, possibly due to the accumulation of large aggregated substrates like phosphorylated tau [91].

To date, there is significant evidence for impaired autophagy in AD. For example, PS1 acts as a chaperone for the vacuolar ATPase which acidifies the lysosomal lumen [92]. PS1 mutations, which cause familial AD, impair lysosomal acidity thereby compromising the autophagy system and leading to the accumulation of autophagosomal vacuoles [81]. However, impaired autophagy was found even in the absence of mutant PS1 [93].

Thus, it appears that under physiological conditions much of tau is degraded by the proteasomal system being soluble and monomeric [82]. In the context of AD, tau undergoes a series of modifications that shift the balance toward autophagy degradation, but autophagy is impaired in AD. Further increases in size and density of tau oligomers as well as folding of these oligomers into filaments and tangles may further directly impair the proteasomal system. Proteases, like caspase-3, may be compensatorily activated but will lead to the accumulation of cleaved, toxic forms of tau [81].

TAUOPATHIES

A heterogenous group of movement disorders and dementias occurring as neurodegenerative diseases are classified as tauopathies. Neuropathologically, they are characterized by intracellular accumulation of tau filaments in the form of neurofibrillary tangles or other tau inclusions in neurons and glial cells [17]. In addition, at least one other amyloidogenic protein such as β-amyloid, α-synuclein, or huntingtin accumulates as well, leading to a spectrum of tauopathies (Fig. **4**). Depending on the associated aggregated protein and the predominant localisation of the inclusions, the clinical symptoms may vary.

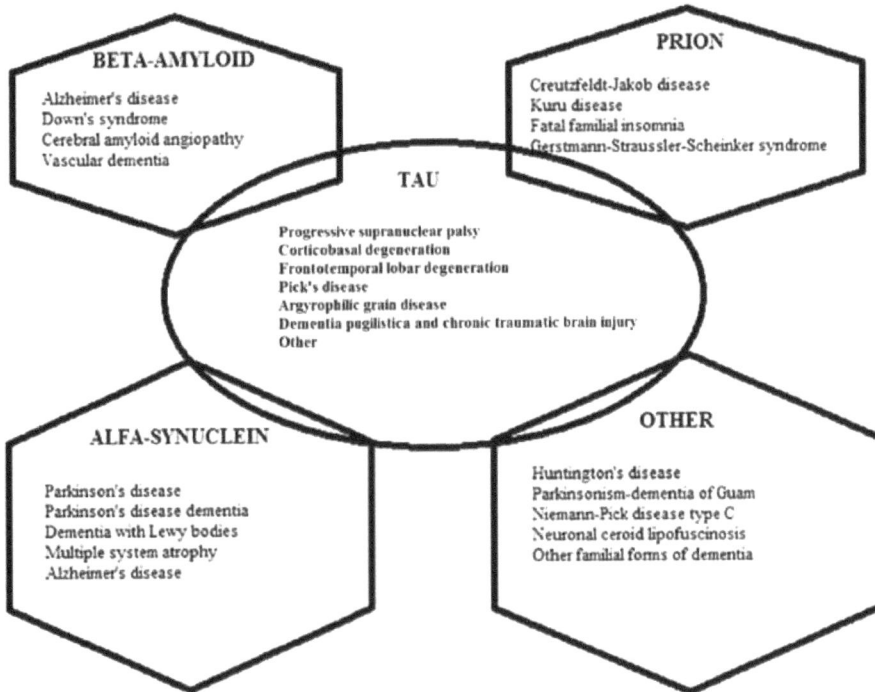

Fig. (4). Tauopathies. Diagram illustrating the wide range of tauopathies. The central panel lists diseases in which tau pathology is the main feature. The overlapping pannels list diseases in which other disease-associated proteins accompany the tau inclusions. (adapted from Guo *et al*, reference 17).

Tauopathies may also be classified according to the 3R- and 4R-tau ratio and according to the presence of 2 or 3 major bands of 60, 64, and 68 kDa in a western blot of sarkosyl-insoluble fractions [94]. For example, Pick's disease is a 3R tauopathy with 60 and 64 kDa bands. Progressive supranuclear palsy, corticobasal degeneration, or argyrophilic grain disease are 4R tauopathies with 64 and 68 kDa bands. Alzheimer's disease is a mixed 3R and 4R tauopathy with 60, 64, and 68 kDa bands [94, 95].

Tau-Mediated Neurodegeneration

A large amount of research focuses on understanding how alterations of the properties of tau protein due to mis-splicing, aggregation, or post-translational modifications transform physiological into pathological tau leading to mislocalisation of tau and synaptic dysfunction, disturbed axonal transport, and impaired protein quality control in the cells of the central nervous system. The findings are summarized below.

Tau Gene Dysfunction

Since the identification of P301L, the first mutation in the MAPT gene leading to frontotemporal lobar degeneration [10, 96], a series of other MAPT mutations have been identified associating frontotemporal dementia, Parkinson's disease dementia, dementia with Lewy bodies, progressive supranuclear palsy, cortico-basal degeneration, Pick's disease, or amyotrophic lateral sclerosis with frontotemporal dementia [17]. Because 4R tau isoforms are more prone to bind to microtubules, mutations that alter the 3R/4R isoform ratio may increase tau phosphorylation and significantly influence the binding of tau to microtubules [97]. However, MAPT gene mutations impact the conformation of tau and interfere with post-translational modifications, with the interaction of tau with other proteins, as well as with a wide variety of intracellular processes.

Tau Aggregation

In all of the tauopathies, tau undergoes misfolding and oligomerisation followed by self-aggregation into insoluble tau deposits [98]. In diseases such as AD, there is a shift in the balance of kinase and phosphatase activity leading to the appearance of hyperphosphorylated tau [99] with decreased affinity for microtubules [100]. Increasing the proportion of tau detached from microtubules allows monomeric hyperphosphorylated tau to form oligomers by binding one another [101]. Two hexapeptide motifs situated in the second and third microtubule-binding repeats of tau are considered the drivers of the protein's self-assembly [21]. These motifs are tau residues 306-311 and 317-355 [102] which can interact with one another promoting tau dimerisation [103]. Additional recruitment of tau monomers and dimers forms a nucleation centre. Once a threshold size is reached, tau oligomerisation proceeds in a dose- and time-dependent manner [104]. The tau oligomers undergo further hyperphosphoryl-ation and change their conformation taking on a beta-sheet structure which makes them insoluble [99]. Fusion of the oligomeric structures results in the appearance of paired helical filaments and culminates with the formation of neurofibrillary tangles [105, 106]. In this process, tau passes through a state in which the protein is hyperphosphorylated, mislocalized, soluble, and conformationally changed but not yet fibrillar, a state which seems to be the most toxic form of tau [99, 107, 108]. Transgenic mice inducibly expressing P301L tau showed memory improvement and reduced neuronal loss when the mutant gene was switched off despite continued neurofibrillary tangle accumulation [109]. In fact, the presence of tangles does not alter electrophysiological functions, the capacity to respond to stimuli, or spine density [110]. It has even been suggested that tangles may sequester toxic soluble tau species, thereby serving as a protective mechanism [111].

The 3R tau isoform is relatively resistant to aggregation. Altering its composition or electric charge may promote aggregation, as happens in deletions of the positively charged Lys 280 residue, phosphorylation of tau on serines, tyrosines and threonines, acetylation of tau, or interaction with fatty acids or polyglutamic acid [17,112, 113]. In addition, if in a normal adult brain the ratio of 3R:4R tau is approximately 1, in AD a shift of this ratio towards 4R tau isoform has been documented [100].

Tau Truncation

Several tau fragments have been described in the different tauopathies. Already in 1993, Novak *et al.* showed that the core of tau within the paired helical filaments in AD consists of tau fragments of 12 and 9.5 kDa [114]. These fragments are resistant to proteases, have an abnormal conformation, and are more prone to misfolding and self-assembly into filaments than full-length tau [115]. In the following years, studies performed on the human AD brain and transgenic animal models of AD identified several sites of truncation which suggest that this post-translational modification of tau facilitates misfolding and may lead to the generation of tangles [116, 117].

Without going into further detail, it appears that cleavage of tau could generate more toxic fragments able to switch on a cell death cascade, or, on the contrary, fragments which drive tau aggregation leading to the loss of its normal function [17, 118]. A number of proteases that can truncate tau have been identified, including caspases, calpains, cathepsins, thrombin, *etc.,* and will be briefly discussed below [119, 120].

<u>Caspases</u>

Although several aspartate residues can be targeted by caspases, to date only Asp421 has been shown *in vitro* and *in vivo* to be involved in the development of tau pathology. This residue is targeted by caspases-1, -3, -6, -7, and -8 which, by removing the C-terminus, generate approximately 5 kDa smaller tau fragments [121, 122]. An increase in caspase activity and caspase-3 truncated tau is positively correlated with cognitive decline and tangle formation in aged wild-type mice [123], possibly related to the induction of apoptosis by the expression of $Tau_{151-421}$ in hippocampal neurons [124]. In addition, $Tau_{151-421}$ may also cause mitochondrial fragmentation and increased oxidative stress in cells [125].

Cleavage of tau at Asp314 by caspase-2 generates an N-terminal fragment able to dislocate glutamate receptors from dendritic spines and cause synaptic dysfunction [126]. Phosphorylation of Ser422 can prevent tau truncation by caspase-3 *in vitro* and might have the same effect *in vivo* [127].

Calpains

Calpains (with the 2 major forms, calpain-1 and calpain-2) are cysteine proteases activated mainly by calcium [17]. Calpain cleavage of tau leads to the formation of a 17 kDa neurotoxic fragment and is an event described in many tauopathies [128]. Activation of calpain (described in the frontal cortex of patients with AD) [129] may be caused by increased calcium influx through the N-methyl-D-aspartate (NMDA) receptors [130] as well as by ERK-mediated phosphorylation, a pathway activated by Aβ in neurons [131]. Recent studies indicate that altered calpain activity may be an early event in AD, preceding tau hyperphosphorylation and loss of post-synaptic markers [132].

Cathepsins

Cathepsins are lysosomal proteases involved in protein quality control through the proteasome system and autophagy, which remove harmful aggregated proteins [133]. Active cathepsins D and B have been found in amyloid plaques in the brain of AD patients [134]. Upregulation of cathepsin D and its gene expression occurs as a response to altered lysosomal function in the hippocampus of cognitively impaired patients [135, 136]. Cleavage of tau by cathepsin D between amino acids 200 and 257 generates 29 kDa tau fragments [137] and it appears that, at least *in vitro*, tau hyperphosphorylation enhances this process, as opposed to tau degradation by calpain, thrombin, or caspase-3 [137]. As lysosomal enzymes, it is still unclear how cathepsins gain access to tau in neurons but it has been suggested that impaired translocation of tau across the lysosomal membrane could lead to its cleavage [138].

Thrombin

This serine protease formed through proteolytic cleavage of prothrombin has been found in tangles and amyloid plaques in brains of AD patients [139], suggesting that thrombin could act on tau in the brain. *In vitro*, tau is cleaved by thrombin on several arginine and lysine residues generating tau fragments which are less able to promote microtubule assembly [140]. Similar to cleavage by caspases and calpains, phosphorylated tau appears more resistant to thrombin cleavage while dephosphorylation of aggregated tau from the AD brain reduces its susceptibility to thrombin degradation [141]. However, the importance of thrombin cleavage of tau in AD or other tauopathies is still questioned [142].

Asparagine Endopeptidase (AEP)

AEP is another lysosomal cysteine protease that translocates from lysosomes into the neuronal cytoplasm where it can cleave tau [143], α-synuclein [144], or Aβ precursor protein [145]. Aβ oligomers can activate AEP resulting in increased levels of tau fragments, mainly tau 1-368 [142], which can reduce microtubule stability and lead to neuronal loss [145].

AEP has recently emerged as a potential therapeutic target in tauopathies [145] after studies have shown that a novel agent, termed compound 11, can inhibit AEP as it crosses the blood-brain barrier, has no observed systemic toxicity, and results in significant reductions of AEP-cleaved tau fragments, synapse loss and improved performance in the Morris water maze in mice [146].

Puromycine-sensitive Aminopeptidase (PSA)

This aminopeptidase, found only in neurons and absent in glial cells or cerebral blood vessels [147], appears to be a protective factor against tau-induced neurodegeneration being able to degrade tau and control the levels of the protein. Its expression correlates inversely with the vulnerability of the tissues to tau pathology, being overexpressed in the cerebellum of patients with frontotemporal lobar degeneration or AD as compared to the frontal cortex [148]. These findings make PSA a possible therapeutic target in AD as well as other tauopathies [149].

Human High-temperature Requirement Serine Protease A1 (HTRA1)

This ATP-independent serine protease has an important role in tau degradation and prevents the accumulation of tau aggregates [150], being able to untangle tau and expose sites for subsequent cleavage in a second step in the region essential for aggregation, namely between residues Asp239 and Val399 [151]. In the AD brain, there is an inverse correlation between HTRA1 and plaque and tangle number, total amount, and phosphorylated tau [151]. In addition, HTRA1 is able to cleave members of the lipid transporting apolipoprotein E (ApoE) family, the ApoE4 variant being a major genetic risk factor for sporadic AD [152]. It appears that ApoE4 is cleaved more efficiently than ApoE3 by HTRA1 [153].

The Ubiquitin-Proteasome System (UPS)

Ubiquitin is a small peptide found in all eukaryotic cells that are bound by ubiquitin-ligase to target proteins which are subsequently recognized by the 26S proteasome for degradation [154, 155]. Growing evidence suggests that UPS dysfunction may be involved in AD pathogenesis. The production of oxidized proteins in the brains of AD patients [156] may further exacerbate this dysfunction [157] leading to the accumulation of protein aggregates and tau filaments [158]. The binding of phosphorylated tau aggregates to the 20S subunit of the proteasome inhibits its activity [91].

A Disintegrin and Metalloprotease 10 (ADAM10)

This zinc metalloprotease is an extracellular, membrane-bound protease that cleaves amyloid β precursor protein into non-amyloidogenic fragments but which can also cleave tau resulting in a $tau_{153-441}$ fragment [159, 160]. However, it is not clear how an extracellular enzyme could interact with intracellular tau, or if the aforementioned tau fragment has any pathophysiological effects [142].

Auto-Proteolysis of Tau

Recent studies have shown that tau is able to self-degrade without the need for a protease, through acetyl coenzyme A-induced autoacetylation of cysteine residues [161]. This process occurs when tau dissociates from microtubules [67]. However, it is not yet clear whether the two resulting fragments, $tau_{282-441}$ and $tau_{341-441}$, appear in the brain or cerebrospinal fluid of AD patients or if any downstream effects of these fragments contribute to AD pathogenesis [142].

Axonal Transport Impairment in Tauopathies

Aside from stabilizing the microtubules, tau has also an important function in regulating the axonal transport of proteins and organelles. There are two major classes of motor proteins that ensure fast axonal transport of organelles such as mitochondria or synaptic vesicles [162]: kinesin, which enables anterograde transport towards the synapse, and dynein, which mediates retrograde transport towards the cell soma [163 - 165].

Axonal transport requires intact microtubules, normally functioning motor proteins, correct attachment of cargoes to these proteins, and ATP stores supplied by mitochondria [17]. Neurofibrillary tangles-containing neurons exhibit impaired anterograde transport along axons and impaired retrograde transport in apical dendrites [166]. Accumulation of organelles in axons and cell bodies is a common finding in neurodegenerative diseases and leads to the appearance of axonal swellings and spheroids [17, 166].

Multiple mechanisms for these impairments have been described [17]:

- Acetylation, truncation, and mainly hyperphosphorylation of tau impair the microtubule-stabilizing function of tau, thereby leading to the loss of the microtubule tracks needed for normal axonal transport [167]. In addition, pathological fragments of tau are less able to promote assembly of the microtubule structure and integrate it into the cytoskeleton [168].
- Overexpression of tau results, at least in mice, in mislocalization of tau in dendrites, a feature which has been described in patients with AD as well even prior to tau aggregation and the emergence of clinical symptoms of dementia [169, 170].
- By interfering with the access of motor proteins to microtubules, tau slows the transport by reducing the velocity and the distance traveled by individual kinesin molecules [171 - 173]. In addition, kinesin light chains (a component that binds cargoes during anterograde transport) and dynein intermediate chain (a component of the dynein complex during retrograde transport) were found reduced in the frontal cortex of patients with Alzheimer's disease [174].

Together, reduced expression of GSK-3β-mediated phosphorylation of the kinesin light chain, and sequestration of available kinesin by tau [26, 175] contribute to the impaired axoplasmic flow and interfere with the synaptic integrity in AD.

Nuclear Tau Dysfunction

Abnormal phosphorylation of tau reduces its ability to translocate into the nucleus and bind to and protect nuclear DNA [176, 177], thereby allowing oxidative stress-induced DNA and chromosomal damage [17]. In addition, Aβ exposure as well, as viral infections may lead to nuclear accumulation of phosphorylated tau [178], which can be recruited by Tia1 cytotoxic granule-associated RNA binding protein (TIA1) to stress granules, sensitizing cells to stress [179]. The downstream events of nuclear tau dysfunction include:

- Altered heterochromatin organisation with cell-cycle re-entry and neuronal loss [180].
- Dysregulation of gene expression and rRNA synthesis leading to altered protein synthesis [181, 182].

Tau-Mediated Dendritic Dysfunction

Exposure to Aβ leads to mislocalisation of tau into the somatodendritic compartment [17]. Here, through phosphorylation of the NR2B subunit of the NMDA receptor, a reaction mediated by the non-receptor-associated tyrosine kinase Fyn, tau causes neuronal excitotoxicity mediated by Aβ [64, 183]. In addition, disturbed functioning of NMDA receptors, calcium entry through L-type calcium channels, and insufficient mitochondrial calcium buffering lead to an increase in intracellular calcium [184] which can activate calcineurin [185]. In turn, calcineurin dephosphorylates the GluR1 S845 subunit of the AMPA receptors [186] and leads to internalization of the AMPA receptors, thereby potentiating synaptic dysfunction and loss of dendritic spines [187].

Another way through which dendritic tau exerts neurotoxic effects is through its interplay with katanin and spastin, enzymes that depolymerize the microtubules [165]. Dendritic tau recruits tubulin tyrosine ligase-like 6 (TTLE6) which polyglutamylates the microtubules, rendering them susceptible to spastin cleavage [188].

Tau and Mitochondrial Dysfunction

To support neuronal function, mitochondria must be delivered from cell bodies, where biogenesis of mitochondria takes place, to distal parts of axons *via* anterograde transport [189, 190], while damaged mitochondria must be returned to the soma *via* retrograde transport to undergo mitophagy [190]. Mitochondrial trafficking is essential for providing ATP to the synapses, for calcium buffering, and to ensure mitochondrial repair and degradation [191] by transporting "young" mitochondria from the soma to distal regions where they will fuse and share their components with other mitochondria promoting their repair [189]. The impairment of mitochondrial transport is mediated, at least partly, by GSK3 and axonal protein phosphatase 1 (PP1) activated by filamentous but not soluble forms of tau [192]. GSK3β phosphorylates and deactivates the motor proteins and hyperphosphorylates tau [193], leading to destabilization of microtubules. In addition, abnormal tau has been shown to trap the kinesin motor protein JIP1 in the soma, and prevent it from loading cargoes onto the kinesin system for anterograde transport along the axon [194, 195].

For retrograde transport, dysfunctional mitochondria must be included in autophagic vacuoles for which autophagosomes fuse with late endosomes and form amphisomes in distal axons [189, 196]. Accumulation of these amphisomes at axonal terminals has been described in AD, presumably due to Aβ mediated interruption of the coupling with motor proteins [197].

In order to meet the energy demands of the cell, mitochondria can rapidly change their size and shape, these morphological changes are being regulated by the fission and fusion proteins [189]. Three GTPases govern fusion: mitofusins 1 and 2 (MFN1 and MFN2), situated on the outer membranes of adjacent mitochondria ensure outer membrane fusion, while the fusion of inner membranes is mediated by OPA1 (optic atrophy 1) which is located in the intermembrane space [179, 195, 198]. Fission is mediated by another GTPase, dynamin-related protein 1 (DRP1), which is recruited and assembled at the outer membrane by additional proteins like mitochondrial fission factor (MFF) and mitochondrial dynamics proteins of 49 and 51 kDa (MiD49 and MiD51) [179, 199].

Altered mitochondrial dynamics has been found in human AD as well as in animal models of AD, with increased fission being described upon exposure to elevated levels of Aβ and hyperphosphorylated tau [200]. Excessive fission of mitochondria affects the integrity of cristae and impacts energy production [201]. In addition, in a mouse model of AD, Zhang and coworkers [202] found even a novel mitochondrial phenotype, termed "mitochondria-on-a-string", consisting of long mitochondria with the bulbous parts of the organelles connected to form a tunnel-like structure. The formation of these structures is attributed to calcium influx and likely represents an attempt to make them more resistant to mitophagy [203]. However, a larger diameter of mitochondria could interfere with their entrance into axons and dendrites and thereby lead to synaptic dysfunction [189].

Research has shown that in AD, impairment of axonal transport of mitochondria precedes the accumulation of protein aggregates and is linked with an imbalance of fission/fusion proteins, increased levels of phosphorylated tau and Aβ, and increased oxidative stress [204, 205].

Tau Seeding and Propagation

The neuropathological staging of AD brain (Braak staging of AD) comprises six stages of increasing severity of amyloid plaque and neurofibrillary tangle accumulation in different brain regions as follows [206]:

- Braak stage I – in the entorhinal and peripheral cortex
- Braak stage II – in the hippocampus, mainly in the CA1 region
- Braak stage III – in the limbic structures
- Braak stage IV – in the amygdala, thalamus, and claustrum
- Braak stage V – in isocortical areas
- Braak stage VI – in the primary motor, sensory, and visual cortical areas

Based on this typical progression in time and space of tau pathology in AD, which correlates with the clinical cognitive decline [207] as opposed to beta-amyloid pathology, it is likely that these affected brain regions are anatomically connected [17] and that tau has prion-like properties being able to propagate trans-synaptically between cells from an initial restricted area [208]. A number of studies showed that various forms of injected tau protein such as fibrils [209], filaments [210], monomers [211], or oligomers [212] can spread from the region injected to anatomically connected brain areas.

Current *in vitro* evidence shows that tau secretion can be mediated by:

1. Release from synaptic vesicles [17, 213],
2. Translocation across the membrane [214],
3. Secretion into extracellular vesicles (exosomes) [17, 215],
4. Transport through tunneling nanotubes [17, 216].

Reuptake of tau from the extracellular space occurs through:

1. Bulk endocytosis [217],
2. Macropinocytosis by heparan sulfate proteoglycans [218], and/or
3. Clathrin-mediated endocytosis [219].

In contrast to the small proportion of tau released in exosomes and ectosomes [214, 220], the majority of released tau is not surrounded by lipid membranes [17], occuring in the absence of cell death [203]. Stimulating neural activity increases tau release, which, in turn, activates neural activity in a positive feedback loop [221, 222]. Structural changes, such as tau fragmentation and/or oligomerisation, appear to increase the propensity of tau to propagate between cells [17]. In addition, depolarisation increases tau release in AD nerve terminals, which may suggest that tau cleavage facilitates its secretion [223].

The demonstration of tau secretion from neurons changed the assumption that tau in cerebrospinal fluid is passively released from dead neurons [224]. This new finding is in line with the time course of tau in the CSF, with maximal CSF-tau levels reached in Braak stages 3-4, when the neurofibrillary tangles are confined to the limbic areas of the temporal lobe which represent less than 10% of brain volume [214, 224], while in later stages it remains stable or may even fall despite widespread neurofibrillary pathology and neuronal loss [206, 214].

In addition, not only does tau spread interneuronal *in vivo*, but the delivery of different strains of tau to the hippocampus or entorhinal cortex leads to the appearance of the same strain in neighboring cell bodies and then in distant brain regions connected to the initial cells, giving birth to pathological phenotypes of different tauopathies [225 - 227]. Thus, it appears that tau toxicity is spread *via* a prion-like mechanism with induction of a specific, "templated" conformation of tau protein in recipient neurons. Evidence for this hypothesis has accumulated from both cellular [224, 228, 229] and transgenic [230 - 232] tauopathy models. There may be additional features that may vary in the various tauopathies because human tauopathy-derived aggregates affect distinct neuronal populations in different brain regions in mice [17].

The propagation of tau can further be accelerated through the activation of inflammatory pathways. Extracellularly released tau can promote microglial activation [233] resulting in high levels of pro-inflammatory cytokines which promote tau phosphorylation [234]. A recently published research showed that tau seeds can activate the inflammasome *via* the NLRP3–ASC axis (Nod-like receptor pyrin-domain containing 3–apoptosis-associated speck-like protein containing a CARD) leading to increased spreading of tau pathology and that inhibition of the NLRP3–ASC axis inhibits tau-seeded tau pathology [235]. In addition, microglia can actively phagocytose soluble and insoluble forms of tau followed by its exocytosis, thereby contributing to tau propagation. Depleting microglia can significantly reduce tau pathology [236].

Amyloid-Beta and Tau

There is a striking neuropathological difference between amyloid deposition and neurofibrillary tangle pathology in the course of Alzheimer's disease. Amyloid deposits start to form in the basal portions of the frontal, temporal, and occipital lobes with sparing of the hippocampus in stage A, followed by all cortex except the motor and sensory one and mild deposits of amyloid in the hippocampus in stage B, and culminate with involvement of the entire neocortex, hippocampus, and subcortical regions like the striatum, thalamus, hypothalamus, cerebellum, subthalamic, and red nucleus in stage C [206]. A more recent temporal pattern of the amyloid deposition describes five phases [237]:

- In phase 1–Aβ deposits in the frontal, parietal, temporal, and occipital neocortex
- In phase 2 – Aβ deposition is present also in the entorhinal region, CA1, the insular cortex, amygdala, the cingulate gyrus, the presubicular region, and the molecular layer of the fascia dentata.
- In phase 3 – deposits appear in subcortical regions: the caudate nucleus, claustrum, basal forebrain nuclei, substantia innominata, thalamus, hypothalamus and mammillary bodies, lateral habenular nucleus, and in the white matter; up to 45% of cases show amyloid deposits also in the central gray matter of the midbrain, the colliculi superior and inferior, CA4, the red nucleus, and the subthalamic nucleus.
- In phase 4 – additional Aβ deposits are visible in the inferior olivary nuclei, the reticular formation of the medulla oblongata, the substantia nigra, while the central gray matter of the midbrain, superior and inferior colliculi, the red nucleus, and CA4 constantly exhibit the amyloid deposits.
- In phase 5 – deposition of Aβ continues in the reticular formation of the pons, the central and dorsal raphe nuclei, the locus coeruleus, pontine nuclei, the parabrachial nuclei, dorsal tegmental nucleus, reticulotegmental nucleus of the pons, and the cerebellum (mainly in the molecular layer and less often in the granular layer and cerebellar nuclei).

However, amyloid burden fails to correlate with the duration or severity of AD [238]. Neurofibrillary tangle pathology has a different temporal course, the first region affected being the limbic cortex while the neocortex exhibits NFTs only in the late stages of AD as discussed earlier [206]. In contrast to plaque burden, the link between NFTs and disease progression is stronger [225, 239].

The link between Aβ and tau is further complicated by findings of studies performed in APP and PS1 transgenic mice which showed that, although beta-amyloid showed significant increases along with increased cortical plaque burden from 2-3 months onward, synaptic plasticity impairment and plaque formation in the hippocampus were delayed until 5-8 months [240]. This is in line with findings of other researchers which found robust Aβ deposits and significant neuropil abnormalities but no hippocampal neuronal loss in Swedish-type APP transgenic mice [241]. Despite the non-significant neuronal or synaptic loss, and even slight facilitation of short term potentiation (explained as a compensatory mechanism) at 6 months, long term potentiation was found completely obliterated in 8 and 15 months old animals as opposed to good LTP in age-matched wild type animals [240]. It appears that soluble oligomers of beta-amyloid can also have detrimental effects on synaptic transmission by decreasing the pool of releasable synaptic vesicles. In addition, although these animals did not show a significant cortical neuronal loss, regions with high neuron density, for example, the granule cell layer of the dentate gyrus, did show loss of neurons [242]. The above-mentioned findings suggest that additional temporal and pathological factors are needed for the massive neuronal loss seen in human AD [243], highlighting our incomplete knowledge of the complex relationship between Aβ deposition, tau aggregation, and neurodegeneration. Other factors, like astroglial and microglial activation and neuroinflammation, must also be taken into account, and these may be altered in transgenic animal models of AD.

This raises the question of what the toxic elements are: amyloid plaques or neurofibrillary tangles? The discovery of the first mutation in the tau gene which caused an autosomal dominant neurodegenerative disease with extensive tau pathology but no Aβ accumulation, namely frontotemporal dementia with parkinsonism linked to chromosome 17, revived the studies focusing on tau [96, 244]. Since then, a number of mutations in tau leading to several neurodegenerative disorders in which there is no Aβ accumulation have been characterized [245, 246]. Knocking out tau attenuates Aβ-induced behavioral deficits without affecting Aβ levels in an APP mouse model of AD [247]. Moreover, in a P301S transgenic model hippocampal synaptic loss, morphological changes in the hippocampal neurons, such as axonal spheroids, and microglial activation appeared 2 months before neurofibrillary tangles could be demonstrated [248]. More recently, Rudinsky and coworkers showed that in a mutant transgenic mouse model, the presence of NFTs was not related to the ability of visual experience to induce expression of early genes linked to synaptic plasticity [249].

From this large amount of research, it became clear that early signs of neurodegeneration, like synapse loss or defective axonal transport, do not require the presence of NFTs despite the fact that the tangles contribute to the later manifestations of the disease [250]. Attenuating the expression of FTDP-17 mutant tau in an inducible tauopathy model improved the behavioral deficits and neuronal loss in spite of NFT numbers being unchanged or even increasing [109]. In other words, NFTs are not the "toxic" tau species but rather soluble oligomers are the culprits. Tau oligomers appear as dimers, tri- or tetramers, are present in neuropil threads, pre-tangles, intraneuronal NFTs as well as in the extracellular space [225], and appear on slices as granular assemblies sized between 4 to 50 nm [225, 251]. When added to hippocampal slices, tau oligomers reduced LTP, an effect blocked by antibodies against these oligomers [225]. Wild-type mice injected with tau oligomers performed worse on object recognition and memory tests followed after months by widespread filamentous tau pathology [252].

To date, it is believed that tau is released as a monomer from normal neurons in response to neuronal activity *in vitro* [213] and *in vivo* [253], as well as from dying cells into the interstitial fluid. Tau oligomers can modify the size and conformation of released tau, the smallest toxic species being trimerized tau which is harmful in nanomolar concentrations [254, 255]. Further, tau oligomers starting with trimeric tau can be taken up by cells through endocytosis [256] into both soma and axons [217] and induce conformational changes to intracellular tau as well as its detachment from microtubules. Tau pathology then spreads as outlined above.

To summarize, it appears that tau and Aβ cooperate synergistically in the pathogenesis of AD [257, 258] and will be discussed further.

TAU IN THE PATHOGENESIS OF ALZHEIMER'S DISEASE

The complex pathogenesis of Alzheimer's disease is yet incompletely elucidated. What we know so far from *in vitro* and animal models of AD, corroborated with pathological findings in human AD is discussed below.

Abnormal or excessive Aβ oligomer formation in cortical neurons including those in the temporal, basal frontal, and occipital regions (due to genetic or predisposing factors) initiates the pathogenetic cascade [225] by entering neurons of the entorhinal cortex either in the presynaptic terminals of cortical axons or through anterograde vesicular axonal transport. It is known that the entorhinal region receives massive input from the cortex and is connected to the hippocampus [259]. Endosomal processing of APP leads to enlargement of endosomes and impairment of their function with disruption of the endosomal and lysosomal pathways [260], endocytosis of surface receptors for neurotransmitters and neurotrophic factors, thereby interfering with synaptic signaling and compromising the trophic support [261]. Downstream events require the presence of a small amount of somatodendritic tau which is not attached to microtubules [64]. Further, although the link between Aβ oligomers and tau oligomers is a missing piece, it seems reasonable to assume that Aβ induces phosphorylation of tau [262] and the production of tau oligomers [225] which then spread to connected neurons as described above from the entorhinal cortex to neighbouring and distant neurons. Basically, Aβ plays a similar role to that of tauopathy mutations in exacerbating the toxicity of tau and results in NFT formation [248]. Research focusing on this issue showed that the likely mechanism is an increase in tau truncation mediated by caspases and calpain [264, 265] causing aggregation of the microtubule binding-containing C-terminal tau fragments. In addition, it appears that tau abnormalities induced by Aβ might in turn increase Aβ production [263, 266]. Studies have confirmed that eliminating tau confers protection against the harmful effects of Aβ accumulation in a mouse model of AD [267, 268].

Abnormal phosphorylation of tau decreases its binding to microtubules and this "loss of function" destabilizes microtubules resulting in cellular damage [269]. However, there are also "gain of function" mechanisms of tau toxicity which do not depend on tau's microtubule-binding ability [244].

Tau filaments at the same physiological concentrations as normal tau inhibit axonal anterograde transport either by inhibiting the interaction of cargoes with kinesin [270, 271] or by disruption of the attachment of kinesin to microtubules [272], an issue not yet settled. Tau-induced impairment of axonal transport leads to mislocalization of proteins and organelles, especially mitochondria, to the soma [171, 244]. Sequestration of mitochondria, the main source of energy and calcium buffering in cells, away from the areas of high energy demand and calcium influx of the axons, such as the synapses and nodes of Ranvier, significantly impacts cellular homeostasis [194].

Calcium homeostasis is dysregulated in AD [273] due to the presence of Aβ which can either create a non-specific cation channel or alter the function of ion channels at synaptic sites such as the NMDA and AMPA receptors [274, 275] or within membranes of intracellular calcium stores. Mitochondria have a crucial role in buffering excess cytosolic calcium. Wang and Schwarz [276] showed that local elevations in calcium can arrest mitochondria along their route of transport thereby securing mitochondria where needed. Mislocalized, perinuclear mitochondria fail in performing this task. In addition, mutations of tau can alter the function of voltage-gated calcium channels [277]. In an FTDP-17 tau mouse model, hyperphosphorylated tau interacted selectively with c-Jun N-terminal kinase interacting protein (JIP1) which controls the association of cargoes with the kinesin motor complex and led to mislocalization of JIP1 to the cell soma and consequently to axonal transport deficits [278].

In addition, mitochondria exhibit abnormal fission/fusion [279], abnormal levels of respiratory proteins [280], and deficient energy metabolism observed on PET scans [281]. In fact, tau synergizes with Aβ at the level of mitochondria to dysregulate several oxidative phosphorylation proteins [258], leading to loss of mitochondrial membrane potential and increased production of reactive oxygen species [282]. Tau also recruits intrinsic mitochondrial proteins to the exosomal proteome, thereby disrupting autophagy and leading to the accumulation of autophagy-associated proteins in granulovacuolar lesions described in AD [283].

The tau N-terminal projection interacts with the plasma membrane and non-receptor tyrosine kinases such as fyn and src [64], especially in the dendrites where tau is redistributed in AD in response to Aβ [284]. Fyn upregulates the NMDA receptor activity flooding the dendrites with calcium which cannot be buffered by dysfunctional mitochondria and results in damage of postsynaptic sites and excitotoxic neuronal death [266]. The activation of "executioner" caspases and the excess generation of oxidative species by the damaged mitochondria are both involved in apoptosis and neurodegeneration in AD [285]. These dysfunctions may be exacerbated by glia-mediated changes of glutamate reuptake and inflammatory changes of microglia caused by Aβ [286].

But these kinases also regulate mitotic activity and neuronal differentiation [263, 287]. The reactivation of the genes associated with cell cycle re-entry and neuronal differentiation [288, 289] leads to the appearance of apoptotic elements and aberrant cellular changes such as neuropil thread formation [290 - 293].

Tau kinases such as GSK3β, PKA, MARK1, and others play important roles in establishing neuronal polarity. Thus, neuronal dedifferentiation and loss of axonal identity induced by the dysregulated cell cycle reentry events could lead to tau mislocalization mainly to the axon and the activation of aberrant developmental events such as the growth of neuropil threads (axon-like neurites) which frequently originate from dendrites [293, 294]. Known predisposing factors for late-ons *et al.* Alzheimer's disease, such as head trauma, could str*etch* axons facilitating calcium entry which destabilizes the axonal microtubules, releasing tau and exposing it to hyperphosphorylation, abnormal cleavage, and abnormal interactions with various signal transduction pathways [295, 296].

Another hallmark of AD, synaptic dysfunction and loss, may be caused by the accumulation of somatodendritic tau due to prior axonal injury, calcium dysregulation, or inhibition of fast axonal transport by dysfunctional interaction of tau with the microtubules [36, 263] enabling the interaction of somatodendritic tau with fyn or src family kinases [263]. Activation of these kinases mediates endocytotic Aβ/APP uptake which potentiates the excitotoxic effects of external Aβ *via* synaptic glutamate receptors and the toxic interaction between Aβ and postsynaptic density associated tau leading to supplemental dendritic tau accumulation, microtubule loss, tau secretion, and progressive degeneration of dendrites [297].

Tau may also undergo toxic interactions with Aβ and APP in the endosome, either directly or through intermediates [298, 299], and thereby contribute to tau seeding and spread of tau pathology since tau oligomerization is favored by the association of the protein with membrane elements [113]. Studies have suggested that oligomerization and enhanced tau cleaving, as seen in tauopathy mutations, lead to the formation of more toxic fragments which can be secreted through multiple pathways [300].

As neuronal axons and dendrites degenerate, they release toxic tau species which activate microglia and are taken up and further spread through exocytosis by these glial cells [30]. The proinflammatory cytokines secreted by microglia, such as IL-1β, TNF-α, IL-6, co-activate astrocytes which will lose their ability to promote neuronal survival, growth, and synaptogenesis and become toxic for neurons [301]. In addition, tau accumulation in astrocytes and displacement of GFAP are followed by induction of heat shock proteins and disruption of the blood-brain barrier [302].

TAU-TARGETED THERAPY IN ALZHEIMER'S DISEASE

With a projected 13.5 million citizens affected by Alzheimer's disease by 2050 in the US and 131 million affected worldwide [303], the need for treatments to delay the onset, slow the progression, and improve symptoms of AD is obvious. The associated costs are rising rapidly – from 1 trillion globally in 2018 and it is estimated to double by 2030 [304]. Currently, four drugs are approved: 3 cholinesterase inhibitors (donepezil, rivastigmine, and galantamine) for mild to moderate AD, memantine, an NMDA receptor antagonist, approved in 2003 for moderate to severe AD [305]. Neither of them is disease-modifying.

Historically, Alzheimer's disease drug development showed a remarkably high failure rate. Of the 244 drugs tested in clinical trials between 2002 and 2012, only memantine has approved by the FDA, which has a success rate of 0.4% [306]. For over 25 years the search for effective drugs was dominated by amyloid-beta centered strategies, with largely disappointing results, despite the fact that pure amyloidosis is asymptomatic [307] and despite the characterization of tau in neurofibrillary tangles already in 1988 by Wischik and coworkers [308]. In addition, tau pathology correlates better with the clinical symptoms and progression of AD [309]. Only the discovery of tau mutations, that are able to cause neurodegenerative diseases on their own [10], has revived the research focusing on tau and led to pursuing tau-targeted therapies.

The main challenge in tau-targeted therapies is the diversity of physiological and pathological tau molecules. Alternative splicing gives rise to 6 tau isoforms; there are over 80 possible phosphorylation sites on the protein, the phosphorylation pattern being established by an interplay between various kinases and phosphatases [310]. A number of other post-translational modifications, like truncation [311], nitration [312], glycation [313], glycosylation [314], or ubiquitination [315] further expand the diversity of tau molecules. These various proteins – the tau proteome – may be different in tauopathies as compared to the physiological tau proteome, but with significant overlap [316]. In addition, the functions of tau may differ depending on its localisation: axonal [317], dendritic

[318], nuclear [167], or extraneuronal [253]. Further, given the multiple facets of its neuronal function, deficits in tau could have serious consequences. Although mice completely lacking tau have been shown to have normal learning, memory, and cognition [319], lack of tau has been associated with iron deposition in the brain [320] and insulin resistance [321]. Furthermore, the full consequences of tau knockdown in the adult brain are yet unknown.

To further complicate the issue, tau may also play an important role in cell signaling: it may serve as a "phosphorylation sink" for the p25-Cdk5 complex, isolating it from other death-inducing substrates [322]; it may interfere with the tyrosine kinase family Src/Fyn signaling at dendrites [64], or modulate insulin signaling by interacting with phosphatases and tensin homolog PTEN [94].

As with other Alzheimer's disease drugs, clinical trials are challenging due to the slow disease progression rate (trials often last over 18 months) and to the cognitive performance measures recognized as endpoints by regulatory agencies, which are less robust than biochemical or physiological measures [305, 323].

In the following section, we provide an overview of the recent developments and various points of intervention with disease-modifying molecules potentially targeting tau protein which are pursued in clinical development for Alzheimer's disease.

Modulating Tau Phosphorylation

Since tau is hyperphosphorylated in neurofibrillary tangles, and phosphorylation of tau is under the tight control of various kinases, modulating the activity of these enzymes could be a potentially useful approach. Whether a specific kinase or a group of kinases should be targeted remained a debate [324].

Tideglusib (NP-12, NP031112) is a non-competitive GSK3β inhibitor [325, 326] belonging to the thiadiazolidinone class, available as an oral formulation. In preclinical studies, the molecule reduced tau phosphorylation, neuronal loss, and gliosis in mouse entorhinal cortex and hippocampus and was able to reverse spatial memory deficits in transgenic mice [327]. In a phase IIa clinical trial (NCT00948259) sponsored by NOSCIRA SA, thirty patients with mild to moderate AD received escalating doses ranging between 400 and 1000 mg/day of tideglusib for 20 weeks and were compared to placebo at 3 sites in Germany. The drug was well-tolerated and researchers reported a trend toward improvement of cognition on the MMSE and ADAS-Cog tests [328]. However, the following phase IIb study conducted in 55 centers across Europe between April 2011 and July 2012 enrolling 308 patients and known as ARGO (NCT01350362), sponsored by NOSCIRA SA and ICON Clinical Research, compared 1000 mg taken once daily, 1000 mg taken every other day, and 500 mg tideglusib once daily against placebo and missed its primary endpoints and some secondary ones [329].

In 2009, the compound received FDA (Food and Drug Administration) and EMA (European Medicines Agency) orphan drug status and was evaluated in 146 patients with another tauopathy, progressive supranuclear palsy, in the TAUROS study (NCT01049399). Patients received 600 mg, 800 mg tideglusib once daily or placebo, and changes were measured on the progressive Supranuclear Palsy Rating Scale of Globe. Unfortunately, tideglusib missed again its primary endpoints [330]. As of 2019, tideglusib in clinical trials for myotonic dystrophy.

Lithium has many effects, among them is being the ability to inhibit GSK-3β. Some observational studies, with a low number of subjects included and low compliance, claimed to have found positive effects of lithium in cognitive impairment [331]. In June 2014, the New York State Psychiatric Institute started recruiting patients to evaluate the efficacy and side effects of low doses of lithium in the treatment of agitation/aggression in 80 patients with AD. It was a phase II randomized, double-blind, placebo-controlled trial (NCT02129348) extending over 12-weeks, which ended in January 2020 but the results are not yet published [321].

Another kinase evaluated as a potential therapeutic target is tyrosine kinase Fyn which phosphorylates the N-terminal domain of dendritic tau. **Saracatinib (AZD0530)** is a Fyn inhibitor developed originally by Astra Zeneca for various types of cancer but for which it lacked efficacy. When evaluated for Alzheimer's disease, it was reported to rescue synaptic depletion, boost hippocampal synaptic density [332], and reduce memory deficits in APP/PS1 transgenic mice [333]. In July 2013, the University of Yale began a phase Ib study evaluating a 1-month course of 50 to 125 mg saracatinib daily in 24 patients with mild to moderate AD (NCT01864655) and reported it to be safe and well-tolerated with a reasonable penetration of the blood-brain barrier [334]. In December 2014, a phase IIa clinical trial (NCT 02167256) began enrolling 159 patients across 22 centers to compare the effect of a 12-month course of 100 and 125 mg saracatinib daily to placebo. The trial ended in March 2018 [321]. The investigators reported that the drug did not slow the course of the disease or the decline in cerebral glucose metabolism compared to placebo despite less shrinkage of the hippocampus and entorhinal cortex detected in the treatment arm [335]. In addition, the drug caused diarrhea and prominent gastrointestinal side effects. As of April 2019 saracatinib is evaluated in a trial at King's College London expected to end in September

2021 involving 30 patients with Parkinson's disease who experience delusions or hallucinations [324].

Activation of protein phosphatases may be an additional strategy to reduce tau phosphorylation [324]. However, protein phosphatase 2A (PP2A), which is the main enzyme that regulates tau phosphorylation in the brain, has several regulatory subunits and broad substrate specificity, making it difficult to be targeted to the right extent [324, 336].

Targeting Tau Acetylation

Lysine acetylation promotes tau aggregation and impairs tau function thus making tau acetylation inhibitors potential therapeutic molecules [324]. The non-steroidal drug **salsalate**, already used for arthritis, has been shown in a frontotemporal dementia mouse model to inhibit acetyltransferase p330-induced tau acetylation and neurodegeneration, thereby improving behavioral impairment [337]. In July 2017, a phase Ib clinical trial has started recruiting 40 patients with mild to moderate AD to evaluate the safety and tolerability of salsalate 1500 mg twice daily for 12 months compared to placebo (NCT03277573). The estimated study end date is July 2021; currently, it is active, but not recruiting [321, 338].

Targeting Tau Glycosylation

Tau glycosylation is a carefully balanced process in which O-linked N-acetylglucosamine (O-GlcNAc) transferase, known as OGT, catalyzes the transfer of GlcNAc to tau, and O-GlcNAcase (OGA) catalyzes the removal of GlcNAc [339]. Since glycosylation has been shown to affect phosphorylation and aggregation of tau, OGA modulators have been studied as disease-modifying therapies for tauopathies [324].

MK-8719 has been evaluated in a phase I study in 16 healthy volunteers and is further pursued in progressive supranuclear palsy, but transgenic mouse models of AD in which MK-8719 has significantly diminished the degree of brain atrophy suggests it may be useful in AD as well [340].

ASN120290 is an OGA inhibitor which, by increasing tau glycosylation, appears to stabilize tau in a soluble form. In animal studies, it has reportedly reduced insoluble paired helical filaments by 40% and neurofibrillary tangles by 80% [341]. Asceneuron conducted a phase I safety and tolerability study in healthy volunteers in 2017 and reported no significant adverse events or toxicity at the 13th International Conference on Alzheimer's and Parkinson's Disease [324]. In

July 2018, ASN120290 received orphan drug designation by the USA FDA for treatment of progressive supranuclear palsy, but clinical studies are not yet listed in Clinicaltrials.gov or other trial registries [338].

Nilotinib is an oral Abl tyrosine kinase inhibitor currently used for the treatment of chronic myeloid leukemia [342]. The molecule has been tested in preclinical models of Parkinson's disease and Lewy body dementia and, aside from other effects, lowered the levels of phosphorylated tau [343]. Thus, a phase II randomized, double-blind, placebo-controlled trial was started by the Georgetown University Medical Center in October 2016, aiming at recruiting 42 participants with mild to moderate AD (NCT02947893) to assess safety and impact on biomarkers and clinical outcome of 150 mg nilotinib taken orally for 6 months followed by 300 mg daily for 6 months *versus* placebo. The estimated end of the study was February 2020. It is listed as being active, not recruiting [338] but no published data are available so far.

Microtubule Stabilizers

Although hyperphosphorylated tau has a reduced affinity for microtubules, leading to its detachment and loss of the microtubule-stabilizing function, it has been shown that Aβ oligomers can also destabilize MTs and impair fast axonal transport through activation of calcineurin in tau-deficient mice [344], suggesting that MT destabilization might generally characterize neurodegenerative diseases.

The first microtubule-stabilizing molecule tested in tau-overexpressing mice was the antimitotic drug paclitaxel, followed by epothilone D (BMS-241027), which is a taxol-derived molecule that crosses the blood-brain barrier. In transgenic animal models, epothilone D was able to reduce tau pathology and hippocampal neuronal loss as well as to restore spatial memory deficits [345]. A phase I multicenter, randomized, double-blind, placebo-controlled clinical study (NCT01492374) was run to assess the safety, tolerability, and pharmacodynamics of the drug in 40 patients with mild AD between February 2012 and October 2013, but no published results are available and the further development of the drug has been halted [324].

Another taxol-derived small molecule, intravenous **abeotaxane or TPI-287**, was recently tested in a phase I trial (NCT01966666) to assess safety and tolerability in 33 patients with mild to moderate AD. The trial recruited only 29 patients between November 2013 and September 2019, tolerance was poor in the patients receiving the drug [346] and cognitive endpoints did not improve significantly [324].

A third microtubule-stabilizing agent is **davunetide** (also known as **NAP** or **AL-108**). This 8 amino acid peptide displayed promising effects in tau transgenic mice by improving memory and behavior [347] and was tested to assess safety, tolerability, and effects on cognition of intranasal administration of AL-108 in 144 patients with mild cognitive impairment in a multicenter, double-blind, placebo-controlled study (NCT00422981). The study started in January 2007 and ran for 12 months. The treatment was safe and well-tolerated but lacked efficacy in slowing cognitive decline [348]. The molecule was further evaluated in progressive supranuclear palsy (NCT 01110720) and other various tauopathies (NCT01056965) but, at least for PSP, missed all endpoints [349].

Tau Aggregation Inhibitors

The various tau species formed during aggregation (monomers, oligomers, prefilaments, granules, fibrils, insoluble aggregates) have been evaluated as targets for therapeutic intervention but the various small molecules tested, although potent *in vitro*, lacked efficacy *in vivo* and failed to significantly improve cognition [350].

The first direct inhibitor of tau protein aggregation is **methylene blue**, also known as methylthionine chloride. It was a thiazine dye known since the end of the nineteenth century [94]. It was first used for malaria, later as an antibacterial and antiviral agent, and even in the treatment of cancer [94]. The tau anti-aggregating effect of phenothiazines was discovered in 1996 by Wischik [351] but at that time, therapy research for AD was focused mainly on amyloid-beta.

Methylene blue occurs in two main forms: the oxidized cationic form, and the reduced form, also known as leucomethylene blue, leucomethylthioninum, or LMT, which is classified as a phenothiazine [94]. Methylene blue is converted into LMT, which penetrates the blood-brain barrier and has been shown to inhibit microtubule assembly and prevent tau interaction *in vitro* as well as in cell and animal models of tauopathy [324]. A phase II randomized, dose-ranging clinical study was initiated in August 2004 by TauRx Therapeutics to compare 30mg, 60 mg, and 100 mg **Trx-0014** (also known as **Rember**) taken three times/day for 24 months to placebo in 321 patients (actually 323 enrolled) with mild to moderate dementia not taking acetylcholinesterase or memantine, the primary endpoint being assessed on the ADAS-Cog scale. The study lasted until December 2007 (NCT00515333) [338]. Although the investigators reported some improvements in symptoms, the drug had many undesirable side effects, such as diarrhea, the urgency of urination, painful urination, and dizziness [352]. The trial continued with a phase II open-label extension study (NCT00684944) recruiting 111 patients between September 2007 and December 2010 to further assess the cognitive effects of 30 mg and 60 mg daily of TRX0014 for 12 months, but no results were published. A second-generation formulation of the methylthioninium moiety, leuco-methylthioninium bis-hydro methanesulfonate, known as **LMTM, LMT-X, or TRx0237**, with better absorption, bioavailability, and tolerance has been developed and subsequently tested. A first study assessed the safety of 250 mg TRx0237 daily taken for 8 weeks in 9 patients already taking other medications for AD (NCT01626391) and reported prolongation of the QT interval and increase in troponin in 20% of patients [338]. A phase 2 safety study of 250 mg/day of TRx0237 administered for 4 weeks to patients with mild to moderate Alzheimer's disease began in September 2012 but was prematurely terminated for administrative reasons in April 2013 [353]. A randomized, controlled, double-blind, parallel-group phase 3 clinical trial (NCT01689233) was started in September 2012, in which 95 centers enrolled 800 participants with either all-cause dementia or mild to moderate AD (but MMSE score above 20) to assess the safety and efficacy of 100 mg TRx0237 twice daily as compared to placebo after an 18-month treatment period. Primary outcome measures were the ADAS-Cog 11 and ADCS-CGIC batteries as well as the number of participants who tolerate the oral form of the drug. The trial was terminated in May 2016 but showed no effect on the primary endpoints, although after posthoc subgroup secondary analysis seemed to show efficacy [354]. Another phase III randomized, double-blind, placebo-controlled, parallel-group clinical trial aimed at evaluating 2 doses of TRx0237 (75 mg and 125 mg twice daily) for 15 months *versus* placebo in patients with mild to moderate Alzheimer's disease [338]. It started in January 2013 and was completed in November 2015 being conducted in 121 locations and enrolling 891 patients. For this study (NCT01689246), the inclusion criteria allowed the MMSE score to be as low as 14. As primary endpoints, the ADAS-Cog 11 and the ADCS-ADL23 batteries were used, as well as the number of patients who tolerated the oral doses of TRx0237. Again, the trial missed its primary endpoints with no improvement in the treatment arm [355]. In May 2013, a phase III multinational, randomized, double-blind, placebo-controlled, parallel-group clinical trial (NCT01626378) started and ran through February 2016 in 67 locations across Europe, North America, Australia, and Singapore, enrolling 220 participants to evaluate the safety and efficacy of TRx0237 200 mg daily for 12 months in patients with probable behavioral variant frontotemporal dementia (bvFTD) with an MMSE score above 20. Primary outcome measures were changed from baseline on Addenbrooke's Cognitive Examination (ACE-R), a psychometric tool widely used in FTD clinical research, change from baseline on Functional Activities Questionnaire, and change from baseline on whole brain volume as assessed by brain MRI [338]. The preliminary results are encouraging [356]. These three Phase 3 trials were not truly placebo-controlled since the placebo tablets included 4 mg of TRx0237 as a urinary and fecal colorant. Thus, the placebo group received a total of 8 mg/day of TRx0237 to help maintain blinding.

The molecule is still pursued, since in January 2018 TauRx started another phase III trial which aims at enrolling 180 people with all-cause dementia and Alzheimer's disease to compare a six-month course of 4 mg of LMTM twice daily to a different kind of placebo (LUCIDITY, NCT 03446001). It was originally intended to be run at 55 sites in North America, Belgium, Poland, and the U.K. Original primary endpoints were safety and 18F-FDG-PET imaging, while measures of cognition and activities of daily living as well as structural MRI were listed as secondary endpoints. In September 2018, the sponsor changed the trial protocol: a third treatment arm of 8 mg LMTM twice daily was added, leading to an increase in enrollment to 375 patients from 156 sites across Europe and North America; the dosing period was extended to 9 months; eligibility criteria were restricted to patients with mild cognitive impairment due to AD, with a positive amyloid PET scan and a Global Clinical Dementia Rating (CDR) of 0.5; primary outcomes were changed to include a composite measure of cognition and function from selected items from the ADAS-Cog and ADCS-ADL scales. Recruitment ended in October 2019 [338]. Subsequently, in November 2019, one of the three listed primary outcomes was changed from 18F FDG PET to ADAS-Cog 11. Further, the inclusion criteria were changed again allowing patients with more advanced disease to enter the study: the range of the MMSE was lowered from 20-27 to 16-27, the CDR was increased up to 2, and patients with a single epileptic episode were allowed to participate as well. Enrollment was again increased to 450, the study duration increased to 12 months and an open-label, 12-months extension phase was added. The estimated study completion date is December 2022 [338].

A number of other compounds belonging to several chemical classes, such as phenothiazines, polyphenols, benzothiazoles, porphyrins, and amino-thienpyrydazines, have been shown to inhibit both tau filament and Aβ fibril formation [357, 358] but no clinical trial was started for neither of these molecules.

Another potential anti-aggregation strategy is the multi-target-directed ligand approach [94]. Multifunctional compounds combine various pharmacophores targeting neurodegenerative processes into a single molecule [359]. **AZP2006** is an N, N'-disubstituted piperazine claimed to block tau phosphorylation and inhibit its aggregation [94, 324] as well as reduce the release of Aβ species [94]. Although it was able to improve cognitive abilities in mouse models of AD and has been tested in phase I clinical trials on AD [94], it is currently evaluated in phase II clinical trial (NCT04008355) sponsored by AlzProtect and coordinated from Lille to assess safety, tolerability, pharmacokinetics, and efficacy of the compound in patients with progressive supranuclear palsy [338].

Curcumin has also been proposed as a possible therapy in AD. A phase II clinical trial (NCT01001637) with curcumin started in October 2009 at Jaslock Hospital in India to evaluate the safety and efficacy of curcumin against placebo in 26 patients with probable AD and MMSE between 5 and 20 but the results have not been published [338]. A phase I/II trial with curcumin and Ginkgo with 3 arms (placebo, 1 g curcumin, and 120 mg Ginkgo, and 4 g curcumin and 120 mg Ginkgo daily for 6 months) was run between October 2004 and July 2006 at the Kwon Wah Hospital in Hong Kong sponsored by the Chinese University (NCT00164749) [338]. Changes in isoprostane level in plasma and changes in Aβ level in serum have been set as primary endpoints. The Aβ levels actually rose in plasma of treated patients, explained as an effect of curcumin to disaggregate amyloid beta in the brain and release it into circulation, while the isoprostane levels were similar. No changes in cognition could be detected, possibly because of the short duration of the trial [360]. Another phase II clinical trial (NCT00099710) was conducted between July 2003 and December 2007 at the UCLA Medical Center sponsored by the John Douglas French Foundation and the Institute for the Study of Aging (ISOA) to examine the safety and tolerability as well as the effects of curcumin dietary supplements in patients with mild to moderate AD. The study enrolled 33 patients randomly assigned to 2 doses of curcumin for the first 6 months, after which patients on placebo were switched to one of the doses of curcumin. The dietary supplement was well tolerated, but the researchers were not able to demonstrate clinical or biochemical evidence of efficacy [361]. Finally, trial NCT01811381 started in January 2014 in Los Angeles, which enrolled 80 participants with mild cognitive impairment being at risk of developing AD, tried to evaluate whether a formulation of curcumin with better intestinal absorption and brain penetration combined with aerobic yoga can alter clinical and biological markers associated with increased risk of AD. The trial is currently active but not enrolling new patients, and results have not yet been published [338, 362].

PE859 is a new tau inhibitor with a structure similar to that of curcumin, promoted by Japanese researchers [363]. Results of preclinical studies were promising both *in vitro* and *in vivo* [364].

Other tau aggregating compounds with pleiotropic effects are:

• SLM and SLOH, two carbazole-based cyanine compounds which inhibit tau aggregation, reduce tau hyperphosphorylation, attenuate neuroinflammation by inhibiting GSK-3β, and were able to improve pathological symptoms and memory deficits in mouse models of AD [365, 366].

- 2,4-thiazolidinedione derivatives were shown to inhibit both GSK-3β activity and tau aggregation [367].
- Rhein-huprine hybrids showed acetylcholinesterase and BACE1 inhibitory activity as well as tau and $A\beta_{1-42}$ anti aggregating properties [94, 368].

It is difficult to foresee further development of these compounds which have shown activity *in vitro* and some have been evaluated *in vivo* as well, but neither has reached the stage of clinical trials. However, given the complex pathogenesis of AD, it seems reasonable to assume that molecules capable of acting on at least two recognized targets, of which one belongs to the tau cascade, could show enhanced clinical benefit as compared to drugs with a single mechanism of action [94].

Antisense Oligonucleotides (ASOs)

This is an exciting approach for degenerative diseases in which a genetic defect could be identified. The efficacy and safety of ASO therapy have been proven for nusinersen for the treatment of spinal muscular dystrophy (NCT02193074), for eteplirsen for the treatment of Duchenne muscular dystrophy (NCT00844597, NCT01396239/NCT01540409, NCT02255552), and recently RG6042 was tested for treating Huntington's disease (NCT02519036) [94, 369].

ASOs are short oligonucleotides that specifically bind to pre-messenger or mature mRNA sequences and mediate degradation of the target mRNA or prevent translation, thereby reducing protein production [94]. Chemical modifications of the DNA or RNA phosphodiesters or ribose sugars, as well as the use of liposomes, cell-penetrating proteins, or viral vectors, can overcome the challenges of cellular uptake, susceptibility to nucleases-mediated degradation, and blood-brain barrier penetrance [94, 370, 371]. Based on the demonstrated 75% reduction of MAPT mRNA in the hippocampus and cortex of primates *via* intrathecally delivered ASOs [372, 373], a phase II randomized, double-blind, placebo-controlled clinical trial sponsored by IONIS Pharmaceuticals was designed to assess safety, tolerability, pharmacokinetics, and pharmacodynamics of multiple ascending doses of intrathecally administered **IONIS-MAPTRx** in patients with mild Alzheimer's disease followed by an open-label extension phase (NCT03186989). The trial started in June 2017, enrolled 46 patients, and is currently active but not recruiting [338]. It is very likely that ASOs will be pursued in other tauopathies, like frontotemporal lobar dementia, progressive supranuclear palsy, or corticobasal degeneration [94, 374].

Phosphodiesterase 4 Inhibitors

The mammalian phosphodiesterases (PDEs) comprise a family of up to 100 enzymes encoded by 21 genes, most of which are expressed as multiple variants due to alternative splicing and multiple transcription start sites [375]. Based on sequence homology, substrate specificity, and pharmacological properties, the enzymes can be divided into 11 families: 3 include enzymes that hydrolyze cAMP, 3 are selective for cGMP, and 5 enzyme families hydrolyze both cyclic nucleotides [376]. Compared with other tissues, the brain expresses the highest level of PDE activity [375].

The PDE4 family comprises 4 genes, each of which can be expressed as multiple variants, resulting in more than 25 individual proteins. Several preclinical studies with PDE4 inhibitors such as **Rolipram** showed that inactivating the enzymes improves memory and learning [377, 378]. Other studies suggested a therapeutic potential for PDE4 inhibition in AD by demonstrating reversal of beta-amyloid induced inhibition of the PKA/p-CREB pathway and LTP in cultured hippocampal neurons [379] as well as in mice expressing amyloid precursor protein and presenilin-1 [380]. Although Rolipram has been used to treat major depression in human patients, it is highly emetic and prevents proper dosing [381].

BPN14770 is an allosteric inhibitor of phosphodiesterase 4D (PDE4D) which binds to a primate-specific, N-terminal region of PDE4D [382]. In preclinical work, BPN14770 enhanced brain cAMP, hippocampal LTP, increased hippocampal phospho-CREB, and BDNF, which are markers of consolidated memory [383], and prevented Aβ42-induced memory deficits and synapse damage [384] in a mouse line expressing a humanized PDE4D sequence.

Between December 2015 and February 2016, a phase I randomized, double-blind, placebo-controlled ascending single-dose clinical trial (NCT02648672) was conducted in Michigan, sponsored by Tetra Discovery Partners, to evaluate the safety, tolerability, and pharmacokinetic profile of BPN14770 in 32 healthy subjects [338]. This study was followed between June and November 2016 by another phase I randomized, double-blind, placebo-controlled, multiple ascending-dose study designed to evaluate the safety, tolerability, and pharmacokinetic profile of BPN14770 in 77 healthy young and elderly subjects and to assess the cognitive effects of BPN14770 in healthy elderly subjects (NCT02840279), a study also conducted in Michigan [338]. Single doses up to 100 mg were well tolerated and had linear pharmacokinetics, while higher doses caused nausea. In the elderly, headache was the most common side effect. The 10 mg and 20 mg dose groups showed an improvement in working memory compared to placebo [385], according to a press release, although no formal results were published in peer-reviewed journals. In January 2017, a phase I randomized, double-blind, placebo-controlled study was initiated in Texas (NCT03030105) to evaluate the effect of BPN14770 10 mg and 50 mg daily in reversing scopolamine-induced cognitive impairment in 38 healthy volunteers [338]. The study design also allowed a comparison with 10 mg of donepezil and was conducted until May 2017 [338]. No results have yet been presented. In 2019, the company started PICASSO, (NCT03817684), a phase II randomized, double-blind, placebo-controlled, 3-arm parallel design study to evaluate the effects of BPN14770 in patients with early Alzheimer's disease [338]. Patients were offered either 10 mg or 25 mg bid of BPN14770 or placebo for 3 months. The primary outcome listed was changed from baseline in the Repeatable Battery for the Assessment of Neurological Status-Delayed Memory Index, while secondary outcomes were standard measures of memory, cognition, and function and RBANS total score. A total of 255 patients clinically diagnosed with early AD who were on stable doses of a cholinesterase inhibitor but not on memantine were enrolled at the 60 sites across the US. The trial ended in February 2020 but no results have yet been published [338]. However, in May 2020, Shionogi bought Tetra Therapeutics and disclosed that no safety issues were observed but PICASSO had missed its primary endpoint despite an above-median CDR-SB score demonstrated by subgroup analysis in patients in the 25 mg bid treatment arm [386]. As such, Shionogi decided to further develop BPN14770 [386]. BPN14770 is also tested with an orphan drug designation for the treatment of Fragile X in a phase II trial started in 2018 (NCT03569631) and completed in July 2020 [338].

Anti-tau Immunotherapy

To counteract tau pathology through immunotherapy, one must [316]:

- Characterize the pathological tau proteome in the target disorder.
- Find the proper epitope(s) that would capture the entire range of pathological tau species.
- Interfere with the propagation of tau pathology.

The AD tau proteome defines the range of tau forms and derivatives present in AD and significantly overlaps with the tau forms present in the normal brain. Still, certain tau moieties may be lost in disease states [316].

Due to the high number of post-translational modifications of tau, the range of epitopes is very wide but epitopes that appear predominantly on pathological tau or are more accessible in pathological tau molecules should be preferred [316]. It has been shown that truncation of tau at the N- and C-terminus promotes transition into pathological forms of tau [387] and that aggregation occurs *via* the microtubule-binding region (MTBR) [388]. However, tau fragments isolated from the CSF of AD patients rarely contain the MTBR [389], suggesting that this form of tau is degraded to prevent the formation of tau oligomers. Other post-translational modifications (phosphorylation, glycation, glycosylation, ubiquitination) create new epitopes on tau but, being highly fluctuating, it is difficult to characterize their distribution across the pool of pathological tau species [316]. Conformational changes of tau, depending on the binding partners, may also unmask some epitopes [311]. In addition, as opposed to Aβ, tau is located intracellularly but may be also extracellular in its seeding process, where it can be accessed by immunotherapy [316].

Human IgG isotypes differ in functions and downstream effects after binding as well as in their propensity to generate immune complexes. For example, IgG1 and IgG3 initiate complement activation and phagocytosis by microglia [390] as opposed to IgG4 [316]. A question, although based on limited evidence [316], seems to be whether engaging the innate immune system of the patient in the immune response would be dangerous [391]. In fact, microglia may be a victim of the neurodegenerative process and actually need help to oppose the progression of AD [311].

In summary, the mechanism of action would be as follows: initially, pre-aggregate tau species are generated intracellularly due to multiple endogenous and environmental factors [392, 393]. For example, psychological stress was shown to promote tau phosphorylation [394]. At this step, antibodies would need to enter neurons, which may not be possible [316]. Once the intracellular tau pathology exceeds a certain threshold, neurons start shedding tau seeds and "tauons" [395]. These extracellular tau species are accessible to immunotherapy. Antibodies may bind and immobilize them [316], disassembly them [396], or change the molecule conformation creating a new template with the antibody acting as template/chaperone [397]. Antibodies may prevent the binding of "tauons" to receptors that mediate the internalization of these seeds [219], or activate immune cells, such as microglia, to degrade and phagocyte pathological tau [398].

Table 1. An overview of the targeted epitopes and molecules in clinical development (adapted from [316]).

Compound	Type and Isotype	Epitope	Targeted Tau Species
AADvac1	Active immunotherapy response is mostly IgG1	Primary: 294-305 Secondary: 268-283, 330-335, 362-367	Conformationally altered extracellular tau, mainly seeding-capable tau

(Table 1) cont.....

Compound	Type and Isotype	Epitope	Targeted Tau Species
ACI-35	Active	pS396	Tau pS396, preferentially multimeric conformers
ABBV-8E12	Passive, igG4	25-30	Extracellular tau
BIIB076	Passive, IgG1	undisclosed	Monomeric and fibrillary forms
BIIB092	Passive, IgG4	undisclosed	Extracellular, N-terminal fragments of tau
JNJ-63733657	Passive, undisclosed isotype	Seeding-capable tau	-
LY3303560	Passive, undisclosed isotype	undisclosed	Soluble tau aggregates with intact N-terminus
RO 7105705	Passive, IgG4	N-terminus of tau	Monomeric and oligomeric tau regardless of its phosphorylation status
UCB0107	Passive, undisclosed isotype	235-246	Seeding-capable tau

Overview of tau-targeted immunotherapies in clinical development.

Anti-tau Active Immunotherapy

Anti-tau vaccines elicit specific antibody responses which can clear pathological tau species and improve neuronal function. Hence, choosing the right epitope is crucial for the success of the approach. Most of the preclinical studies demonstrated a reduction in the extent of tau pathology [399, 400] and improvement in cognitive or sensorimotor abilities in animal models [401, 402]. However, the antibody response in terms of quantity and quality has not been completely characterized [316].

Currently, there are 2 vaccines in clinical trials: AADvac1 and ACI-35.

AADvac1 for Alzheimer's Disease

The AADvac1 vaccine is a synthetic peptide derived from aminoacids 294 to 305 of tau [403]. In animal models, vaccination induced a robust immune response, the antibodies being able to discriminate between physiological and pathological tau, reduced the levels of tau oligomers and the levels of hyperphosphorylated tau by about 95%, reduced the extent of neurofibrillary pathology in the brain and improved the clinical phenotype of the transgenic animals [404]. Between May 2013 and March 2015, a phase I trial sponsored by Axon Neuroscience (NCT01850238) at 3 sites of Austria was conducted to evaluate the safety, tolerability, and immunogenicity as well as efficacy on cognition of AADvac1 in patients with mild to moderate Alzheimer's disease with an MMSE between 15 and 26 [338]. Three monthly subcutaneous injections of a single dose were administered to 30 patients. The double-blind, placebo-controlled phase was followed by a further three months of open-label monthly dosing. Thus, the treatment group received six injections and placebo participants received three. Of the 30 patients enrolled, 29 developed an IgG immune response. The most serious side effects were local reactions at the injection site (53%) but 2 patients developed more serious side effects and withdrew from the trial; one patient developed a viral infection followed by epileptic seizures, a condition which was considered as possibly related to the study medication [405]. After that, 26 patients who completed the first study enrolled in a follow-up open-label extension trial (FUNDAMANT, NCT02031198) [338], lasting a further 18 months between January 2014 and August 2016. All responders maintained an IgG antibody response against the tau peptide component of AADvac1 over 6 months, although booster doses at 6 months were required to restore the previous IgG levels. There were no significant safety issues or serious adverse events. The degree of hippocampal atrophy was lower in patients with high IgG levels, which also scored better on the cognitive assessment tools [406].

In March 2016, a 24-month randomized, placebo-controlled, double-blinded, multicenter, phase II study (ADAMANT, NCT02579252) began enrolling 185 patients (actually 196 enrolled) with mild to moderate AD (MMSE ≥ 20 and ≤ 26) and an MRI consistent with this diagnosis to assess safety and efficacy of AADvac1 at 42 locations across Europe [338]. The initial inclusion criteria required demonstration of medial temporal lobe atrophy but during the study, this criterion was changed to medial temporal lobe atrophy and/or a positive AD biomarker profile of amyloid and tau in CSF. It compared six-monthly subcutaneously administered doses of 40 micrograms of vaccine followed by five boosters of 40 micrograms of vaccine given every 3 months to placebo. The primary outcome was safety. Secondary outcomes included immunogenicity as well as cognitive and clinical efficacy measures. Exploratory outcomes listed were FDG PET, MRI volumetry, and CSF biochemistry. The study was completed in June 2019. No peer-reviewed publication of the results is yet available, but at the 2020 virtual AAT-AD/PD Focus Meeting [407], the research team reported similar incidence and type of adverse event between the treatment and placebo groups. Of 196 participants enrolled, 163 completed the study and more than 80 percent of patients receiving the vaccine developed a robust immune response. In addition, AADvac1 slowed the increase in blood neurofibrils. As a whole, the treatment group showed no cognitive benefit, but a preplanned age subgroup analysis showed a trend toward slower cognitive decline in younger participants, which also had greater reduction in plasma neurofibrils and a slower rate of cortical atrophy [407].

The vaccine is still under trial for nonfluent/agrammatic variant primary progressive aphasia in a phase I open-label pilot study that started in July 2017 and expected to be completed in November 2020 (AIDA, NCT03174886) [338].

ACI-35

ACI-35 is a liposome-based vaccine containing 16 copies of a synthetic tau fragment that is phosphorylated at the S396 and S404 residues and is anchored into a lipid bilayer [408]. In both wild-type and transgenic mouse models of AD, ACI-35 generated a robust immune response with polyclonal IgG antibodies directed mainly against phosphorylated tau and was able to lower the levels of soluble and aggregated tau in brain extracts [409].

In July 2013, AC Immune and Janssen Pharmaceuticals began a phase Ib multi-center double-blind randomized placebo-controlled study to compare safety and effects of low, medium, and high (undisclosed) doses of ACI-35 to placebo in 24 patients with mild to moderate AD with MMSE between 18 and 28, on a stable dose of cholinesterase inhibitors in the prior 3 months (ISRCTN13033912) [410]. The initial vaccine was followed by a booster shot and a subsequent six-month safety observation period. Primary outcomes listed were antibody titers, biochemical measures from the CSF, and MRI changes during the study, while the secondary outcomes included a battery of neuropsychological tests such as the Neuropsychiatric Inventory Scale, MMSE, ADAS-Cog, trail-making and fluency tests, and the Clinical Global Impression of Change Disability Assessment in Dementia [338]. The trial was completed in June 2017 and the results were presented at the virtual AAT-AD/PD Focus Meeting [407]. The vaccine was safe but elicited a weak immune response with boosters having little effect. A second generation vaccine was designed, to include a second epitope and an adjuvant to activate T helper cells, named **ACI-35.030**. In monkeys, the new compound induced a stronger immune response with specific antibodies for phosphorylated tau which recognized paired helical filaments [411].

In July 2019, the sponsoring companies started a phase Ib/IIa multicenter, double-blind, randomized, placebo-controlled trial (NCT04445831) to test the safety and immunogenicity of ACI-35.030 in people with early AD (MMSE score 22 and above) [338]. The trial aims at enrolling 24 people who will be sequentially tested after receiving up to 3 dose levels of the vaccine multiple times over 48 weeks and compared to placebo-receivers. Primary outcomes listed are safety, the occurrence of adverse events, and anti-phosphoTau IgG titers in blood up to 74 weeks after baseline. The trial will also monitor cognition and behavior changes using the CDR-SB, RBANS, and NPI. The study is running at 6 centers in

Finland, the Netherlands, and the UK and is expected to be completed in October 2022 [338].

Other Vaccines

A number of other compounds are in preclinical studies targeting various epitopes, as shown in the Table **2** below (adapted from Jadhav *et al.* [94]).

Table 2. Epitopes targeted in preclinical studies of anti-tau vaccines.

Epitope	Improvement		Efficacy		Reference
	Cognitive	Sensorimotor	Neurofibrillary tangles	Insoluble tau	
Tau379-408	No change	Improved	Decreased	Decreased	[399]
Tau 417-426	Improved	Not disclosed	Reduced	Reduced	[402]
Tau393-408	Not defined	Improved	No change	Reduced	[409]
Tau379-408;	Improved	No change	Reduced	Reduced	[401]
Tau195-213; Tau207-220; Tau224-238	Not defined	Not defined	Reduced	Not defined	[412]
Tau aa395-406	Not defined	Not defined	Reduced	Not defined	[413]
Human paired helical filaments	Not defined	Not defined	Reduced	Reduced	[414]
Tau199-208	Not defined	Improved	No change	No change	[415]
Tau294-305	Not defined	Improved	Reduced	Reduced	[404]
Tau379-408	No change	Not defined	Reduced	Reduced	[400]
Tau294-305	Improved	Not defined	Reduced	Reduced	[416]

Passive Anti-tau Immunotherapy

As opposed to active immunization, which is long-lasting and polyclonal but depends on the host's immune system, passive immunization implies a repeated transfer of antibodies to the host which neutralize and/or eliminate monomeric, aggregated, or conformationally changed forms of tau protein, thereby preventing the formation of neurofibrillary tangles [94, 417]. They differ in their binding site recognizing the N-terminus, the proline-rich region, the MTBR, or the C-terminus [94]. Currently, several humanized tau antibodies are being investigated in clinical trials, as discussed below:

Passive anti-tau Immunotherapy Targeting the N-Terminus of Tau

The N-terminus of tau protein is targeted by many antibodies due to several reasons [94]:

- The conformational changes in the N-terminal of tau occur early in AD pathogenesis [418]
- Only N-terminal tau fragments were detected in the CSF of AD patients [419]
- The N-terminal fragment of tau increases amyloid-beta production and impairs mitochondrial function and synaptic plasticity [222, 420, 421].

On the other hand, most of the tau in the AD brain is truncated, mainly at the N-terminus, thereby tau oligomers and fibrils in AD lack the N-terminal fragment [94, 422] which means that N-terminal tau antibodies do not recognize the whole spectrum of pathological tau in AD brain [94, 423]. In addition, these antibodies are not specific to diseased tau and may reduce the level of physiological tau as well [94].

Although several antibodies from this subcategory are in preclinical tests, humanized versions of N-terminal specific antibodies in clinical tests are BIIB092 (also known as BMS-986168 or IPN007) and RO 7105705 (also known as RG 6100).

BIIB092, or Gosuranemab is a humanized IgG4 monoclonal antibody against extracellular N-terminal tau fragments [324]. In November 2014, Biogen started a phase I randomized, double-blind, placebo-controlled, single-ascending-dose study in 65 healthy volunteers aged 21 to 65 to assess safety, tolerability, pharmacokinetics, and pharmacodynamics as well as immunogenicity of an intravenous infusion of BIIB092 (NCT02294851) [338]. It was completed in April 2016 and according to published results [424], there were no serious adverse effects recorded. Following the infusion, a dose-dependent increase in blood and CSF BIIB092 was recorded, which reduced the levels of unbound cerebrospinal fluid N-terminal tau by 67 to 97%, 4 weeks after the dose, and the effect was lasted for over 12 weeks with doses of ≥ 210 mg. BIIB092 was safe and well-tolerated after single dose of up to 4200 mg (2100 mg in Japanese participants). In May 2018 Biogen initiated a phase II randomized, double-blind, placebo-controlled, parallel-group study (TANGO, NCT03352557) to assess the safety, tolerability, and efficacy of BIIB092 in patients with mild cognitive impairment due to Alzheimer's disease or with mild Alzheimer's disease. The trial enrolled 654 participants from 105 locations who are receiving low doses of BIIB092 IV every 4 weeks or every 12 weeks with placebo infusions on the other 4-week visits, and medium or high doses every 4 weeks for about 1.5 years were given followed by 3 years of extended dosing [338]. Inclusion criteria required an MMSE from 22 to 30 and amyloid positivity at screening. It is currently active, but not recruiting and is estimated to be completed in March 2024. It looks at the number of adverse events as a primary outcome, while secondary outcomes listed are anti-BIIB092 antibodies developed by the participants and change on the CDR-SB after 1.5 years of treatment [338]. The antibody has also been evaluated in progressive supranuclear palsy between October 2015 and October 2016 in a phase I trial which raised no safety issues [425], followed by a Phase II study (PASSPORT, NCT03068468) and an extension trial (NCT02658916) for PSP, but which unfortunately missed primary and secondary endpoints in terms of efficacy [426]. Another study, NCT03658135, conducted between September 2018 and December 2019 in 22 patients with various other tauopathies (corticobasal degeneration, frontotemporal lobar degeneration with MAPT mutations, traumatic encephalopathy, or non-fluent primary progressive aphasia) was also ended due to lack of efficacy [427]. Currently, Biogen runs only the TANGO trial in patients with mild cognitive impairment due to AD [338].

RO07105705 (RG6100) or Semorinemab is an anti-tau IgG4 antibody that binds the N-terminus of all six monomeric and oligomeric isoforms of human tau regardless of phosphorylation status [94, 391]. It appeared as a result of research collaboration between AC Immune and Genentech, belonging to the Roche Group. Between June 2016 and June 2017, Genentech ran at Tennessee a phase I, randomized, placebo-controlled, double-blind study to evaluate the safety, tolerability, pharmacokinetics, and preliminary activity of RO7105705 in 74 volunteers, both healthy participants and patients with mild to moderate Alzheimer's disease (NCT02820896). It combined single-dose, dose-escalation, and multiple dosing given intravenously and subcutaneously [338]. The results are not published, but at the 2017 AD/PD and AAIC conferences, the sponsor reported that there were no serious adverse events and that plasma and CSF concentrations increased with dose, which went as high as 16,800 mg in healthy volunteers [428, 429]. Thus, in October 2017, Genentech started TAURIEL (NCT03828747), a phase II, randomized, placebo-controlled, double-blind study to evaluate the efficacy and safety of Semorinemab in participants with prodromal to mild Alzheimer's disease [338]. Patients who complete the study and who, in the opinion of the investigator could benefit, will be offered the choice to enroll in a 96-week open-label extension period. The 457 patients enrolled at the 133 locations were randomized to an 18-month course of placebo or one of three doses of RO7105705 infusion. Positive amyloid PET or amyloid Aβ42 CSF findings were required in the inclusion criteria. Primary outcomes are safety and change on the CDR-sum of boxes. As secondary outcomes, the trial will assess serum drug concentration, anti-drug antibodies, and neuropsychological changes using the

Repeatable Battery for Assessment of Neuropsychological Status (RBANS), the new Amsterdam Instrumental Activity of Daily Living questionnaire, the ADAS-Cog13, and ADCS-ADL Inventory. Enrollment was completed, and the trial is now active in the extension phase [338]. In January 2019, Genentech started NCT03289143, another phase II, randomized, placebo-controlled, double-blind study to evaluate the efficacy and safety of Semorinemab in participants with prodromal to moderate Alzheimer's disease. Again, the possibility to enroll in a 96-week open-label extension period will present to patients who complete the double-blind treatment period and who could benefit from Semorinemab treatment in the opinion of the investigator. AD must be confirmed by amyloid positivity *via* PET or CSF testing and the MMSE between 16 and 21. The primary endpoints listed are changes from baseline to week 49 on ADAS-Cog1 and ADCS-ADL. Secondary endpoints include serum concentration of RO7105705, antibody levels to RO7105705, adverse events, as well as changes in MMSE and CDR-SB. The trial plans to enroll 260 patients from 50 locations and is estimated to be completed in August 2023 [338].

C2N or 8E12 is a humanized IgG4 antibody developed by C2N Diagnostics and AbbVie to be used in the treatment of tauopathies. It recognizes an extracellular aggregated form of tau implicated in the seeding of tau pathology [430]. In P301S tau-transgenic mice the mouse version of this antibody was reported to reduce brain neurofibrillary pathology, insoluble tau, microgliosis, seeding activity, and brain atrophy [431]. In October 2016, AbbVie started NCT02880956, a phase II multiple-dose, multicenter, randomized, double-blind, placebo-controlled study to evaluate the safety and efficacy of ABBV-8E12 in subjects with early Alzheimer's disease [338]. Inclusion required a CDR rating of 0.5, an MMSE of 22 or higher, and an RBANS of 85 or lower as well as a positive amyloid PET scan. The study compares three doses of 8E12 to placebo, to be infused over a period of 96 weeks plus a 16-week follow-up. The primary outcomes are a decline on the CDR Sum of Boxes and in adverse events. Secondary outcomes include blood-based pharmacokinetic parameters for the 8E12 antibody, and a range of clinical and functional measures such as the ADAS-Cog14, RBANS, FAQ, ADCS-MCI-ADL-24 [338]. It aimed at enrolling 400 patients until June 2021, but in March 2019, the trial stopped enrolling at 453 participants. A long-term safety extension (NCT03712787) with dosing extended up to 5.5 years was offered to patients who completed the study [338]. The antibody was also tested in progressive supranuclear palsy but missed its primary and secondary endpoints, which led AbbVie to stop these studies in July 2019 [432].

Passive Anti-tau Immunotherapy Targeting the Mid-domain of Tau Protein

Phosphorylation of tau at pS202 and pT205 is an early intra- and extracellular marker of tau pathology in AD [94] but antibodies targeting this region, although effective in the neutralization of tau seeding and in reduction of tau uptake *in vitro*, showed inconsistent results in preclinical *in vivo* experiments [94, 433].

UCB0107 is a monoclonal antibody that binds to the mid-region of tau. It recognizes amino acids 235–246 near tau's MTBR. Supposedly, the binding of antibodies to this region could more efficiently interfere with the seeding of pathogenic, aggregated tau. UCB0107 also binds tau monomers.

From February 2018 until December 2018, UCB Biopharma ran a phase I randomized, double-blind, placebo-controlled study of single ascending doses of UCB0107 (NCT03464227) to evaluate the safety, tolerability, pharmacokinetics, and pharmacodynamics in healthy male subjects. The trial enrolled 52 volunteers assigned to up to seven doses to be received for 20 weeks. The results were presented at the 2019 Movement Disorders Conference [434]. All participants completed the study. There were no drug-related adverse events, no evidence of anti-drug antibodies, and serum and CSF concentrations increased with dose [435]. Safety, tolerability, pharmacokinetics, and pharmacodynamics were also assessed in Japanese subjects (NCT03605082) and a study is ongoing to test safety and tolerability in patients with progressive supranuclear palsy (NCT04185415). In July 2020, UCB announced an agreement with Roche/Genentech in which, in exchange for receiving $120 million, the company committed to conducting a proof-of-concept study on AD [436].

JNJ-63733657 is a monoclonal antibody that recognizes the mid-region of tau developed by Janssen Pharmaceutical. It was shown to eliminate pathogenic tau seeds in a cell-based assay and was also able to inhibit the spread of tau pathology in a mouse model of AD [437]. In December 2017, a phase I randomized, placebo-controlled, double-blind, single, and multiple ascending dose study (NCT03375697) was initiated to evaluate the safety, tolerability, pharmacokinetics, and pharmacodynamics of JNJ-63733657 in 72 volunteers, either healthy subjects or patients with prodromal or mild Alzheimer's disease [338]. In part 1 of the trial, a single, ascending dose intravenous infusion was administered to healthy volunteers, followed in part 2 by multiple ascending intravenous doses administered to patients with prodromal or mild AD based on the findings from part 1. The study was completed in December 2019 but no results have yet been published [338]. Another phase I randomized, placebo-controlled, double-blind, single ascending dose study was conducted between September 2018 and July 2019 in Japan on 24 participants to assess safety, tolerability, pharmacokinetics, and pharmacodynamics in healthy Japanese subjects (NCT03689153) [338]. No results have been published yet.

Passive Anti-tau Immunotherapy Targeting the Microtubule-Binding Region

This region plays an important role in the polymerization and stability of microtubules and is also responsible for the interaction of pathological tau [379]. Preclinical studies have shown promising results with passive immunotherapy targeting the MTBR of tau [396, 417].

LY3303560, also known as **Zagotenemab**, developed by Eli Lilly & Co, is a modified MC1 antibody recognizing the N-terminus and the microtubule-binding domain of tau [94]. In 2017 at the AAIC meeting, Lilly further characterized the antibody as having selectivity to tau aggregates over monomers and high affinity for soluble tau aggregates [438]. In April 2016, Lilly started a phase I single-dose, dose-escalation, placebo-controlled study (NCT02754830) to evaluate the safety, tolerability, and pharmacokinetics of LY3303560 in 110 healthy volunteers or patients with mild cognitive impairment due to Alzheimer's disease or mild to moderate AD diagnosed with a positive amyloid PET scan. The study evaluated a single, escalating intravenous infusion or subcutaneous injection of the antibody. It was completed in July 2018 but no results were yet published [338]. Another phase I study conducted at 12 sites in the UK, US, and Japan between January 2017 and June 2019 enrolled 24 participants with mild cognitive impairment due to Alzheimer's disease or mild to moderate Alzheimer's disease (NCT03019536) [338]. Participants received escalating multiple doses of LY3303560 or placebo for 48 weeks, followed by a 16-week follow-up period. Concomitantly with the drug, patients also received amyloid and tau PET tracers. No results have yet been published [338]. A phase II study (NCT03518073) started in April 2018 enrolled 285 patients with early symptomatic AD (gradual and progressive memory decline for at least 6 months) and compared 2 intravenous doses (not disclosed) to placebo. The primary outcome is change from baseline on Lilly's integrated Alzheimer's Disease Rating Scale (iADRS), while secondary outcome measures include the levels of antibodies against LY3303560, tau PET, volumetric MRI as well as assessments on a series of scales including ADAS-Cog13, ADCS-iADL, CDR-SB, MMSE, and the CogState Brief Battery (CBB) [338]. The duration of the treatment has not been disclosed, but it is currently active, not recruiting, and estimated to end in October 2021 [338].

Other Passive Tau-targeting Immunotherapies

The C-terminus of tau enhances its microtubule-binding capacity but also influences pathological tau aggregation [439]. Several studies investigated the effect of C-terminal tau-specific antibodies in animal models [440, 441]. However, no antibody has escalated to clinical trials.

BIIB076 (also known as **NI-105, 6C5, or huIgG1/1**) is a human recombinant monoclonal anti-tau IgG1 developed by Neuroimmune's reverse translational medicine platform and reported in 2017 at AD/PD and AAIC conferences to recognize monomeric and fibrillar forms of tau to which it binds with high affinity [442]. In animal models, it increased total plasma tau while CSF total tau remained unchanged but free tau unbound to BIIB076 dropped by 75% after 24 hours [443].

In February 2017, Biogen initiated a phase I randomized, blinded, placebo-controlled, single ascending dose study on 56 healthy volunteers and subjects with mild Alzheimer's disease (MMSE 18-30 and amyloid positivity) to evaluate the safety, tolerability, and pharmacokinetics of BIIB076 (NCT03056729) [338]. A single intravenous infusion was administered. Healthy volunteers were grouped in five dosing cohorts, whereas AD patients received 2 different doses [444]. In June 2019, the sponsor changed the trial protocol, reduced participant number to 46 by dropping the more advanced AD cohort, and set the number of adverse events as the only primary outcome [444]. The study was run in the US and completed in March 2020 but no results are yet available [338].

Other molecules were also developed as passive anti-tautherapies, such as **RO 6926496** or **RG7345**. They are humanized monoclonal antibodies targeting the tau phosphoepitope pS422. Tau phosphorylated at this site has been shown to relocalize tau away from microtubules and toward the somatodendritic compartment of the neuron [445]. In January 2015, Roche started a randomized, double-blind, single ascending dose, placebo-controlled, parallel study to evaluate the safety, tolerability, and pharmacokinetics of RO6926496 in healthy male participants in the UK (NCT02281786), but the study was terminated in October 2015 and further development of the drug was discontinued [338, 446].

Lu AF87908 is a humanized mouse IgG1 monoclonal antibody to phosphorylated tau protein. In animal models, the mouse version of this antibody, C10.2, reduced aggregated tau seeding in cultured neurons and rTg4510 tau transgenic mice [447]. In September 2019, Lundbeck initiated a phase 1 interventional, randomized, double-blind, placebo-controlled, single-ascending-dose trial to investigate the safety, tolerability, and pharmacokinetics of Lu AF87908 in healthy subjects and patients with Alzheimer's disease (NCT04149860). The study was conducted in Los Angeles and enrolled 96 adults in three sequential cohorts: healthy, healthy Japanese and Chinese, and patients with Alzheimer's disease. Outcomes were safety and plasma antibody concentrations, which was monitored for three months following infusion. The study was estimated to run through March 2021 [338].

PNT001 is a monoclonal antibody to the cis isomer of tau phosphorylated at threonine 231. This tau species has been detected in brain tissue from people with AD and was shown to be resistant to dephosphorylation and degradation and to promote aggregation [448]. In preclinical models of brain injury, the mouse version of the antibody prevented tau oligomerization and tangle formation as well as axonal pathology and astrogliosis [449]. In September 2019, Pinteon began a phase 1 randomized, double-blind, placebo-controlled, single-ascending - dose trial to evaluate the safety, tolerability, immunogenicity, and pharma-cokinetics of intravenous PNT001 in 48 healthy volunteers (NCT04096287). Participants received single doses of 33, 100, 300, 900, or 2,700 mg antibody or placebo. The primary outcomes listed were adverse events and abnormalities on clinical and laboratory measures 16 weeks after the infusion. The trial also assessed antibody concentration in serum and CSF, levels of neurofilament light chains in CSF as well as levels of total tau, cis pT231 tau, and pT231 tau in the CSF. The trial was estimated to run through November 2020 [338].

Active tau vaccines and therapeutic monoclonal antibodies are among the most promising therapeutic strategies for preventing, slowing, or ameliorating the production, oligomerization, aggregation, and accumulation of pathological tau protein. However, to maximize the gain of knowledge from the trials described and pursue more efficiently on this aspect, certain methodological and reporting standards should be implemented [316]. The targeted epitopes and mechanism of action of the immunotherapeutic agent should be well characterized. In addition, a great benefit would be a complete characterization of the diseased tau proteome in terms of molecular properties of individual tau species, their location (cellular or extracellular), and effects [316]. As for active immunotherapy, to avoid T-cell responses aimed at self-antigens, as happened with AN-1792, the first anti-amyloid vaccine (NCT00021723) [450], a T-cell independent antibody response should be induced [409].

CONCLUSION

The search for disease-modifying therapies for AD is a very fluctuating field, with repeated rises and falls, shifting from agony to ecstasy. For example, in March 2019 the announcement of the futility of aducanumab [451] left companies that have invested in amyloid-targeting treatments to reconsider their position. In October 2019, when the sponsors of EMERGE announced that their data were positive for clinical efficacy and anticipated filing in 2020 with the FDA for a clinical indication of aducanumab for AD [452], research investigating anti-amyloid methods were again encouraged [453].

However, the World Health Organization registry indicates that at the moment there are only 170 drugs in development for AD worldwide, as compared to 6833 drugs for malignancies and 433 for diabetes [303]. These figures reflect the greater expenses, higher risk of failure of AD drugs, longer trial durations, and less well-defined target biology [303]. Of the 29 agents in phase 3 clinical trials, only 17 (59%) are disease-modifying ones and only 1 (6%) is tau-targeted. In phase 2 trials are 65 agents, of which 55 (85%) are DMT and 6 (11%) are tau-targeted, 4 being monoclonal antibodies, 1 – an antisense nucleotide and 1 aims at decreasing phosphorylation [303]. Overall, there are fewer drugs in the AD pipeline than in 2019: 121 as compared to 134 in 2019, of which 29 agents in phase III trials in 2020 as compared to 29 in 2019 and 28 in 2018, 65 agents in phase II trials as compared to 75 in 2019 and 63 in 2018, and 27 agents in phase I studies as compared to 30 in 2019 and 23 in 2018 [303, 454]. Over the last 5 years, there is a trend toward increased target diversity, enrollment of participants in earlier stages of cognitive decline, and increased use of biomarkers to define trial populations [303]. However, tau-based therapies may prove useful in other tauopathies as well. As for AD treatment, it is more likely that none of the monotherapies will be the answer, but rather combined approaches, similar to chemotherapy, will be more successful.

CONSENT FOR PUBLICATION

Not applicable.

CONFLICT OF INTEREST

The authors declare no conflict of interest, financial or otherwise.

ACKNOWLEDGEMENTS

Declared none.

REFERENCES

[1]   Alzheimer A. Über eine eigenartige Erkrankung der Hirnrinde. Allgemeine Zeitschrift für Psychiatrie und Psychisch-gerichtliche Medizin 1907; 64: 146-8.

[2]   Maurer K, Volk S, Gerbaldo H. Auguste D and Alzheimer's disease. Lancet 1997; 349(9064): 1546-9.
       [http://dx.doi.org/10.1016/S0140-6736(96)10203-8] [PMID: 9167474]

[3]   DeTure MA, Dickson DW. The neuropathological diagnosis of Alzheimer's disease. Mol Neurodegener 2019; 14(1): 32.
       [http://dx.doi.org/10.1186/s13024-019-0333-5] [PMID: 31375134]

[4]   Kidd M. Paired helical filaments in electron microscopy of Alzheimer's disease. Nature 1963; 197:

192-3.
[http://dx.doi.org/10.1038/197192b0] [PMID: 14032480]

[5]   Terry RD. The fine structure of neurofibrillary tangles in Alzheimer's disease. J Neuropathol Exp Neurol 1963; 22: 629-42.
[http://dx.doi.org/10.1097/00005072-196310000-00005] [PMID: 14069842]

[6]   Weingarten MD, Lockwood AH, Hwo SY, Kirschner MW. A protein factor essential for microtubule assembly. Proc Natl Acad Sci USA 1975; 72(5): 1858-62.
[http://dx.doi.org/10.1073/pnas.72.5.1858] [PMID: 1057175]

[7]   Mandelkow E-M, Mandelkow E. Biochemistry and cell biology of tau protein in neurofibrillary degeneration. Cold Spring Harb Perspect Med 2012; 2(7): a006247.
[http://dx.doi.org/10.1101/cshperspect.a006247] [PMID: 22762014]

[8]   Cleveland DW, Hwo SY, Kirschner MW. Physical and chemical properties of purified tau factor and the role of tau in microtubule assembly. J Mol Biol 1977; 116(2): 227-47.
[http://dx.doi.org/10.1016/0022-2836(77)90214-5] [PMID: 146092]

[9]   Cleveland DW, Hwo SY, Kirschner MW. Purification of tau, a microtubule-associated protein that induces assembly of microtubules from purified tubulin. J Mol Biol 1977; 116(2): 207-25.
[http://dx.doi.org/10.1016/0022-2836(77)90213-3] [PMID: 599557]

[10]   Poorkaj P, Bird TD, Wijsman E, *et al.* Tau is a candidate gene for chromosome 17 frontotemporal dementia. Ann Neurol 1998; 43(6): 815-25.
[http://dx.doi.org/10.1002/ana.410430617] [PMID: 9629852]

[11]   Spillantini MG, Goedert M. Tau pathology and neurodegeneration. Lancet Neurol 2013; 12(6): 609-22.
[http://dx.doi.org/10.1016/S1474-4422(13)70090-5] [PMID: 23684085]

[12]   Lee VMY, Brunden KR, Hutton M, Trojanowski JQ. Developing therapeutic approaches to tau, selected kinases, and related neuronal protein targets. Cold Spring Harb Perspect Med 2011; 1(1): a006437.
[http://dx.doi.org/10.1101/cshperspect.a006437] [PMID: 22229117]

[13]   Schneider A, Mandelkow E. Tau-based treatment strategies in neurodegenerative diseases. Neurotherapeutics 2008; 5(3): 443-57.
[http://dx.doi.org/10.1016/j.nurt.2008.05.006] [PMID: 18625456]

[14]   Neve RL, Harris P, Kosik KS, Kurnit DM, Donlon TA. Identification of cDNA clones for the human microtubule-associated protein tau and chromosomal localization of the genes for tau and microtubule-associated protein 2. Brain Res 1986; 387(3): 271-80.
[PMID: 3103857]

[15]   Andreadis A. Misregulation of tau alternative splicing in neurodegeneration and dementia. Prog Mol Subcell Biol 2006; 44: 89-107.
[http://dx.doi.org/10.1007/978-3-540-34449-0_5] [PMID: 17076266]

[16]   Couchie D, Mavilia C, Georgieff IS, Liem RK, Shelanski ML, Nunez J. Primary structure of high molecular weight tau present in the peripheral nervous system. Proc Natl Acad Sci USA 1992; 89(10): 4378-81.
[http://dx.doi.org/10.1073/pnas.89.10.4378] [PMID: 1374898]

[17]   Guo T, Noble W, Hanger DP. Roles of tau protein in health and disease. Acta Neuropathol 2017; 133(5): 665-704.
[http://dx.doi.org/10.1007/s00401-017-1707-9] [PMID: 28386764]

[18]   Goedert M, Wischik CM, Crowther RA, Walker JE, Klug A. Cloning and sequencing of the cDNA encoding a core protein of the paired helical filament of Alzheimer disease: identification as the microtubule-associated protein tau. Proc Natl Acad Sci USA 1988; 85(11): 4051-5.
[http://dx.doi.org/10.1073/pnas.85.11.4051] [PMID: 3131773]

[19]   Avila J, Jiménez JS, Sayas CL, *et al.* Tau structures. Front Aging Neurosci 2016; 8: 262.
         [http://dx.doi.org/10.3389/fnagi.2016.00262] [PMID: 27877124]

[20]   Schweers O, Schönbrunn-Hanebeck E, Marx A, Mandelkow E. Structural studies of tau protein and
         Alzheimer paired helical filaments show no evidence for β-structure. J Biol Chem 1994; 269(39):
         24290-7.
         [http://dx.doi.org/10.1016/S0021-9258(19)51080-8] [PMID: 7929085]

[21]   Mukrasch MD, Bibow S, Korukottu J, *et al.* Structural polymorphism of 441-residue tau at single
         residue resolution. PLoS Biol 2009; 7(2): e34.
         [http://dx.doi.org/10.1371/journal.pbio.1000034] [PMID: 19226187]

[22]   Jeganathan S, von Bergen M, Brutlach H, Steinhoff HJ, Mandelkow E. Global hairpin folding of tau in
         solution. Biochemistry 2006; 45(7): 2283-93.
         [http://dx.doi.org/10.1021/bi0521543] [PMID: 16475817]

[23]   Morris M, Maeda S, Vossel K, Mucke L. The many faces of tau. Neuron 2011; 70(3): 410-26.
         [http://dx.doi.org/10.1016/j.neuron.2011.04.009] [PMID: 21555069]

[24]   Li XC, Hu Y, Wang ZH, *et al.* Human wild-type full-length tau accumulation disrupts mitochondrial
         dynamics and the functions *via* increasing mitofusins. Sci Rep 2016; 6: 24756.
         [http://dx.doi.org/10.1038/srep24756] [PMID: 27099072]

[25]   Li X, Kumar Y, Zempel H, Mandelkow EM, Biernat J, Mandelkow E. Novel diffusion barrier for
         axonal retention of Tau in neurons and its failure in neurodegeneration. EMBO J 2011; 30(23): 4825-
         37.
         [http://dx.doi.org/10.1038/emboj.2011.376] [PMID: 22009197]

[26]   Utton MA, Noble WJ, Hill JE, Anderton BH, Hanger DP. Molecular motors implicated in the axonal
         transport of tau and α-synuclein. J Cell Sci 2005; 118(Pt 20): 4645-54.
         [http://dx.doi.org/10.1242/jcs.02558] [PMID: 16176937]

[27]   Aronov S, Aranda G, Behar L, Ginzburg I. Axonal tau mRNA localization coincides with tau protein
         in living neuronal cells and depends on axonal targeting signal. J Neurosci 2001; 21(17): 6577-87.
         [http://dx.doi.org/10.1523/JNEUROSCI.21-17-06577.2001] [PMID: 11517247]

[28]   Morita T, Sobue K. Specification of neuronal polarity regulated by local translation of CRMP2 and
         Tau *via* the mTOR-p70S6K pathway. J Biol Chem 2009; 284(40): 27734-45.
         [http://dx.doi.org/10.1074/jbc.M109.008177] [PMID: 19648118]

[29]   Kahlson MA, Colodner KJ. Glial tau pathology in Tauopathies: functional consequences. J Exp
         Neurosci 2015; 5: 943-50.
         [http://dx.doi.org/10.4137/JEN.S25515] [PMID: 26884683]

[30]   Leyns CEG, Holtzman DM. Glial contributions to neurodegeneration in tauopathies. Mol
         Neurodegener 2017; 12(1): 50.
         [http://dx.doi.org/10.1186/s13024-017-0192-x] [PMID: 28662669]

[31]   Togo T, Dickson DW. Tau accumulation in astrocytes in progressive supranuclear palsy is a
         degenerative rather than a reactive process. Acta Neuropathol 2002; 104(4): 398-402.
         [http://dx.doi.org/10.1007/s00401-002-0569-x] [PMID: 12200627]

[32]   Ikeda K, Akiyama H, Kondo H, *et al.* Thorn-shaped astrocytes: possibly secondarily induced tau-
         positive glial fibrillary tangles. Acta Neuropathol 1995; 90(6): 620-5.
         [http://dx.doi.org/10.1007/BF00318575] [PMID: 8615083]

[33]   Arima K. Ultrastructural characteristics of tau filaments in tauopathies: immuno-electron microscopic
         demonstration of tau filaments in tauopathies. Neuropathology 2006; 26(5): 475-83.
         [http://dx.doi.org/10.1111/j.1440-1789.2006.00669.x] [PMID: 17080728]

[34]   Ferrer I, López-González I, Carmona M, *et al.* Glial and neuronal tau pathology in tauopathies:
         characterization of disease-specific phenotypes and tau pathology progression. J Neuropathol Exp

Neurol 2014; 73(1): 81-97.
[http://dx.doi.org/10.1097/NEN.0000000000000030] [PMID: 24335532]

[35] Odawara T, Iseki E, Kosaka K, Akiyama H, Ikeda K, Yamamoto T. Investigation of tau-2 positive microglia-like cells in the subcortical nuclei of human neurodegenerative disorders. Neurosci Lett 1995; 192(3): 145-8.
[http://dx.doi.org/10.1016/0304-3940(95)11595-N] [PMID: 7566636]

[36] LaPointe NE, Morfini G, Pigino G, *et al.* The amino terminus of tau inhibits kinesin-dependent axonal transport: implications for filament toxicity. J Neurosci Res 2009; 87(2): 440-51.
[http://dx.doi.org/10.1002/jnr.21850] [PMID: 18798283]

[37] Mitchison T, Kirschner M. Cytoskeletal dynamics and nerve growth. Neuron 1988; 1(9): 761-72.
[http://dx.doi.org/10.1016/0896-6273(88)90124-9] [PMID: 3078414]

[38] Panda D, Samuel JC, Massie M, Feinstein SC, Wilson L. Differential regulation of microtubule dynamics by three- and four-repeat tau: implications for the onset of neurodegenerative disease. Proc Natl Acad Sci USA 2003; 100(16): 9548-53.
[http://dx.doi.org/10.1073/pnas.1633508100] [PMID: 12886013]

[39] Wilson L, Jordan MA. New microtubule/tubulin-targeted anticancer drugs and novel chemotherapeutic strategies. J Chemother 2004; 16 (Suppl. 4): 83-5.
[http://dx.doi.org/10.1179/joc.2004.16.Supplement-1.83] [PMID: 15688618]

[40] Chen J, Kanai Y, Cowan NJ, Hirokawa N. Projection domains of MAP2 and tau determine spacings between microtubules in dendrites and axons. Nature 1992; 360(6405): 674-7.
[http://dx.doi.org/10.1038/360674a0] [PMID: 1465130]

[41] Chen Q, Zhou Z, Zhang L, *et al.* Tau protein is involved in morphological plasticity in hippocampal neurons in response to BDNF. Neurochem Int 2012; 60(3): 233-42.
[http://dx.doi.org/10.1016/j.neuint.2011.12.013] [PMID: 22226842]

[42] Gauthier-Kemper A, Weissmann C, Golovyashkina N, *et al.* The frontotemporal dementia mutation R406W blocks tau's interaction with the membrane in an annexin A2-dependent manner. J Cell Biol 2011; 192(4): 647-61.
[http://dx.doi.org/10.1083/jcb.201007161] [PMID: 21339331]

[43] Kim W, Lee S, Jung C, Ahmed A, Lee G, Hall GF. Interneuronal transfer of human tau between Lamprey central neurons *in situ.* J Alzheimers Dis 2010; 19(2): 647-64.
[http://dx.doi.org/10.3233/JAD-2010-1273] [PMID: 20110609]

[44] Pooler AM, Usardi A, Evans CJ, Philpott KL, Noble W, Hanger DP. Dynamic association of tau with neuronal membranes is regulated by phosphorylation. Neurobiol Aging 2012; 33(2): 431.e27-38.
[http://dx.doi.org/10.1016/j.neurobiolaging.2011.01.005] [PMID: 21388709]

[45] Jeganathan S, Hascher A, Chinnathambi S, Biernat J, Mandelkow EM, Mandelkow E. Proline-directed pseudo-phosphorylation at AT8 and PHF1 epitopes induces a compaction of the paperclip folding of Tau and generates a pathological (MC-1) conformation. J Biol Chem 2008; 283(46): 32066-76.
[http://dx.doi.org/10.1074/jbc.M805300200] [PMID: 18725412]

[46] Wang Y, Loomis PA, Zinkowski RP, Binder LI. A novel tau transcript in cultured human neuroblastoma cells expressing nuclear tau. J Cell Biol 1993; 121(2): 257-67.
[http://dx.doi.org/10.1083/jcb.121.2.257] [PMID: 8468346]

[47] Camero S, Benítez MJ, Barrantes A, *et al.* Tau protein provides DNA with thermodynamic and structural features which are similar to those found in histone-DNA complex. J Alzheimers Dis 2014; 39(3): 649-60.
[http://dx.doi.org/10.3233/JAD-131415] [PMID: 24254705]

[48] Violet M, Delattre L, Tardivel M, *et al.* A major role for Tau in neuronal DNA and RNA protection *in vivo* under physiological and hyperthermic conditions. Front Cell Neurosci 2014; 8: 84.
[http://dx.doi.org/10.3389/fncel.2014.00084] [PMID: 24672431]

[49]   Bewley CA, Gronenborn AM, Clore GM. Minor groove-binding architectural proteins: structure, function, and DNA recognition. Annu Rev Biophys Biomol Struct 1998; 27: 105-31.
[http://dx.doi.org/10.1146/annurev.biophys.27.1.105] [PMID: 9646864]

[50]   de Barreda EG, Dawson HN, Vitek MP, Avila J. Tau deficiency leads to the upregulation of BAF-57, a protein involved in neuron-specific gene repression. FEBS Lett 2010; 584(11): 2265-70.
[http://dx.doi.org/10.1016/j.febslet.2010.03.032] [PMID: 20338169]

[51]   Kimura T, Whitcomb DJ, Jo J, *et al.* Microtubule-associated protein tau is essential for long-term depression in the hippocampus. Philos Trans R Soc Lond B Biol Sci 2013; 369(1633): 20130144.
[http://dx.doi.org/10.1098/rstb.2013.0144] [PMID: 24298146]

[52]   Fuster-Matanzo A, de Barreda EG, Dawson HN, Vitek MP, Avila J, Hernández F. Function of tau protein in adult newborn neurons. FEBS Lett 2009; 583(18): 3063-8.
[http://dx.doi.org/10.1016/j.febslet.2009.08.017] [PMID: 19695252]

[53]   Pallas-Bazarra N, Jurado-Arjona J, Navarrete M, *et al.* Novel function of Tau in regulating the effects of external stimuli on adult hippocampal neurogenesis. EMBO J 2016; 35(13): 1417-36.
[http://dx.doi.org/10.15252/embj.201593518] [PMID: 27198172]

[54]   Klein C, Kramer EM, Cardine AM, Schraven B, Brandt R, Trotter J. Process outgrowth of oligodendrocytes is promoted by interaction of fyn kinase with the cytoskeletal protein tau. J Neurosci 2002; 22(3): 698-707.
[http://dx.doi.org/10.1523/JNEUROSCI.22-03-00698.2002] [PMID: 11826099]

[55]   LoPresti P, Szuchet S, Papasozomenos SC, Zinkowski RP, Binder LI. Functional implications for the microtubule-associated protein tau: localization in oligodendrocytes. Proc Natl Acad Sci USA 1995; 92(22): 10369-73.
[http://dx.doi.org/10.1073/pnas.92.22.10369] [PMID: 7479786]

[56]   Ilschner S, Brandt R. The transition of microglia to a ramified phenotype is associated with the formation of stable acetylated and detyrosinated microtubules. Glia 1996; 18(2): 129-40.
[http://dx.doi.org/10.1002/(SICI)1098-1136(199610)18:2<129::AID-GLIA5>3.0.CO;2-W]   [PMID: 8913776]

[57]   Hanger DP, Anderton BH, Noble W. Tau phosphorylation: the therapeutic challenge for neurodegenerative disease. Trends Mol Med 2009; 15(3): 112-9.
[http://dx.doi.org/10.1016/j.molmed.2009.01.003] [PMID: 19246243]

[58]   Liu F, Grundke-Iqbal I, Iqbal K, Gong CX. Contributions of protein phosphatases PP1, PP2A, PP2B and PP5 to the regulation of tau phosphorylation. Eur J Neurosci 2005; 22(8): 1942-50.
[http://dx.doi.org/10.1111/j.1460-9568.2005.04391.x] [PMID: 16262633]

[59]   Liu F, Iqbal K, Grundke-Iqbal I, Rossie S, Gong CX. Dephosphorylation of tau by protein phosphatase 5: impairment in Alzheimer's disease. J Biol Chem 2005; 280(3): 1790-6.
[http://dx.doi.org/10.1074/jbc.M410775200] [PMID: 15546861]

[60]   Lee YI, Seo M, Kim Y, *et al.* Membrane depolarization induces the undulating phosphorylation/dephosphorylation of glycogen synthase kinase 3β, and this dephosphorylation involves protein phosphatases 2A and 2B in SH-SY5Y human neuroblastoma cells. J Biol Chem 2005; 280(23): 22044-52.
[http://dx.doi.org/10.1074/jbc.M413987200] [PMID: 15799972]

[61]   Yao XQ, Zhang XX, Yin YY, *et al.* Glycogen synthase kinase-3β regulates Tyr307 phosphorylation of protein phosphatase-2A *via* protein tyrosine phosphatase 1B but not Src. Biochem J 2011; 437(2): 335-44.
[http://dx.doi.org/10.1042/BJ20110347] [PMID: 21554241]

[62]   Ksiezak-Reding H, Pyo HK, Feinstein B, Pasinetti GM. Akt/PKB kinase phosphorylates separately Thr212 and Ser214 of tau protein *in vitro.* Biochim Biophys Acta 2003; 1639(3): 159-68.
[http://dx.doi.org/10.1016/j.bbadis.2003.09.001] [PMID: 14636947]

[63] von Bergen M, Friedhoff P, Biernat J, Heberle J, Mandelkow EM, Mandelkow E. Assembly of τ protein into Alzheimer paired helical filaments depends on a local sequence motif ((306)VQIVYK(311)) forming β structure. Proc Natl Acad Sci USA 2000; 97(10): 5129-34.
[http://dx.doi.org/10.1073/pnas.97.10.5129] [PMID: 10805776]

[64] Ittner LM, Ke YD, Delerue F, *et al.* Dendritic function of tau mediates amyloid-β toxicity in Alzheimer's disease mouse models. Cell 2010; 142(3): 387-97.
[http://dx.doi.org/10.1016/j.cell.2010.06.036] [PMID: 20655099]

[65] Hoover BR, Reed MN, Su J, *et al.* Tau mislocalization to dendritic spines mediates synaptic dysfunction independently of neurodegeneration. Neuron 2010; 68(6): 1067-81.
[http://dx.doi.org/10.1016/j.neuron.2010.11.030] [PMID: 21172610]

[66] Min SW, Cho SH, Zhou Y, *et al.* Acetylation of tau inhibits its degradation and contributes to tauopathy. Neuron 2010; 67(6): 953-66.
[http://dx.doi.org/10.1016/j.neuron.2010.08.044] [PMID: 20869593]

[67] Cohen TJ, Friedmann D, Hwang AW, Marmorstein R, Lee VMY. The microtubule-associated tau protein has intrinsic acetyltransferase activity. Nat Struct Mol Biol 2013; 20(6): 756-62.
[http://dx.doi.org/10.1038/nsmb.2555] [PMID: 23624859]

[68] Cook C, Carlomagno Y, Gendron TF, *et al.* Acetylation of the KXGS motifs in tau is a critical determinant in modulation of tau aggregation and clearance. Hum Mol Genet 2014; 23(1): 104-16.
[http://dx.doi.org/10.1093/hmg/ddt402] [PMID: 23962722]

[69] Morris M, Knudsen GM, Maeda S, *et al.* Tau post-translational modifications in wild-type and human amyloid precursor protein transgenic mice. Nat Neurosci 2015; 18(8): 1183-9.
[http://dx.doi.org/10.1038/nn.4067] [PMID: 26192747]

[70] Tracy TE, Sohn PD, Minami SS, *et al.* Acetylated tau obstructs KIBRA-mediated signaling in synaptic plasticity and promotes tauopathy-related memory loss. Neuron 2016; 90(2): 245-60.
[http://dx.doi.org/10.1016/j.neuron.2016.03.005] [PMID: 27041503]

[71] Heitz FD, Farinelli M, Mohanna S, *et al.* The memory gene KIBRA is a bidirectional regulator of synaptic and structural plasticity in the adult brain. Neurobiol Learn Mem 2016; 135: 100-14.
[http://dx.doi.org/10.1016/j.nlm.2016.07.028] [PMID: 27498008]

[72] Sohn PD, Tracy TE, Son HI, *et al.* Acetylated tau destabilizes the cytoskeleton in the axon initial segment and is mislocalized to the somatodendritic compartment. Mol Neurodegener 2016; 11(1): 47.
[http://dx.doi.org/10.1186/s13024-016-0109-0] [PMID: 27356871]

[73] Liu F, Zaidi T, Iqbal K, Grundke-Iqbal I, Gong CX. Aberrant glycosylation modulates phosphorylation of tau by protein kinase A and dephosphorylation of tau by protein phosphatase 2A and 5. Neuroscience 2002; 115(3): 829-37.
[http://dx.doi.org/10.1016/S0306-4522(02)00510-9] [PMID: 12435421]

[74] Liu F, Iqbal K, Grundke-Iqbal I, Hart GW, Gong CX. O-GlcNAcylation regulates phosphorylation of tau: a mechanism involved in Alzheimer's disease. Proc Natl Acad Sci USA 2004; 101(29): 10804-9.
[http://dx.doi.org/10.1073/pnas.0400348101] [PMID: 15249677]

[75] Yuzwa SA, Cheung AH, Okon M, McIntosh LP, Vocadlo DJ. O-GlcNAc modification of tau directly inhibits its aggregation without perturbing the conformational properties of tau monomers. J Mol Biol 2014; 426(8): 1736-52.
[http://dx.doi.org/10.1016/j.jmb.2014.01.004] [PMID: 24444746]

[76] Wang AC, Jensen EH, Rexach JE, Vinters HV, Hsieh-Wilson LC. Loss of O-GlcNAc glycosylation in forebrain excitatory neurons induces neurodegeneration. Proc Natl Acad Sci USA 2016; 113(52): 15120-5.
[http://dx.doi.org/10.1073/pnas.1606899113] [PMID: 27956640]

[77] Petrucelli L, Dickson D, Kehoe K, *et al.* CHIP and Hsp70 regulate tau ubiquitination, degradation and aggregation. Hum Mol Genet 2004; 13(7): 703-14.

[http://dx.doi.org/10.1093/hmg/ddh083] [PMID: 14962978]

[78] Luo HB, Xia YY, Shu XJ, *et al*. SUMOylation at K340 inhibits tau degradation through deregulating its phosphorylation and ubiquitination. Proc Natl Acad Sci USA 2014; 111(46): 16586-91.
[http://dx.doi.org/10.1073/pnas.1417548111] [PMID: 25378699]

[79] Funk KE, Thomas SN, Schafer KN, *et al*. Lysine methylation is an endogenous post-translational modification of tau protein in human brain and a modulator of aggregation propensity. Biochem J 2014; 462(1): 77-88.
[http://dx.doi.org/10.1042/BJ20140372] [PMID: 24869773]

[80] Watanabe A, Hong WK, Dohmae N, Takio K, Morishima-Kawashima M, Ihara Y. Molecular aging of tau: disulfide-independent aggregation and non-enzymatic degradation *in vitro* and *in vivo*. J Neurochem 2004; 90(6): 1302-11.
[http://dx.doi.org/10.1111/j.1471-4159.2004.02611.x] [PMID: 15341514]

[81] Chesser AS, Pritchard SM, Johnson GVW. Tau clearance mechanisms and their possible role in the pathogenesis of Alzheimer disease. Front Neurol 2013; 4: 122.
[http://dx.doi.org/10.3389/fneur.2013.00122] [PMID: 24027553]

[82] David DC, Layfield R, Serpell L, Narain Y, Goedert M, Spillantini MG. Proteasomal degradation of tau protein. J Neurochem 2002; 83(1): 176-85.
[http://dx.doi.org/10.1046/j.1471-4159.2002.01137.x] [PMID: 12358741]

[83] Babu JR, Geetha T, Wooten MW. Sequestosome 1/p62 shuttles polyubiquitinated tau for proteasomal degradation. J Neurochem 2005; 94(1): 192-203.
[http://dx.doi.org/10.1111/j.1471-4159.2005.03181.x] [PMID: 15953362]

[84] Johansen T, Lamark T. Selective autophagy mediated by autophagic adapter proteins. Autophagy 2011; 7(3): 279-96.
[http://dx.doi.org/10.4161/auto.7.3.14487] [PMID: 21189453]

[85] Watson RO, Manzanillo PS, Cox JS. Extracellular *M. tuberculosis* DNA targets bacteria for autophagy by activating the host DNA-sensing pathway. Cell 2012; 150(4): 803-15.
[http://dx.doi.org/10.1016/j.cell.2012.06.040] [PMID: 22901810]

[86] Dice JF. Chaperone-mediated autophagy. Autophagy 2007; 3(4): 295-9.
[http://dx.doi.org/10.4161/auto.4144] [PMID: 17404494]

[87] Knaevelsrud H, Simonsen A. Fighting disease by selective autophagy of aggregate-prone proteins. FEBS Lett 2010; 584(12): 2635-45.
[http://dx.doi.org/10.1016/j.febslet.2010.04.041] [PMID: 20412801]

[88] Dolan PJ, Johnson GV. A caspase cleaved form of tau is preferentially degraded through the autophagy pathway. J Biol Chem 2010; 285(29): 21978-87.
[http://dx.doi.org/10.1074/jbc.M110.110940] [PMID: 20466727]

[89] Zhang JY, Liu SJ, Li HL, Wang JZ. Microtubule-associated protein tau is a substrate of ATP/Mg$^{(2+)}$-dependent proteasome protease system. J Neural Transm (Vienna) 2005; 112(4): 547-55.
[http://dx.doi.org/10.1007/s00702-004-0196-x] [PMID: 15372326]

[90] Korolchuk VI, Menzies FM, Rubinsztein DC. Mechanisms of cross-talk between the ubiquitin-proteasome and autophagy-lysosome systems. FEBS Lett 2010; 584(7): 1393-8.
[http://dx.doi.org/10.1016/j.febslet.2009.12.047] [PMID: 20040365]

[91] Keck S, Nitsch R, Grune T, Ullrich O. Proteasome inhibition by paired helical filament-tau in brains of patients with Alzheimer's disease. J Neurochem 2003; 85(1): 115-22.
[http://dx.doi.org/10.1046/j.1471-4159.2003.01642.x] [PMID: 12641733]

[92] Nixon RA, Yang DS. Autophagy failure in Alzheimer's disease--locating the primary defect. Neurobiol Dis 2011; 43(1): 38-45.
[http://dx.doi.org/10.1016/j.nbd.2011.01.021] [PMID: 21296668]

[93]   Yang DS, Stavrides P, Mohan PS, *et al.* Reversal of autophagy dysfunction in the TgCRND8 mouse model of Alzheimer's disease ameliorates amyloid pathologies and memory deficits. Brain 2011; 134(Pt 1): 258-77.
[http://dx.doi.org/10.1093/brain/awq341] [PMID: 21186265]

[94]   Jadhav S, Avila J, Schöll M, *et al.* A walk through tau therapeutic strategies. Acta Neuropathol Commun 2019; 7(1): 22.
[http://dx.doi.org/10.1186/s40478-019-0664-z] [PMID: 30767766]

[95]   Kovacs GG. Invited review: Neuropathology of tauopathies: principles and practice. Neuropathol Appl Neurobiol 2015; 41(1): 3-23.
[http://dx.doi.org/10.1111/nan.12208] [PMID: 25495175]

[96]   Hutton M, Lendon CL, Rizzu P, *et al.* Association of missense and 5'-splice-site mutations in tau with the inherited dementia FTDP-17. Nature 1998; 393(6686): 702-5.
[http://dx.doi.org/10.1038/31508] [PMID: 9641683]

[97]   Alonso AdelC, Mederlyova A, Novak M, Grundke-Iqbal I, Iqbal K. Promotion of hyperphosphorylation by frontotemporal dementia tau mutations. J Biol Chem 2004; 279(33): 34873-81.
[http://dx.doi.org/10.1074/jbc.M405131200] [PMID: 15190058]

[98]   Ward SM, Himmelstein DS, Lancia JK, Binder LI. Tau oligomers and tau toxicity in neurodegenerative disease. Biochem Soc Trans 2012; 40(4): 667-71.
[http://dx.doi.org/10.1042/BST20120134] [PMID: 22817713]

[99]   Kopeikina KJ, Hyman BT, Spires-Jones TL. Soluble forms of tau are toxic in Alzheimer's disease. Transl Neurosci 2012; 3(3): 223-33.
[http://dx.doi.org/10.2478/s13380-012-0032-y] [PMID: 23029602]

[100]   Spires-Jones TL, Stoothoff WH, de Calignon A, Jones PB, Hyman BT. Tau pathophysiology in neurodegeneration: a tangled issue. Trends Neurosci 2009; 32(3): 150-9.
[http://dx.doi.org/10.1016/j.tins.2008.11.007] [PMID: 19162340]

[101]   Alonso AC, Grundke-Iqbal I, Iqbal K. Alzheimer's disease hyperphosphorylated tau sequesters normal tau into tangles of filaments and disassembles microtubules. Nat Med 1996; 2(7): 783-7.
[http://dx.doi.org/10.1038/nm0796-783] [PMID: 8673924]

[102]   Pérez M, Valpuesta JM, Medina M, Montejo de Garcini E, Avila J. Polymerization of tau into filaments in the presence of heparin: the minimal sequence required for tau-tau interaction. J Neurochem 1996; 67(3): 1183-90.
[http://dx.doi.org/10.1046/j.1471-4159.1996.67031183.x] [PMID: 8752125]

[103]   Peterson DW, Zhou H, Dahlquist FW, Lew J. A soluble oligomer of tau associated with fiber formation analyzed by NMR. Biochemistry 2008; 47(28): 7393-404.
[http://dx.doi.org/10.1021/bi702466a] [PMID: 18558718]

[104]   Barghorn S, Mandelkow E. Toward a unified scheme for the aggregation of tau into Alzheimer paired helical filaments. Biochemistry 2002; 41(50): 14885-96.
[http://dx.doi.org/10.1021/bi026469j] [PMID: 12475237]

[105]   Meraz-Ríos MA, Lira-De León KI, Campos-Peña V, De Anda-Hernández MA, Mena-López R. Tau oligomers and aggregation in Alzheimer's disease. J Neurochem 2010; 112(6): 1353-67.
[http://dx.doi.org/10.1111/j.1471-4159.2009.06511.x] [PMID: 19943854]

[106]   Mocanu MM, Nissen A, Eckermann K, *et al.* The potential for beta-structure in the repeat domain of tau protein determines aggregation, synaptic decay, neuronal loss, and coassembly with endogenous Tau in inducible mouse models of tauopathy. J Neurosci 2008; 28(3): 737-48.
[http://dx.doi.org/10.1523/JNEUROSCI.2824-07.2008] [PMID: 18199773]

[107]   Cowan CM, Mudher A. Are tau aggregates toxic or protective in tauopathies? Front Neurol 2013; 4: 114.

[http://dx.doi.org/10.3389/fneur.2013.00114] [PMID: 23964266]

[108] Lasagna-Reeves CA, Castillo-Carranza DL, Sengupta U, Clos AL, Jackson GR, Kayed R. Tau oligomers impair memory and induce synaptic and mitochondrial dysfunction in wild-type mice. Mol Neurodegener 2011; 6: 39.
[http://dx.doi.org/10.1186/1750-1326-6-39] [PMID: 21645391]

[109] Santacruz K, Lewis J, Spires T, *et al.* Tau suppression in a neurodegenerative mouse model improves memory function. Science 2005; 309(5733): 476-81.
[http://dx.doi.org/10.1126/science.1113694] [PMID: 16020737]

[110] Fox LM, William CM, Adamowicz DH, *et al.* Soluble tau species, not neurofibrillary aggregates, disrupt neural system integration in a tau transgenic model. J Neuropathol Exp Neurol 2011; 70(7): 588-95.
[http://dx.doi.org/10.1097/NEN.0b013e318220a658] [PMID: 21666499]

[111] Spires-Jones TL, Kopeikina KJ, Koffie RM, de Calignon A, Hyman BT. Are tangles as toxic as they look? J Mol Neurosci 2011; 45(3): 438-44.
[http://dx.doi.org/10.1007/s12031-011-9566-7] [PMID: 21638071]

[112] von Bergen M, Barghorn S, Li L, *et al.* Mutations of tau protein in frontotemporal dementia promote aggregation of paired helical filaments by enhancing local beta-structure. J Biol Chem 2001; 276(51): 48165-74.
[http://dx.doi.org/10.1074/jbc.M105196200] [PMID: 11606569]

[113] Wilson DM, Binder LI. Free fatty acids stimulate the polymerization of tau and amyloid beta peptides. *In vitro* evidence for a common effector of pathogenesis in Alzheimer's disease. Am J Pathol 1997; 150(6): 2181-95.
[PMID: 9176408]

[114] Novak M, Kabat J, Wischik CM. Molecular characterization of the minimal protease resistant tau unit of the Alzheimer's disease paired helical filament. EMBO J 1993; 12(1): 365-70.
[http://dx.doi.org/10.1002/j.1460-2075.1993.tb05665.x] [PMID: 7679073]

[115] Novák M. Truncated tau protein as a new marker for Alzheimer's disease. Acta Virol 1994; 38(3): 173-89.
[PMID: 7817900]

[116] García-Sierra F, Mondragón-Rodríguez S, Basurto-Islas G. Truncation of tau protein and its pathological significance in Alzheimer's disease. J Alzheimers Dis 2008; 14(4): 401-9.
[http://dx.doi.org/10.3233/JAD-2008-14407] [PMID: 18688090]

[117] Horowitz PM, Patterson KR, Guillozet-Bongaarts AL, *et al.* Early N-terminal changes and caspase-6 cleavage of tau in Alzheimer's disease. J Neurosci 2004; 24(36): 7895-902.
[http://dx.doi.org/10.1523/JNEUROSCI.1988-04.2004] [PMID: 15356202]

[118] Zilka N, Kazmerova Z, Jadhav S, *et al.* Who fans the flames of Alzheimer's disease brains? Misfolded tau on the crossroad of neurodegenerative and inflammatory pathways. J Neuroinflammation 2012; 9: 47.
[http://dx.doi.org/10.1186/1742-2094-9-47] [PMID: 22397366]

[119] Hanger DP, Wray S. Tau cleavage and tau aggregation in neurodegenerative disease. Biochem Soc Trans 2010; 38(4): 1016-20.
[http://dx.doi.org/10.1042/BST0381016] [PMID: 20658996]

[120] Wang Y, Garg S, Mandelkow EM, Mandelkow E. Proteolytic processing of tau. Biochem Soc Trans 2010; 38(4): 955-61.
[http://dx.doi.org/10.1042/BST0380955] [PMID: 20658984]

[121] Gamblin TC, Chen F, Zambrano A, *et al.* Caspase cleavage of tau: linking amyloid and neurofibrillary tangles in Alzheimer's disease. Proc Natl Acad Sci USA 2003; 100(17): 10032-7.
[http://dx.doi.org/10.1073/pnas.1630428100] [PMID: 12888622]

[122] Rissman RA, Poon WW, Blurton-Jones M, *et al.* Caspase-cleavage of tau is an early event in Alzheimer disease tangle pathology. J Clin Invest 2004; 114(1): 121-30.
[http://dx.doi.org/10.1172/JCI200420640] [PMID: 15232619]

[123] Means JC, Gerdes BC, Kaja S, *et al.* Caspase-3-dependent proteolytic cleavage of tau causes neurofibrillary tangles and results in cognitive impairment during normal aging. Neurochem Res 2016; 41(9): 2278-88.
[http://dx.doi.org/10.1007/s11064-016-1942-9] [PMID: 27220334]

[124] Fasulo L, Ugolini G, Visintin M, *et al.* The neuronal microtubule-associated protein tau is a substrate for caspase-3 and an effector of apoptosis. J Neurochem 2000; 75(2): 624-33.
[http://dx.doi.org/10.1046/j.1471-4159.2000.0750624.x] [PMID: 10899937]

[125] Quintanilla RA, Matthews-Roberson TA, Dolan PJ, Johnson GV. Caspase-cleaved tau expression induces mitochondrial dysfunction in immortalized cortical neurons: implications for the pathogenesis of Alzheimer disease. J Biol Chem 2009; 284(28): 18754-66.
[http://dx.doi.org/10.1074/jbc.M808908200] [PMID: 19389700]

[126] Zhao X, Kotilinek LA, Smith B, *et al.* Caspase-2 cleavage of tau reversibly impairs memory. Nat Med 2016; 22(11): 1268-76.
[http://dx.doi.org/10.1038/nm.4199] [PMID: 27723722]

[127] Guillozet-Bongaarts AL, Cahill ME, Cryns VL, Reynolds MR, Berry RW, Binder LI. Pseudophosphorylation of tau at serine 422 inhibits caspase cleavage: *in vitro* evidence and implications for tangle formation *in vivo*. J Neurochem 2006; 97(4): 1005-14.
[http://dx.doi.org/10.1111/j.1471-4159.2006.03784.x] [PMID: 16606369]

[128] Ferreira A, Bigio EH. Calpain-mediated tau cleavage: a mechanism leading to neurodegeneration shared by multiple tauopathies. Mol Med 2011; 17(7-8): 676-85.
[http://dx.doi.org/10.2119/molmed.2010.00220] [PMID: 21442128]

[129] Saito K, Elce JS, Hamos JE, Nixon RA. Widespread activation of calcium-activated neutral proteinase (calpain) in the brain in Alzheimer disease: a potential molecular basis for neuronal degeneration. Proc Natl Acad Sci USA 1993; 90(7): 2628-32.
[http://dx.doi.org/10.1073/pnas.90.7.2628] [PMID: 8464868]

[130] Kelly BL, Ferreira A. beta-Amyloid-induced dynamin 1 degradation is mediated by N-methyl-D-aspartate receptors in hippocampal neurons. J Biol Chem 2006; 281(38): 28079-89.
[http://dx.doi.org/10.1074/jbc.M605081200] [PMID: 16864575]

[131] Ferreira A, Lu Q, Orecchio L, Kosik KS. Selective phosphorylation of adult tau isoforms in mature hippocampal neurons exposed to fibrillar A beta. Mol Cell Neurosci 1997; 9(3): 220-34.
[http://dx.doi.org/10.1006/mcne.1997.0615] [PMID: 9245504]

[132] Kurbatskaya K, Phillips EC, Croft CL, *et al.* Upregulation of calpain activity precedes tau phosphorylation and loss of synaptic proteins in Alzheimer's disease brain. Acta Neuropathol Commun 2016; 4: 34.
[http://dx.doi.org/10.1186/s40478-016-0299-2] [PMID: 27036949]

[133] Di Domenico F, Tramutola A, Perluigi M. Cathepsin D as a therapeutic target in Alzheimer's disease. Expert Opin Ther Targets 2016; 20(12): 1393-5.
[http://dx.doi.org/10.1080/14728222.2016.1252334] [PMID: 27805462]

[134] Cataldo AM, Thayer CY, Bird ED, Wheelock TR, Nixon RA. Lysosomal proteinase antigens are prominently localized within senile plaques of Alzheimer's disease: evidence for a neuronal origin. Brain Res 1990; 513(2): 181-92.
[http://dx.doi.org/10.1016/0006-8993(90)90456-L] [PMID: 2350688]

[135] Perez SE, He B, Nadeem M, *et al.* Hippocampal endosomal, lysosomal, and autophagic dysregulation in mild cognitive impairment: correlation with aβ and tau pathology. J Neuropathol Exp Neurol 2015; 74(4): 345-58.

[http://dx.doi.org/10.1097/NEN.0000000000000179] [PMID: 25756588]

[136] Vidoni C, Follo C, Savino M, Melone MA, Isidoro C. The role of cathepsin D in the pathogenesis of human neurodegenerative disorders. Med Res Rev 2016; 36(5): 845-70.
[http://dx.doi.org/10.1002/med.21394] [PMID: 27114232]

[137] Kenessey A, Nacharaju P, Ko LW, Yen SH. Degradation of tau by lysosomal enzyme cathepsin D: implication for Alzheimer neurofibrillary degeneration. J Neurochem 1997; 69(5): 2026-38.
[http://dx.doi.org/10.1046/j.1471-4159.1997.69052026.x] [PMID: 9349548]

[138] Wang Y, Martinez-Vicente M, Krüger U, *et al.* Tau fragmentation, aggregation and clearance: the dual role of lysosomal processing. Hum Mol Genet 2009; 18(21): 4153-70.
[http://dx.doi.org/10.1093/hmg/ddp367] [PMID: 19654187]

[139] Arai T, Miklossy J, Klegeris A, Guo JP, McGeer PL. Thrombin and prothrombin are expressed by neurons and glial cells and accumulate in neurofibrillary tangles in Alzheimer disease brain. J Neuropathol Exp Neurol 2006; 65(1): 19-25.
[http://dx.doi.org/10.1097/01.jnen.0000196133.74087.cb] [PMID: 16410745]

[140] Olesen OF. Proteolytic degradation of microtubule associated protein tau by thrombin. Biochem Biophys Res Commun 1994; 201(2): 716-21.
[http://dx.doi.org/10.1006/bbrc.1994.1759] [PMID: 8003007]

[141] Arai T, Guo JP, McGeer PL. Proteolysis of non-phosphorylated and phosphorylated tau by thrombin. J Biol Chem 2005; 280(7): 5145-53.
[http://dx.doi.org/10.1074/jbc.M409234200] [PMID: 15542598]

[142] Quinn JP, Corbett NJ, Kellett KAB, Hooper NM. Tau proteolysis in the pathogenesis of tauopathies: neurotoxic fragments and novel biomarkers. J Alzheimers Dis 2018; 63(1): 13-33.
[http://dx.doi.org/10.3233/JAD-170959] [PMID: 29630551]

[143] Zhang Z, Song M, Liu X, *et al.* Cleavage of tau by asparagine endopeptidase mediates the neurofibrillary pathology in Alzheimer's disease. Nat Med 2014; 20(11): 1254-62.
[http://dx.doi.org/10.1038/nm.3700] [PMID: 25326800]

[144] Zhang Z, Kang SS, Liu X, *et al.* Asparagine endopeptidase cleaves α-synuclein and mediates pathologic activities in Parkinson's disease. Nat Struct Mol Biol 2017; 24(8): 632-42.
[http://dx.doi.org/10.1038/nsmb.3433] [PMID: 28671665]

[145] Zhang Z, Xie M, Ye K. Asparagine endopeptidase is an innovative therapeutic target for neurodegenerative diseases. Expert Opin Ther Targets 2016; 20(10): 1237-45.
[http://dx.doi.org/10.1080/14728222.2016.1182990] [PMID: 27115710]

[146] Zhang Z, Obianyo O, Dall E, *et al.* Inhibition of delta-secretase improves cognitive functions in mouse models of Alzheimer's disease. Nat Commun 2017; 8: 14740.
[http://dx.doi.org/10.1038/ncomms14740] [PMID: 28345579]

[147] Tobler AR, Constam DB, Schmitt-Gräff A, Malipiero U, Schlapbach R, Fontana A. Cloning of the human puromycin-sensitive aminopeptidase and evidence for expression in neurons. J Neurochem 1997; 68(3): 889-97.
[http://dx.doi.org/10.1046/j.1471-4159.1997.68030889.x] [PMID: 9048733]

[148] Karsten SL, Sang TK, Gehman LT, *et al.* A genomic screen for modifiers of tauopathy identifies puromycin-sensitive aminopeptidase as an inhibitor of tau-induced neurodegeneration. Neuron 2006; 51(5): 549-60.
[http://dx.doi.org/10.1016/j.neuron.2006.07.019] [PMID: 16950154]

[149] Kudo LC, Parfenova L, Ren G, *et al.* Puromycin-sensitive aminopeptidase (PSA/NPEPPS) impedes development of neuropathology in hPSA/TAU(P301L) double-transgenic mice. Hum Mol Genet 2011; 20(9): 1820-33.
[http://dx.doi.org/10.1093/hmg/ddr065] [PMID: 21320871]

[150] Poepsel S, Sprengel A, Sacca B, *et al.* Determinants of amyloid fibril degradation by the PDZ protease

HTRA1. Nat Chem Biol 2015; 11(11): 862-9.
[http://dx.doi.org/10.1038/nchembio.1931] [PMID: 26436840]

[151] Tennstaedt A, Pöpsel S, Truebestein L, *et al.* Human high temperature requirement serine protease A1 (HTRA1) degrades tau protein aggregates. J Biol Chem 2012; 287(25): 20931-41.
[http://dx.doi.org/10.1074/jbc.M111.316232] [PMID: 22535953]

[152] Yu JT, Tan L, Hardy J. Apolipoprotein E in Alzheimer's disease: an update. Annu Rev Neurosci 2014; 37: 79-100.
[http://dx.doi.org/10.1146/annurev-neuro-071013-014300] [PMID: 24821312]

[153] Chu Q, Diedrich JK, Vaughan JM, *et al.* HtrA1 proteolysis of ApoE *in vitro* is allele selective. J Am Chem Soc 2016; 138(30): 9473-8.
[http://dx.doi.org/10.1021/jacs.6b03463] [PMID: 27379525]

[154] Gregori L, Poosch MS, Cousins G, Chau V. A uniform isopeptide-linked multiubiquitin chain is sufficient to target substrate for degradation in ubiquitin-mediated proteolysis. J Biol Chem 1990; 265(15): 8354-7.
[http://dx.doi.org/10.1016/S0021-9258(19)38890-8] [PMID: 2160452]

[155] Deveraux Q, Ustrell V, Pickart C, Rechsteiner M. A 26 S protease subunit that binds ubiquitin conjugates. J Biol Chem 1994; 269(10): 7059-61.
[http://dx.doi.org/10.1016/S0021-9258(17)37244-7] [PMID: 8125911]

[156] Forero DA, Casadesus G, Perry G, Arboleda H. Synaptic dysfunction and oxidative stress in Alzheimer's disease: emerging mechanisms. J Cell Mol Med 2006; 10(3): 796-805.
[http://dx.doi.org/10.1111/j.1582-4934.2006.tb00439.x] [PMID: 16989739]

[157] Bence NF, Sampat RM, Kopito RR. Impairment of the ubiquitin-proteasome system by protein aggregation. Science 2001; 292(5521): 1552-5.
[http://dx.doi.org/10.1126/science.292.5521.1552] [PMID: 11375494]

[158] Keller JN, Hanni KB, Markesbery WR. Impaired proteasome function in Alzheimer's disease. J Neurochem 2000; 75(1): 436-9.
[http://dx.doi.org/10.1046/j.1471-4159.2000.0750436.x] [PMID: 10854289]

[159] Henriksen K, Wang Y, Sørensen MG, *et al.* An enzyme-generated fragment of tau measured in serum shows an inverse correlation to cognitive function. PLoS One 2013; 8(5): e64990.
[http://dx.doi.org/10.1371/journal.pone.0064990] [PMID: 23717682]

[160] Andrew RJ, Kellett KA, Thinakaran G, Hooper NM. A Greek tragedy: the growing complexity of Alzheimer amyloid precursor protein proteolysis. J Biol Chem 2016; 291(37): 19235-44.
[http://dx.doi.org/10.1074/jbc.R116.746032] [PMID: 27474742]

[161] Cohen TJ, Constance BH, Hwang AW, James M, Yuan CX. Intrinsic tau acetylation is coupled to auto-proteolytic tau fragmentation. PLoS One 2016; 11(7): e0158470.
[http://dx.doi.org/10.1371/journal.pone.0158470] [PMID: 27383765]

[162] Morfini GA, Burns M, Binder LI, *et al.* Axonal transport defects in neurodegenerative diseases. J Neurosci 2009; 29(41): 12776-86.
[http://dx.doi.org/10.1523/JNEUROSCI.3463-09.2009] [PMID: 19828789]

[163] Hirokawa N. Kinesin and dynein superfamily proteins and the mechanism of organelle transport. Science 1998; 279(5350): 519-26.
[http://dx.doi.org/10.1126/science.279.5350.519] [PMID: 9438838]

[164] De Vos KJ, Grierson AJ, Ackerley S, Miller CC. Role of axonal transport in neurodegenerative diseases. Annu Rev Neurosci 2008; 31: 151-73.
[http://dx.doi.org/10.1146/annurev.neuro.31.061307.090711] [PMID: 18558852]

[165] Matamoros AJ, Baas PW. Microtubules in health and degenerative disease of the nervous system. Brain Res Bull 2016; 126(Pt 3): 217-25.
[http://dx.doi.org/10.1016/j.brainresbull.2016.06.016] [PMID: 27365230]

[166]   Millecamps S, Julien JP. Axonal transport deficits and neurodegenerative diseases. Nat Rev Neurosci 2013; 14(3): 161-76.
[http://dx.doi.org/10.1038/nrn3380] [PMID: 23361386]

[167]   Paholikova K, Salingova B, Opattova A, *et al.* N-terminal truncation of microtubule associated protein tau dysregulates its cellular localization. J Alzheimers Dis 2015; 43(3): 915-26.
[http://dx.doi.org/10.3233/JAD-140996] [PMID: 25147106]

[168]   Goedert M, Jakes R. Expression of separate isoforms of human tau protein: correlation with the tau pattern in brain and effects on tubulin polymerization. EMBO J 1990; 9(13): 4225-30.
[http://dx.doi.org/10.1002/j.1460-2075.1990.tb07870.x] [PMID: 2124967]

[169]   Braak E, Braak H, Mandelkow EM. A sequence of cytoskeleton changes related to the formation of neurofibrillary tangles and neuropil threads. Acta Neuropathol 1994; 87(6): 554-67.
[http://dx.doi.org/10.1007/BF00293315] [PMID: 7522386]

[170]   Chiasseu M, Cueva Vargas JL, Destroismaisons L, Vande Velde C, Leclerc N, Di Polo A. Tau accumulation, altered phosphorylation, and missorting promote neurodegeneration in glaucoma. J Neurosci 2016; 36(21): 5785-98.
[http://dx.doi.org/10.1523/JNEUROSCI.3986-15.2016] [PMID: 27225768]

[171]   Ebneth A, Godemann R, Stamer K, Illenberger S, Trinczek B, Mandelkow E. Overexpression of tau protein inhibits kinesin-dependent trafficking of vesicles, mitochondria, and endoplasmic reticulum: implications for Alzheimer's disease. J Cell Biol 1998; 143(3): 777-94.
[http://dx.doi.org/10.1083/jcb.143.3.777] [PMID: 9813097]

[172]   Dixit R, Ross JL, Goldman YE, Holzbaur EL. Differential regulation of dynein and kinesin motor proteins by tau. Science 2008; 319(5866): 1086-9.
[http://dx.doi.org/10.1126/science.1152993] [PMID: 18202255]

[173]   Stamer K, Vogel R, Thies E, Mandelkow E, Mandelkow EM. Tau blocks traffic of organelles, neurofilaments, and APP vesicles in neurons and enhances oxidative stress. J Cell Biol 2002; 156(6): 1051-63.
[http://dx.doi.org/10.1083/jcb.200108057] [PMID: 11901170]

[174]   Morel M, Héraud C, Nicaise C, Suain V, Brion JP. Levels of kinesin light chain and dynein intermediate chain are reduced in the frontal cortex in Alzheimer's disease: implications for axoplasmic transport. Acta Neuropathol 2012; 123(1): 71-84.
[http://dx.doi.org/10.1007/s00401-011-0901-4] [PMID: 22094641]

[175]   Konzack S, Thies E, Marx A, Mandelkow EM, Mandelkow E. Swimming against the tide: mobility of the microtubule-associated protein tau in neurons. J Neurosci 2007; 27(37): 9916-27.
[http://dx.doi.org/10.1523/JNEUROSCI.0927-07.2007] [PMID: 17855606]

[176]   Camero S, Benítez MJ, Cuadros R, Hernández F, Avila J, Jiménez JS. Thermodynamics of the interaction between Alzheimer's disease related tau protein and DNA. PLoS One 2014; 9(8): e104690.
[http://dx.doi.org/10.1371/journal.pone.0104690] [PMID: 25126942]

[177]   Qi H, Cantrelle FX, Benhelli-Mokrani H, *et al.* Nuclear magnetic resonance spectroscopy characterization of interaction of Tau with DNA and its regulation by phosphorylation. Biochemistry 2015; 54(7): 1525-33.
[http://dx.doi.org/10.1021/bi5014613] [PMID: 25623359]

[178]   Mondragón-Rodríguez S, Trillaud-Doppia E, Dudilot A, *et al.* Interaction of endogenous tau protein with synaptic proteins is regulated by N-methyl-D-aspartate receptor-dependent tau phosphorylation. J Biol Chem 2012; 287(38): 32040-53.
[http://dx.doi.org/10.1074/jbc.M112.401240] [PMID: 22833681]

[179]   Brunello CA, Yan X, Huttunen HJ. Internalized Tau sensitizes cells to stress by promoting formation and stability of stress granules. Sci Rep 2016; 6: 30498.
[http://dx.doi.org/10.1038/srep30498] [PMID: 27460788]

[180]   Seward ME, Swanson E, Norambuena A, *et al.* Amyloid-β signals through tau to drive ectopic neuronal cell cycle re-entry in Alzheimer's disease. J Cell Sci 2013; 126(Pt 5): 1278-86.
[http://dx.doi.org/10.1242/jcs.1125880] [PMID: 23345405]

[181]   Hernández-Ortega K, Garcia-Esparcia P, Gil L, Lucas JJ, Ferrer I. Altered machinery of protein synthesis in Alzheimer's: from the nucleolus to the ribosome. Brain Pathol 2016; 26(5): 593-605.
[http://dx.doi.org/10.1111/bpa.12335] [PMID: 26512942]

[182]   Frost B, Hemberg M, Lewis J, Feany MB. Tau promotes neurodegeneration through global chromatin relaxation. Nat Neurosci 2014; 17(3): 357-66.
[http://dx.doi.org/10.1038/nn.3639] [PMID: 24464041]

[183]   Roberson ED, Halabisky B, Yoo JW, *et al.* Amyloid-β/Fyn-induced synaptic, network, and cognitive impairments depend on tau levels in multiple mouse models of Alzheimer's disease. J Neurosci 2011; 31(2): 700-11.
[http://dx.doi.org/10.1523/JNEUROSCI.4152-10.2011] [PMID: 21228179]

[184]   Green KN. Calcium in the initiation, progression and as an effector of Alzheimer's disease pathology. J Cell Mol Med 2009; 13(9A): 2787-99.
[http://dx.doi.org/10.1111/j.1582-4934.2009.00861.x] [PMID: 19650832]

[185]   Dell'Acqua ML, Smith KE, Gorski JA, Horne EA, Gibson ES, Gomez LL. Regulation of neuronal PKA signaling through AKAP targeting dynamics. Eur J Cell Biol 2006; 85(7): 627-33.
[http://dx.doi.org/10.1016/j.ejcb.2006.01.010] [PMID: 16504338]

[186]   Lee HK, Kameyama K, Huganir RL, Bear MF. NMDA induces long-term synaptic depression and dephosphorylation of the GluR1 subunit of AMPA receptors in hippocampus. Neuron 1998; 21(5): 1151-62.
[http://dx.doi.org/10.1016/S0896-6273(00)80632-7] [PMID: 9856470]

[187]   Hsieh H, Boehm J, Sato C, *et al.* AMPAR removal underlies Abeta-induced synaptic depression and dendritic spine loss. Neuron 2006; 52(5): 831-43.
[http://dx.doi.org/10.1016/j.neuron.2006.10.035] [PMID: 17145504]

[188]   Valenstein ML, Roll-Mecak A. Graded control of microtubule severing by tubulin glutamylation. Cell 2016; 164(5): 911-21.
[http://dx.doi.org/10.1016/j.cell.2016.01.019] [PMID: 26875866]

[189]   Flannery PJ, Trushina E. Mitochondrial dynamics and transport in Alzheimer's disease. Mol Cell Neurosci 2019; 98: 109-20.
[http://dx.doi.org/10.1016/j.mcn.2019.06.009] [PMID: 31216425]

[190]   Sheng ZH, Cai Q. Mitochondrial transport in neurons: impact on synaptic homeostasis and neurodegeneration. Nat Rev Neurosci 2012; 13(2): 77-93.
[http://dx.doi.org/10.1038/nrn3156] [PMID: 22218207]

[191]   Lin MY, Sheng ZH. Regulation of mitochondrial transport in neurons. Exp Cell Res 2015; 334(1): 35-44.
[http://dx.doi.org/10.1016/j.yexcr.2015.01.004] [PMID: 25612908]

[192]   Kanaan NM, Morfini GA, LaPointe NE, *et al.* Pathogenic forms of tau inhibit kinesin-dependent axonal transport through a mechanism involving activation of axonal phosphotransferases. J Neurosci 2011; 31(27): 9858-68.
[http://dx.doi.org/10.1523/JNEUROSCI.0560-11.2011] [PMID: 21734277]

[193]   Shahpasand K, Uemura I, Saito T, *et al.* Regulation of mitochondrial transport and inter-microtubule spacing by tau phosphorylation at the sites hyperphosphorylated in Alzheimer's disease. J Neurosci 2012; 32(7): 2430-41.
[http://dx.doi.org/10.1523/JNEUROSCI.5927-11.2012] [PMID: 22396417]

[194]   Ittner LM, Ke YD, Götz J. Phosphorylated Tau interacts with c-Jun N-terminal kinase-interacting protein 1 (JIP1) in Alzheimer disease. J Biol Chem 2009; 284(31): 20909-16.

[http://dx.doi.org/10.1074/jbc.M109.014472] [PMID: 19491104]

[195] Eckert A, Nisbet R, Grimm A, Götz J. March separate, strike together--role of phosphorylated TAU in mitochondrial dysfunction in Alzheimer's disease. Biochim Biophys Acta 2014; 1842(8): 1258-66.
[http://dx.doi.org/10.1016/j.bbadis.2013.08.013] [PMID: 24051203]

[196] Cheng XT, Zhou B, Lin MY, Cai Q, Sheng ZH. Axonal autophagosomes use the ride-on service for retrograde transport toward the soma. Autophagy 2015; 11(8): 1434-6.
[http://dx.doi.org/10.1080/15548627.2015.1062203] [PMID: 26102591]

[197] Tammineni P, Ye X, Feng T, Aikal D, Cai Q. Impaired retrograde transport of axonal autophagosomes contributes to autophagic stress in Alzheimer's disease neurons. eLife 2017; 6: e21776.
[http://dx.doi.org/10.7554/eLife.21776] [PMID: 28085665]

[198] Song Z, Ghochani M, McCaffery JM, Frey TG, Chan DC. Mitofusins and OPA1 mediate sequential steps in mitochondrial membrane fusion. Mol Biol Cell 2009; 20(15): 3525-32.
[http://dx.doi.org/10.1091/mbc.e09-03-0252] [PMID: 19477917]

[199] Palmer CS, Osellame LD, Laine D, Koutsopoulos OS, Frazier AE, Ryan MT. MiD49 and MiD51, new components of the mitochondrial fission machinery. EMBO Rep 2011; 12(6): 565-73.
[http://dx.doi.org/10.1038/embor.2011.54] [PMID: 21508961]

[200] Manczak M, Calkins MJ, Reddy PH. Impaired mitochondrial dynamics and abnormal interaction of amyloid beta with mitochondrial protein Drp1 in neurons from patients with Alzheimer's disease: implications for neuronal damage. Hum Mol Genet 2011; 20(13): 2495-509.
[http://dx.doi.org/10.1093/hmg/ddr139] [PMID: 21459773]

[201] Darshi M, Mendiola VL, Mackey MR, *et al.* ChChd3, an inner mitochondrial membrane protein, is essential for maintaining crista integrity and mitochondrial function. J Biol Chem 2011; 286(4): 2918-32.
[http://dx.doi.org/10.1074/jbc.M110.171975] [PMID: 21081504]

[202] Zhang L, Trushin S, Christensen TA, *et al.* Altered brain energetics induces mitochondrial fission arrest in Alzheimer's Disease. Sci Rep 2016; 6: 18725.
[http://dx.doi.org/10.1038/srep18725] [PMID: 26729583]

[203] Vincent AE, Turnbull DM, Eisner V, Hajnóczky G, Picard M. Mitochondrial Nanotunnels. Trends Cell Biol 2017; 27(11): 787-99.
[http://dx.doi.org/10.1016/j.tcb.2017.08.009] [PMID: 28935166]

[204] Stokin GB, Lillo C, Falzone TL, *et al.* Axonopathy and transport deficits early in the pathogenesis of Alzheimer's disease. Science 2005; 307(5713): 1282-8.
[http://dx.doi.org/10.1126/science.1105681] [PMID: 15731448]

[205] Cai Q, Tammineni P. Mitochondrial aspects of synaptic dysfunction in Alzheimer's disease. J Alzheimers Dis 2017; 57(4): 1087-103.
[http://dx.doi.org/10.3233/JAD-160726] [PMID: 27767992]

[206] Braak H, Braak E. Neuropathological stageing of Alzheimer-related changes. Acta Neuropathol 1991; 82(4): 239-59.
[http://dx.doi.org/10.1007/BF00308809] [PMID: 1759558]

[207] Jucker M, Walker LC. Self-propagation of pathogenic protein aggregates in neurodegenerative diseases. Nature 2013; 501(7465): 45-51.
[http://dx.doi.org/10.1038/nature12481] [PMID: 24005412]

[208] Fraser PE. Prions and prion-like proteins. J Biol Chem 2014; 289(29): 19839-40.
[http://dx.doi.org/10.1074/jbc.R114.583492] [PMID: 24860092]

[209] Peeraer E, Bottelbergs A, Van Kolen K, *et al.* Intracerebral injection of preformed synthetic tau fibrils initiates widespread tauopathy and neuronal loss in the brains of tau transgenic mice. Neurobiol Dis 2015; 73: 83-95.
[http://dx.doi.org/10.1016/j.nbd.2014.08.032] [PMID: 25220759]

[210]   Ahmed Z, Cooper J, Murray TK, *et al.* A novel *in vivo* model of tau propagation with rapid and progressive neurofibrillary tangle pathology: the pattern of spread is determined by connectivity, not proximity. Acta Neuropathol 2014; 127(5): 667-83.
        [http://dx.doi.org/10.1007/s00401-014-1254-6] [PMID: 24531916]

[211]   Michel CH, Kumar S, Pinotsi D, *et al.* Extracellular monomeric tau protein is sufficient to initiate the spread of tau protein pathology. J Biol Chem 2014; 289(2): 956-67.
        [http://dx.doi.org/10.1074/jbc.M113.515445] [PMID: 24235150]

[212]   Levarska L, Zilka N, Jadhav S, Neradil P, Novak M. Of rodents and men: the mysterious interneuronal pilgrimage of misfolded protein tau in Alzheimer's disease. J Alzheimers Dis 2013; 37(3): 569-77.
        [http://dx.doi.org/10.3233/JAD-131106] [PMID: 23948940]

[213]   Pooler AM, Phillips EC, Lau DH, Noble W, Hanger DP. Physiological release of endogenous tau is stimulated by neuronal activity. EMBO Rep 2013; 14(4): 389-94.
        [http://dx.doi.org/10.1038/embor.2013.15] [PMID: 23412472]

[214]   Katsinelos T, Zeitler M, Dimou E, *et al.* Unconventional secretion mediates the trans-cellular spreading of tau. Cell Rep 2018; 23(7): 2039-55.
        [http://dx.doi.org/10.1016/j.celrep.2018.04.056] [PMID: 29768203]

[215]   Saman S, Kim W, Raya M, *et al.* Exosome-associated tau is secreted in tauopathy models and is selectively phosphorylated in cerebrospinal fluid in early Alzheimer disease. J Biol Chem 2012; 287(6): 3842-9.
        [http://dx.doi.org/10.1074/jbc.M111.277061] [PMID: 22057275]

[216]   Tardivel M, Bégard S, Bousset L, *et al.* Tunneling nanotube (TNT)-mediated neuron-to neuron transfer of pathological Tau protein assemblies. Acta Neuropathol Commun 2016; 4(1): 117.
        [http://dx.doi.org/10.1186/s40478-016-0386-4] [PMID: 27809932]

[217]   Wu JW, Herman M, Liu L, *et al.* Small misfolded Tau species are internalized *via* bulk endocytosis and anterogradely and retrogradely transported in neurons. J Biol Chem 2013; 288(3): 1856-70.
        [http://dx.doi.org/10.1074/jbc.M112.394528] [PMID: 23188818]

[218]   Holmes BB, DeVos SL, Kfoury N, *et al.* Heparan sulfate proteoglycans mediate internalization and propagation of specific proteopathic seeds. Proc Natl Acad Sci USA 2013; 110(33): E3138-47.
        [http://dx.doi.org/10.1073/pnas.1301440110] [PMID: 23898162]

[219]   Evans LD, Wassmer T, Fraser G, *et al.* Extracellular monomeric and aggregated Tau efficiently enter human neurons through overlapping but distinct pathways. Cell Rep 2018; 22(13): 3612-24.
        [http://dx.doi.org/10.1016/j.celrep.2018.03.021] [PMID: 29590627]

[220]   Dujardin S, Bégard S, Caillierez R, *et al.* Ectosomes: a new mechanism for non-exosomal secretion of tau protein. PLoS One 2014; 9(6): e100760.
        [http://dx.doi.org/10.1371/journal.pone.0100760] [PMID: 24971751]

[221]   Yamada K, Patel TK, Hochgräfe K, *et al.* Analysis of *in vivo* turnover of tau in a mouse model of tauopathy. Mol Neurodegener 2015; 10: 55.
        [http://dx.doi.org/10.1186/s13024-015-0052-5] [PMID: 26502977]

[222]   Bright J, Hussain S, Dang V, *et al.* Human secreted tau increases amyloid-beta production. Neurobiol Aging 2015; 36(2): 693-709.
        [http://dx.doi.org/10.1016/j.neurobiolaging.2014.09.007] [PMID: 25442111]

[223]   Sokolow S, Henkins KM, Bilousova T, *et al.* Pre-synaptic C-terminal truncated tau is released from cortical synapses in Alzheimer's disease. J Neurochem 2015; 133(3): 368-79.
        [http://dx.doi.org/10.1111/jnc.12991] [PMID: 25393609]

[224]   Hall GF, Saman S. Death or secretion? The demise of a plausible assumption about CSF-tau in Alzheimer Disease? Commun Integr Biol 2012; 5(6): 623-6.
        [http://dx.doi.org/10.4161/cib.21437] [PMID: 23740221]

[225] Chen XQ, Mobley WC. Alzheimer disease pathogenesis: insights from molecular and cellular biology studies of oligomeric Aβ and tau species. Front Neurosci 2019; 13: 659.
       [http://dx.doi.org/10.3389/fnins.2019.00659] [PMID: 31293377]

[226] Sanders DW, Kaufman SK, DeVos SL, *et al.* Distinct tau prion strains propagate in cells and mice and define different tauopathies. Neuron 2014; 82(6): 1271-88.
       [http://dx.doi.org/10.1016/j.neuron.2014.04.047] [PMID: 24857020]

[227] Sharma AM, Thomas TL, Woodard DR, Kashmer OM, Diamond MI. Tau monomer encodes strains. eLife 2018; 7: e37813.
       [http://dx.doi.org/10.7554/eLife.37813] [PMID: 30526844]

[228] Guo JL, Lee VMY. Seeding of normal Tau by pathological Tau conformers drives pathogenesis of Alzheimer-like tangles. J Biol Chem 2011; 286(17): 15317-31.
       [http://dx.doi.org/10.1074/jbc.M110.209296] [PMID: 21372138]

[229] Kfoury N, Holmes BB, Jiang H, Holtzman DM, Diamond MI. Trans-cellular propagation of Tau aggregation by fibrillar species. J Biol Chem 2012; 287(23): 19440-51.
       [http://dx.doi.org/10.1074/jbc.M112.346072] [PMID: 22461630]

[230] Liu L, Drouet V, Wu JW, *et al.* Trans-synaptic spread of tau pathology *in vivo.* PLoS One 2012; 7(2): e31302.
       [http://dx.doi.org/10.1371/journal.pone.0031302] [PMID: 22312444]

[231] de Calignon A, Polydoro M, Suárez-Calvet M, *et al.* Propagation of tau pathology in a model of early Alzheimer's disease. Neuron 2012; 73(4): 685-97.
       [http://dx.doi.org/10.1016/j.neuron.2011.11.033] [PMID: 22365544]

[232] Iba M, Guo JL, McBride JD, Zhang B, Trojanowski JQ, Lee VM. Synthetic tau fibrils mediate transmission of neurofibrillary tangles in a transgenic mouse model of Alzheimer's-like tauopathy. J Neurosci 2013; 33(3): 1024-37.
       [http://dx.doi.org/10.1523/JNEUROSCI.2642-12.2013] [PMID: 23325240]

[233] Zilka N, Stozicka Z, Kovac A, Pilipcinec E, Bugos O, Novak M. Human misfolded truncated tau protein promotes activation of microglia and leukocyte infiltration in the transgenic rat model of tauopathy. J Neuroimmunol 2009; 209(1-2): 16-25.
       [http://dx.doi.org/10.1016/j.jneuroim.2009.01.013] [PMID: 19232747]

[234] Ghosh S, Wu MD, Shaftel SS, *et al.* Sustained interleukin-1β overexpression exacerbates tau pathology despite reduced amyloid burden in an Alzheimer's mouse model. J Neurosci 2013; 33(11): 5053-64.
       [http://dx.doi.org/10.1523/JNEUROSCI.4361-12.2013] [PMID: 23486975]

[235] Stancu I-C, Cremers N, Vanrusselt H, *et al.* Aggregated Tau activates NLRP3-ASC inflammasome exacerbating exogenously seeded and non-exogenously seeded Tau pathology *in vivo.* Acta Neuropathol 2019; 137(4): 599-617.
       [http://dx.doi.org/10.1007/s00401-018-01957-y] [PMID: 30721409]

[236] Asai H, Ikezu S, Tsunoda S, *et al.* Depletion of microglia and inhibition of exosome synthesis halt tau propagation. Nat Neurosci 2015; 18(11): 1584-93.
       [http://dx.doi.org/10.1038/nn.4132] [PMID: 26436904]

[237] Thal DR, Rüb U, Orantes M, Braak H. Phases of A beta-deposition in the human brain and its relevance for the development of AD. Neurology 2002; 58(12): 1791-800.
       [http://dx.doi.org/10.1212/WNL.58.12.1791] [PMID: 12084879]

[238] Giannakopoulos P, Herrmann FR, Bussière T, *et al.* Tangle and neuron numbers, but not amyloid load, predict cognitive status in Alzheimer's disease. Neurology 2003; 60(9): 1495-500.
       [http://dx.doi.org/10.1212/01.WNL.0000063311.58879.01] [PMID: 12743238]

[239] Bierer LM, Hof PR, Purohit DP, *et al.* Neocortical neurofibrillary tangles correlate with dementia severity in Alzheimer's disease. Arch Neurol 1995; 52(1): 81-8.

[http://dx.doi.org/10.1001/archneur.1995.00540250089017] [PMID: 7826280]

[240] Gengler S, Hamilton A, Hölscher C. Synaptic plasticity in the hippocampus of a APP/PS1 mouse model of Alzheimer's disease is impaired in old but not young mice. PLoS One 2010; 5(3): e9764.
[http://dx.doi.org/10.1371/journal.pone.0009764] [PMID: 20339537]

[241] Irizarry MC, McNamara M, Fedorchak K, Hsiao K, Hyman BT. APPSw transgenic mice develop age-related A beta deposits and neuropil abnormalities, but no neuronal loss in CA1. J Neuropathol Exp Neurol 1997; 56(9): 965-73.
[http://dx.doi.org/10.1097/00005072-199709000-00002] [PMID: 9291938]

[242] Rupp NJ, Wegenast-Braun BM, Radde R, Calhoun ME, Jucker M. Early onset amyloid lesions lead to severe neuritic abnormalities and local, but not global neuron loss in APPPS1 transgenic mice. Neurobiol Aging 2011; 32(12): 2324.e1-6.
[http://dx.doi.org/10.1016/j.neurobiolaging.2010.08.014] [PMID: 20970889]

[243] Takeuchi A, Irizarry MC, Duff K, *et al.* Age-related amyloid β deposition in transgenic mice overexpressing both Alzheimer mutant presenilin 1 and amyloid β precursor protein Swedish mutant is not associated with global neuronal loss. Am J Pathol 2000; 157(1): 331-9.
[http://dx.doi.org/10.1016/S0002-9440(10)64544-0] [PMID: 10880403]

[244] Pritchard SM, Dolan PJ, Vitkus A, Johnson GVW. The toxicity of tau in Alzheimer disease: turnover, targets and potential therapeutics. J Cell Mol Med 2011; 15(8): 1621-35.
[http://dx.doi.org/10.1111/j.1582-4934.2011.01273.x] [PMID: 21348938]

[245] Selkoe DJ, Hardy J. The amyloid hypothesis of Alzheimer's disease at 25 years. EMBO Mol Med 2016; 8(6): 595-608.
[http://dx.doi.org/10.15252/emmm.201606210] [PMID: 27025652]

[246] Lee VM, Goedert M, Trojanowski JQ. Neurodegenerative tauopathies. Annu Rev Neurosci 2001; 24: 1121-59.
[http://dx.doi.org/10.1146/annurev.neuro.24.1.1121] [PMID: 11520930]

[247] Roberson ED, Scearce-Levie K, Palop JJ, *et al.* Reducing endogenous tau ameliorates amyloid beta-induced deficits in an Alzheimer's disease mouse model. Science 2007; 316(5825): 750-4.
[http://dx.doi.org/10.1126/science.1141736] [PMID: 17478722]

[248] Yoshiyama Y, Higuchi M, Zhang B, *et al.* Synapse loss and microglial activation precede tangles in a P301S tauopathy mouse model. Neuron 2007; 53(3): 337-51.
[http://dx.doi.org/10.1016/j.neuron.2007.01.010] [PMID: 17270732]

[249] Rudinskiy N, Hawkes JM, Wegmann S, *et al.* Tau pathology does not affect experience-driven single-neuron and network-wide Arc/Arg3.1 responses. Acta Neuropathol Commun 2014; 2: 63.
[http://dx.doi.org/10.1186/2051-5960-2-63] [PMID: 24915991]

[250] Brunden KR, Trojanowski JQ, Lee VM. Evidence that non-fibrillar tau causes pathology linked to neurodegeneration and behavioral impairments. J Alzheimers Dis 2008; 14(4): 393-9.
[http://dx.doi.org/10.3233/JAD-2008-14406] [PMID: 18688089]

[251] Maeda S, Sahara N, Saito Y, Murayama S, Ikai A, Takashima A. Increased levels of granular tau oligomers: an early sign of brain aging and Alzheimer's disease. Neurosci Res 2006; 54(3): 197-201.
[http://dx.doi.org/10.1016/j.neures.2005.11.009] [PMID: 16406150]

[252] Lasagna-Reeves CA, Castillo-Carranza DL, Sengupta U, *et al.* Alzheimer brain-derived tau oligomers propagate pathology from endogenous tau. Sci Rep 2012; 2: 700.
[http://dx.doi.org/10.1038/srep00700] [PMID: 23050084]

[253] Yamada K, Holth JK, Liao F, *et al.* Neuronal activity regulates extracellular tau *in vivo.* J Exp Med 2014; 211(3): 387-93.
[http://dx.doi.org/10.1084/jem.20131685] [PMID: 24534188]

[254] Tian H, Davidowitz E, Lopez P, Emadi S, Moe J, Sierks M. Trimeric tau is toxic to human neuronal cells at low nanomolar concentrations. Int J Cell Biol 2013; 2013: 260787.

[http://dx.doi.org/10.1155/2013/260787] [PMID: 24159335]

[255] Kaniyappan S, Chandupatla RR, Mandelkow EM, Mandelkow E. Extracellular low-n oligomers of tau cause selective synaptotoxicity without affecting cell viability. Alzheimers Dement 2017; 13(11): 1270-91.
[http://dx.doi.org/10.1016/j.jalz.2017.04.002] [PMID: 28528849]

[256] Mirbaha H, Holmes BB, Sanders DW, Bieschke J, Diamond MI. Tau trimers are the minimal propagation unit spontaneously internalized to seed intracellular aggregation. J Biol Chem 2015; 290(24): 14893-903.
[http://dx.doi.org/10.1074/jbc.M115.652693] [PMID: 25887395]

[257] Small SA, Duff K. Linking Abeta and tau in late-on set Alzheimer's disease: a dual pathway hypothesis. Neuron 2008; 60(4): 534-42.
[http://dx.doi.org/10.1016/j.neuron.2008.11.007] [PMID: 19038212]

[258] Rhein V, Song X, Wiesner A, *et al.* Amyloid-beta and tau synergistically impair the oxidative phosphorylation system in triple transgenic Alzheimer's disease mice. Proc Natl Acad Sci USA 2009; 106(47): 20057-62.
[http://dx.doi.org/10.1073/pnas.0905529106] [PMID: 19897719]

[259] Witter MP, Doan TP, Jacobsen B, Nilssen ES, Ohara S. Architecture of the entorhinal cortex. A review of entorhinal anatomy in rodents with some comparative notes. Front Syst Neurosci 2017; 11: 46.
[http://dx.doi.org/10.3389/fnsys.2017.00046] [PMID: 28701931]

[260] Nixon RA. Amyloid precursor protein and endosomal-lysosomal dysfunction in Alzheimer's disease: inseparable partners in a multifactorial disease. FASEB J 2017; 31(7): 2729-43.
[http://dx.doi.org/10.1096/fj.201700359] [PMID: 28663518]

[261] Chen XQ, Sawa M, Mobley WC. Dysregulation of neurotrophin signaling in the pathogenesis of Alzheimer disease and of Alzheimer disease in Down syndrome. Free Radic Biol Med 2018; 114: 52-61.
[http://dx.doi.org/10.1016/j.freeradbiomed.2017.10.341] [PMID: 29031834]

[262] Zheng WH, Bastianetto S, Mennicken F, Ma W, Kar S. Amyloid beta peptide induces tau phosphorylation and loss of cholinergic neurons in rat primary septal cultures. Neuroscience 2002; 115(1): 201-11.
[http://dx.doi.org/10.1016/S0306-4522(02)00404-9] [PMID: 12401334]

[263] Bhatia N, Hall GF. Untangling the role of tau in Alzheimer's disease: a unifying hypothesis. Transl Neurosci 2013; 4(2): 115-33.
[http://dx.doi.org/10.2478/s13380-013-0114-5]

[264] Amadoro G, Ciotti MT, Costanzi M, Cestari V, Calissano P, Canu N. NMDA receptor mediates tau-induced neurotoxicity by calpain and ERK/MAPK activation. Proc Natl Acad Sci USA 2006; 103(8): 2892-7.
[http://dx.doi.org/10.1073/pnas.0511065103] [PMID: 16477009]

[265] Corsetti V, Amadoro G, Gentile A, *et al.* Identification of a caspase-derived N-terminal tau fragment in cellular and animal Alzheimer's disease models. Mol Cell Neurosci 2008; 38(3): 381-92.
[http://dx.doi.org/10.1016/j.mcn.2008.03.011] [PMID: 18511295]

[266] Bloom GS. Amyloid-β and tau: the trigger and bullet in Alzheimer disease pathogenesis. JAMA Neurol 2014; 71(4): 505-8.
[http://dx.doi.org/10.1001/jamaneurol.2013.5847] [PMID: 24493463]

[267] Leroy K, Ando K, Laporte V, *et al.* Lack of tau proteins rescues neuronal cell death and decreases amyloidogenic processing of APP in APP/PS1 mice. Am J Pathol 2012; 181(6): 1928-40.
[http://dx.doi.org/10.1016/j.ajpath.2012.08.012] [PMID: 23026200]

[268] Rapoport M, Dawson HN, Binder LI, Vitek MP, Ferreira A. Tau is essential to beta -amyloid-induced

neurotoxicity. Proc Natl Acad Sci USA 2002; 99(9): 6364-9.
[http://dx.doi.org/10.1073/pnas.092136199] [PMID: 11959919]

[269] Gustke N, Steiner B, Mandelkow EM, *et al.* The Alzheimer-like phosphorylation of tau protein reduces microtubule binding and involves Ser-Pro and Thr-Pro motifs. FEBS Lett 1992; 307(2): 199-205.
[http://dx.doi.org/10.1016/0014-5793(92)80767-B] [PMID: 1644173]

[270] Cuchillo-Ibanez I, Seereeram A, Byers HL, *et al.* Phosphorylation of tau regulates its axonal transport by controlling its binding to kinesin. FASEB J 2008; 22(9): 3186-95.
[http://dx.doi.org/10.1096/fj.08-109181] [PMID: 18511549]

[271] Dubey M, Chaudhury P, Kabiru H, Shea TB. Tau inhibits anterograde axonal transport and perturbs stability in growing axonal neurites in part by displacing kinesin cargo: neurofilaments attenuate tau-mediated neurite instability. Cell Motil Cytoskeleton 2008; 65(2): 89-99.
[http://dx.doi.org/10.1002/cm.20243] [PMID: 18000878]

[272] Trinczek B, Ebneth A, Mandelkow EM, Mandelkow E. Tau regulates the attachment/detachment but not the speed of motors in microtubule-dependent transport of single vesicles and organelles. J Cell Sci 1999; 112(Pt 14): 2355-67.
[http://dx.doi.org/10.1242/jcs.112.14.2355] [PMID: 10381391]

[273] Swerdlow RH, Burns JM, Khan SM. The Alzheimer's disease mitochondrial cascade hypothesis. J Alzheimers Dis 2010; 20 (Suppl. 2): S265-79.
[http://dx.doi.org/10.3233/JAD-2010-100339] [PMID: 20442494]

[274] Bezprozvanny I, Mattson MP. Neuronal calcium mishandling and the pathogenesis of Alzheimer's disease. Trends Neurosci 2008; 31(9): 454-63.
[http://dx.doi.org/10.1016/j.tins.2008.06.005] [PMID: 18675468]

[275] Hermes M, Eichhoff G, Garaschuk O. Intracellular calcium signalling in Alzheimer's disease. J Cell Mol Med 2010; 14(1-2): 30-41.
[http://dx.doi.org/10.1111/j.1582-4934.2009.00976.x] [PMID: 19929945]

[276] Wang X, Schwarz TL. The mechanism of Ca^{2+} -dependent regulation of kinesin-mediated mitochondrial motility. Cell 2009; 136(1): 163-74.
[http://dx.doi.org/10.1016/j.cell.2008.11.046] [PMID: 19135897]

[277] Furukawa K, Wang Y, Yao PJ, *et al.* Alteration in calcium channel properties is responsible for the neurotoxic action of a familial frontotemporal dementia tau mutation. J Neurochem 2003; 87(2): 427-36.
[http://dx.doi.org/10.1046/j.1471-4159.2003.02020.x] [PMID: 14511120]

[278] Ittner LM, Fath T, Ke YD, *et al.* Parkinsonism and impaired axonal transport in a mouse model of frontotemporal dementia. Proc Natl Acad Sci USA 2008; 105(41): 15997-6002.
[http://dx.doi.org/10.1073/pnas.0808084105] [PMID: 18832465]

[279] Wang X, Su B, Lee HG, *et al.* Impaired balance of mitochondrial fission and fusion in Alzheimer's disease. J Neurosci 2009; 29(28): 9090-103.
[http://dx.doi.org/10.1523/JNEUROSCI.1357-09.2009] [PMID: 19605646]

[280] Chandrasekaran K, Hatanpää K, Rapoport SI, Brady DR. Decreased expression of nuclear and mitochondrial DNA-encoded genes of oxidative phosphorylation in association neocortex in Alzheimer disease. Brain Res Mol Brain Res 1997; 44(1): 99-104.
[http://dx.doi.org/10.1016/S0169-328X(96)00191-X] [PMID: 9030703]

[281] Grady CL, Haxby JV, Horwitz B, *et al.* Longitudinal study of the early neuropsychological and cerebral metabolic changes in dementia of the Alzheimer type. J Clin Exp Neuropsychol 1988; 10(5): 576-96.
[http://dx.doi.org/10.1080/01688638808402796] [PMID: 3265710]

[282] Eckert A, Schulz KL, Rhein V, Götz J. Convergence of amyloid-beta and tau pathologies on

mitochondria *in vivo*. Mol Neurobiol 2010; 41(2-3): 107-14.
[http://dx.doi.org/10.1007/s12035-010-8109-5] [PMID: 20217279]

[283] Funk KE, Mrak RE, Kuret J. Granulovacuolar degeneration (GVD) bodies of Alzheimer's disease (AD) resemble late-stage autophagic organelles. Neuropathol Appl Neurobiol 2011; 37(3): 295-306.
[http://dx.doi.org/10.1111/j.1365-2990.2010.01135.x] [PMID: 20946470]

[284] Zempel H, Thies E, Mandelkow E, Mandelkow EM. Abeta oligomers cause localized Ca(2+) elevation, missorting of endogenous Tau into dendrites, Tau phosphorylation, and destruction of microtubules and spines. J Neurosci 2010; 30(36): 11938-50.
[http://dx.doi.org/10.1523/JNEUROSCI.2357-10.2010] [PMID: 20826658]

[285] Müller WE, Eckert A, Kurz C, Eckert GP, Leuner K. Mitochondrial dysfunction: common final pathway in brain aging and Alzheimer's disease--therapeutic aspects. Mol Neurobiol 2010; 41(2-3): 159-71.
[http://dx.doi.org/10.1007/s12035-010-8141-5] [PMID: 20461558]

[286] Li S, Mallory M, Alford M, Tanaka S, Masliah E. Glutamate transporter alterations in Alzheimer disease are possibly associated with abnormal APP expression. J Neuropathol Exp Neurol 1997; 56(8): 901-11.
[http://dx.doi.org/10.1097/00005072-199708000-00008] [PMID: 9258260]

[287] Stone JG, Siedlak SL, Tabaton M, *et al*. The cell cycle regulator phosphorylated retinoblastoma protein is associated with tau pathology in several tauopathies. J Neuropathol Exp Neurol 2011; 70(7): 578-87.
[http://dx.doi.org/10.1097/NEN.0b013e3182204414] [PMID: 21666500]

[288] Yasojima K, Kuret J, DeMaggio AJ, McGeer E, McGeer PL. Casein kinase 1 delta mRNA is upregulated in Alzheimer disease brain. Brain Res 2000; 865(1): 116-20.
[http://dx.doi.org/10.1016/S0006-8993(00)02200-9] [PMID: 10814741]

[289] de la Monte SM, Ng SC, Hsu DW. Aberrant GAP-43 gene expression in Alzheimer's disease. Am J Pathol 1995; 147(4): 934-46.
[PMID: 7573369]

[290] Andorfer C, Acker CM, Kress Y, Hof PR, Duff K, Davies P. Cell-cycle reentry and cell death in transgenic mice expressing nonmutant human tau isoforms. J Neurosci 2005; 25(22): 5446-54.
[http://dx.doi.org/10.1523/JNEUROSCI.4637-04.2005] [PMID: 15930395]

[291] Yamaguchi H, Nakazato Y, Shoji M, Ihara Y, Hirai S. Ultrastructure of the neuropil threads in the Alzheimer brain: their dendritic origin and accumulation in the senile plaques. Acta Neuropathol 1990; 80(4): 368-74.
[http://dx.doi.org/10.1007/BF00307689] [PMID: 2239149]

[292] Perry G, Kawai M, Tabaton M, *et al*. Neuropil threads of Alzheimer's disease show a marked alteration of the normal cytoskeleton. J Neurosci 1991; 11(6): 1748-55.
[http://dx.doi.org/10.1523/JNEUROSCI.11-06-01748.1991] [PMID: 1904481]

[293] Ihara Y. Massive somatodendritic sprouting of cortical neurons in Alzheimer's disease. Brain Res 1988; 459(1): 138-44.
[http://dx.doi.org/10.1016/0006-8993(88)90293-4] [PMID: 3139259]

[294] McKee AC, Kowall NW, Kosik KS. Microtubular reorganization and dendritic growth response in Alzheimer's disease. Ann Neurol 1989; 26(5): 652-9.
[http://dx.doi.org/10.1002/ana.410260511] [PMID: 2817839]

[295] Maxwell WL, McCreath BJ, Graham DI, Gennarelli TA. Cytochemical evidence for redistribution of membrane pump calcium-ATPase and ecto-Ca-ATPase activity, and calcium influx in myelinated nerve fibres of the optic nerve after stretch injury. J Neurocytol 1995; 24(12): 925-42.
[http://dx.doi.org/10.1007/BF01215643] [PMID: 8719820]

[296] Jafari SS, Maxwell WL, Neilson M, Graham DI. Axonal cytoskeletal changes after non-disruptive

axonal injury. J Neurocytol 1997; 26(4): 207-21.
[http://dx.doi.org/10.1023/A:1018588114648] [PMID: 9192287]

[297] Lee S, Kim W, Li Z, Hall GF. Accumulation of vesicle-associated human tau in distal dendrites drives degeneration and tau secretion in an *in situ* cellular tauopathy model. Int J Alzheimers Dis 2012; 2012: 172837.
[http://dx.doi.org/10.1155/2012/172837] [PMID: 22315694]

[298] Rank KB, Pauley AM, Bhattacharya K, *et al.* Direct interaction of soluble human recombinant tau protein with Abeta 1-42 results in tau aggregation and hyperphosphorylation by tau protein kinase II. FEBS Lett 2002; 514(2-3): 263-8.
[http://dx.doi.org/10.1016/S0014-5793(02)02376-1] [PMID: 11943163]

[299] Pérez M, Cuadros R, Benítez MJ, Jiménez JS. Interaction of Alzheimer's disease amyloid β peptide fragment 25-35 with tau protein, and with a tau peptide containing the microtubule binding domain. J Alzheimers Dis 2004; 6(5): 461-7.
[http://dx.doi.org/10.3233/JAD-2004-6501] [PMID: 15505366]

[300] Le MN, Kim W, Lee S, McKee AC, Hall GF. Multiple mechanisms of extracellular tau spreading in a non-transgenic tauopathy model. Am J Neurodegener Dis 2012; 1(3): 316-33.
[PMID: 23383401]

[301] Liddelow SA, Guttenplan KA, Clarke LE, *et al.* Neurotoxic reactive astrocytes are induced by activated microglia. Nature 2017; 541(7638): 481-7.
[http://dx.doi.org/10.1038/nature21029] [PMID: 28099414]

[302] Forman MS, Lal D, Zhang B, *et al.* Transgenic mouse model of tau pathology in astrocytes leading to nervous system degeneration. J Neurosci 2005; 25(14): 3539-50.
[http://dx.doi.org/10.1523/JNEUROSCI.0081-05.2005] [PMID: 15814784]

[303] Cummings J, Lee G, Ritter A, Sabbagh M, Zhong K. Alzheimer's disease drug development pipeline: 2020. Alzheimers Dement (N Y) 2020; 6(1): e12050.
[http://dx.doi.org/10.1002/trc2.12050] [PMID: 32695874]

[304] Alzheimer's Disease International. World Alzheimer's Report 2015: the global impact of dementia. London, UK: Alzheimer's Disease International 2015.

[305] Fish PV, Steadman D, Bayle ED, Whiting P. New approaches for the treatment of Alzheimer's disease. Bioorg Med Chem Lett 2019; 29(2): 125-33.
[http://dx.doi.org/10.1016/j.bmcl.2018.11.034] [PMID: 30501965]

[306] Cummings JL, Morstorf T, Zhong K. Alzheimer's disease drug-development pipeline: few candidates, frequent failures. Alzheimers Res Ther 2014; 6(4): 37.
[http://dx.doi.org/10.1186/alzrt269] [PMID: 25024750]

[307] Murray ME, Lowe VJ, Graff-Radford NR, *et al.* Clinicopathologic and 11C-Pittsburgh compound B implications of Thal amyloid phase across the Alzheimer's disease spectrum. Brain 2015; 138(Pt 5): 1370-81.
[http://dx.doi.org/10.1093/brain/awv050] [PMID: 25805643]

[308] Wischik CM, Novak M, Edwards PC, Klug A, Tichelaar W, Crowther RA. Structural characterization of the core of the paired helical filament of Alzheimer disease. Proc Natl Acad Sci USA 1988; 85(13): 4884-8.
[http://dx.doi.org/10.1073/pnas.85.13.4884] [PMID: 2455299]

[309] Bejanin A, Schonhaut DR, La Joie R, *et al.* Tau pathology and neurodegeneration contribute to cognitive impairment in Alzheimer's disease. Brain 2017; 140(12): 3286-300.
[http://dx.doi.org/10.1093/brain/awx243] [PMID: 29053874]

[310] Iqbal K, Liu F, Gong CX. Tau and neurodegenerative disease: the story so far. Nat Rev Neurol 2016; 12(1): 15-27.
[http://dx.doi.org/10.1038/nrneurol.2015.225] [PMID: 26635213]

[311] Novak P, Cehlar O, Skrabana R, Novak M. Tau conformation a s a target for disease-modifying therapy: the role of truncation. J Alzheimers Dis 2018; 64(s1): S535-46.
[http://dx.doi.org/10.3233/JAD-179942] [PMID: 29865059]

[312] Horiguchi T, Uryu K, Giasson BI, *et al.* Nitration of tau protein is linked to neurodegeneration in tauopathies. Am J Pathol 2003; 163(3): 1021-31.
[http://dx.doi.org/10.1016/S0002-9440(10)63462-1] [PMID: 12937143]

[313] Ledesma MD, Bonay P, Colaço C, Avila J. Analysis of microtubule-associated protein tau glycation in paired helical filaments. J Biol Chem 1994; 269(34): 21614-9.
[http://dx.doi.org/10.1016/S0021-9258(17)31849-5] [PMID: 8063802]

[314] Wang JZ, Grundke-Iqbal I, Iqbal K. Glycosylation of microtubule-associated protein tau: an abnormal posttranslational modification in Alzheimer's disease. Nat Med 1996; 2(8): 871-5.
[http://dx.doi.org/10.1038/nm0896-871] [PMID: 8705855]

[315] Mori H, Kondo J, Ihara Y. Ubiquitin is a component of paired helical filaments in Alzheimer's disease. Science 1987; 235(4796): 1641-4.
[http://dx.doi.org/10.1126/science.3029875] [PMID: 3029875]

[316] Novak P, Kontsekova E, Zilka N, Novak M. Then years of tau-targeted immunotherapy: the path walked and the road ahead. Front Neurosci 2018; 12: 798.
[http://dx.doi.org/10.3389/fnins.2018.00798] [PMID: 30450030]

[317] Méphon-Gaspard A, Boca M, Pioche-Durieu C, *et al.* Role of tau in the spatial organization of axonal microtubules: keeping parallel microtubules evenly distributed despite macromolecular crowding. Cell Mol Life Sci 2016; 73(19): 3745-60.
[http://dx.doi.org/10.1007/s00018-016-2216-z] [PMID: 27076215]

[318] Ittner A, Ittner LM. Dendritic tau in Alzheimer's disease. Neuron 2018; 99(1): 13-27.
[http://dx.doi.org/10.1016/j.neuron.2018.06.003] [PMID: 30001506]

[319] Morris M, Hamto P, Adame A, Devidze N, Masliah E, Mucke L. Age-appropriate cognition and subtle dopamine-independent motor deficits in aged tau knockout mice. Neurobiol Aging 2013; 34(6): 1523-9.
[http://dx.doi.org/10.1016/j.neurobiolaging.2012.12.003] [PMID: 23332171]

[320] Lei P, Ayton S, Finkelstein DI, *et al.* Tau deficiency induces parkinsonism with dementia by impairing APP-mediated iron export. Nat Med 2012; 18(2): 291-5.
[http://dx.doi.org/10.1038/nm.2613] [PMID: 22286308]

[321] Marciniak E, Leboucher A, Caron E, *et al.* Tau deletion promotes brain insulin resistance. J Exp Med 2017; 214(8): 2257-69.
[http://dx.doi.org/10.1084/jem.20161731] [PMID: 28652303]

[322] Hamdane M, Bretteville A, Sambo AV, *et al.* p25/Cdk5-mediated retinoblastoma phosphorylation is an early event in neuronal cell death. J Cell Sci 2005; 118(Pt 6): 1291-8.
[http://dx.doi.org/10.1242/jcs.01724] [PMID: 15741232]

[323] Webster L, Groskreutz D, Grinbergs-Saull A, *et al.* Core outcome measures for interventions to prevent or slow the progress of dementia for people living with mild to moderate dementia: Systematic review and consensus recommendations. PLoS One 2017; 12(6): e0179521.
[http://dx.doi.org/10.1371/journal.pone.0179521] [PMID: 28662127]

[324] Medina M. An overview on the clinical development of tau-based therapies. Int J Mol Sci 2018; 19(4): 1160.
[http://dx.doi.org/10.3390/ijms19041160] [PMID: 29641484]

[325] Medina M, Avila J. New insights into the role of glycogen synthase kinase-3 in Alzheimer's disease. Expert Opin Ther Targets 2014; 18(1): 69-77.
[http://dx.doi.org/10.1517/14728222.2013.843670] [PMID: 24099155]

[326] Domínguez JM, Fuertes A, Orozco L, del Monte-Millán M, Delgado E, Medina M. Evidence for irreversible inhibition of glycogen synthase kinase-3β by tideglusib. J Biol Chem 2012; 287(2): 893-904.
[http://dx.doi.org/10.1074/jbc.M111.306472] [PMID: 22102280]

[327] Serenó L, Coma M, Rodríguez M, *et al.* A novel GSK-3β inhibitor reduces Alzheimer's pathology and rescues neuronal loss *in vivo.* Neurobiol Dis 2009; 35(3): 359-67.
[http://dx.doi.org/10.1016/j.nbd.2009.05.025] [PMID: 19523516]

[328] del Ser T, Steinwachs KC, Gertz HJ, *et al.* Treatment of Alzheimer's disease with the GSK-3 inhibitor tideglusib: a pilot study. J Alzheimers Dis 2013; 33(1): 205-15.
[http://dx.doi.org/10.3233/JAD-2012-120805] [PMID: 22936007]

[329] Lovestone S, Boada M, Dubois B, *et al.* ARGO investigators. A phase II trial of tideglusib in Alzheimer's disease. J Alzheimers Dis 2015; 45(1): 75-88.
[http://dx.doi.org/10.3233/JAD-141959] [PMID: 25537011]

[330] Tolosa E, Litvan I, Höglinger GU, *et al.* TAUROS Investigators. A phase 2 trial of the GSK-3 inhibitor tideglusib in progressive supranuclear palsy. Mov Disord 2014; 29(4): 470-8.
[http://dx.doi.org/10.1002/mds.25824] [PMID: 24532007]

[331] Forlenza OV, Diniz BS, Radanovic M, Santos FS, Talib LL, Gattaz WF. Disease-modifying properties of long-term lithium treatment for amnestic mild cognitive impairment: randomised controlled trial. Br J Psychiatry 2011; 198(5): 351-6.
[http://dx.doi.org/10.1192/bjp.bp.110.080044] [PMID: 21525519]

[332] Toyonaga T, Smith LM, Finnema SJ, *et al. In vivo* synaptic density imaging with 11C-UCB-J detects treatment effects of saracatinib (AZD0530) in a mouse model of Alzheimer's disease. J Nucl Med 2019; 60(12): 1780-6.
[http://dx.doi.org/10.2967/jnumed.118.223867] [PMID: 31101744]

[333] Kaufman AC, Salazar SV, Haas LT, *et al.* Fyn inhibition rescues established memory and synapse loss in Alzheimer mice. Ann Neurol 2015; 77(6): 953-71.
[http://dx.doi.org/10.1002/ana.24394] [PMID: 25707991]

[334] Nygaard HB, Wagner AF, Bowen GS, *et al.* A phase Ib multiple ascending dose study of the safety, tolerability, and central nervous system availability of AZD0530 (saracatinib) in Alzheimer's disease. Alzheimers Res Ther 2015; 7(1): 35.
[http://dx.doi.org/10.1186/s13195-015-0119-0] [PMID: 25874001]

[335] van Dyck CH, Nygaard HB, Chen K, *et al.* Effect of AZD0530 on cerebral metabolic decline in alzheimer disease: A Randomized Clinical Trial. JAMA Neurol 2019; 76(10): 1219-29.
[http://dx.doi.org/10.1001/jamaneurol.2019.2050] [PMID: 31329216]

[336] Sontag JM, Sontag E. Protein phosphatase 2A dysfunction in Alzheimer's disease. Front Mol Neurosci 2014; 7: 16.
[http://dx.doi.org/10.3389/fnmol.2014.00016] [PMID: 24653673]

[337] Min SW, Chen X, Tracy TE, *et al.* Critical role of acetylation in tau-mediated neurodegeneration and cognitive deficits. Nat Med 2015; 21(10): 1154-62.
[http://dx.doi.org/10.1038/nm.3951] [PMID: 26390242]

[338] ClinicalTrials.gov [homepage on the internet]. US National Library of Medicine [accessed August 2020] Available at: http://www.clinicaltrials.gov

[339] Yu Y, Zhang L, Li X, *et al.* Differential effects of an O-GlcNAcase inhibitor on tau phosphorylation. PLoS One 2012; 7(4): e35277.
[http://dx.doi.org/10.1371/journal.pone.0035277] [PMID: 22536363]

[340] Wang X, Li W, Marcus J, *et al.* MK-8719, a novel and selective O-GlcNAcase inhibitor that reduces the formation of pathological tau and ameliorates neurodegeneration in a mouse model of tauopathy. J Pharmacol Exp Ther 2020; 374(2): 252-63.

[http://dx.doi.org/10.1124/jpet.120.266122] [PMID: 32493725]

[341]   alzforum.org [homepage on the internet]. Apr 2017 conference news. AD/PD 2017 draws record number of scientists to Vienna, 2017. available at : https://www.alzforum.org/news/conference-coverage/treating-tau-finally-clinical-candidates-are-stepping-ring

[342]   Ursan ID, Jiang R, Pickard EM, Lee TA, Ng D, Pickard AS. Emergence of BCR-ABL kinase domain mutations associated with newly diagnosed chronic myeloid leukemia: a meta-analysis of clinical trials of tyrosine kinase inhibitors. J Manag Care Spec Pharm 2015; 21(2): 114-22.
[http://dx.doi.org/10.18553/jmcsp.2015.21.2.114] [PMID: 25615000]

[343]   Pagan F, Hebron M, Valadez EH, *et al.* Nilotinib Effects in Parkinson's disease and Dementia with Lewy bodies. J Parkinsons Dis 2016; 6(3): 503-17.
[http://dx.doi.org/10.3233/JPD-160867] [PMID: 27434297]

[344]   Ramser EM, Gan KJ, Decker H, *et al.* Amyloid-β oligomers induce tau-independent disruption of BDNF axonal transport *via* calcineurin activation in cultured hippocampal neurons. Mol Biol Cell 2013; 24(16): 2494-505.
[http://dx.doi.org/10.1091/mbc.e12-12-0858] [PMID: 23783030]

[345]   Brunden KR, Zhang B, Carroll J, *et al.* Epothilone D improves microtubule density, axonal integrity, and cognition in a transgenic mouse model of tauopathy. J Neurosci 2010; 30(41): 13861-6.
[http://dx.doi.org/10.1523/JNEUROSCI.3059-10.2010] [PMID: 20943926]

[346]   Tsai RM, Miller Z, Koestler M, *et al.* Reactions to multiple ascending doses of the microtubule stabilizer TPI-287 in patients with Alzheimer disease, progressive supranuclear palsy, and corticobasal syndrome: a randomized clinical trial. JAMA Neurol 2020; 77(2): 215-24.
[http://dx.doi.org/10.1001/jamaneurol.2019.3812] [PMID: 31710340]

[347]   Matsuoka Y, Jouroukhin Y, Gray AJ, *et al.* A neuronal microtubule-interacting agent, NAPVSIPQ, reduces tau pathology and enhances cognitive function in a mouse model of Alzheimer's disease. J Pharmacol Exp Ther 2008; 325(1): 146-53.
[http://dx.doi.org/10.1124/jpet.107.130526] [PMID: 18199809]

[348]   Morimoto BH, Schmechel D, Hirman J, Blackwell A, Keith J, Gold M. AL-108-211 Study. A double-blind, placebo-controlled, ascending-dose, randomized study to evaluate the safety, tolerability and effects on cognition of AL-108 after 12 weeks of intranasal administration in subjects with mild cognitive impairment. Dement Geriatr Cogn Disord 2013; 35(5-6): 325-36.
[http://dx.doi.org/10.1159/000348347] [PMID: 23594991]

[349]   Boxer AL, Lang AE, Grossman M, *et al.* AL-108-231 Investigators. Davunetide in patients with progressive supranuclear palsy: a randomised, double-blind, placebo-controlled phase 2/3 trial. Lancet Neurol 2014; 13(7): 676-85.
[http://dx.doi.org/10.1016/S1474-4422(14)70088-2] [PMID: 24873720]

[350]   Bulic B, Pickhardt M, Mandelkow E. Progress and developments in tau aggregation inhibitors for Alzheimer disease. J Med Chem 2013; 56(11): 4135-55.
[http://dx.doi.org/10.1021/jm3017317] [PMID: 23484434]

[351]   Wischik CM, Edwards PC, Lai RY, Roth M, Harrington CR. Selective inhibition of Alzheimer disease-like tau aggregation by phenothiazines. Proc Natl Acad Sci USA 1996; 93(20): 11213-8.
[http://dx.doi.org/10.1073/pnas.93.20.11213] [PMID: 8855335]

[352]   Wischik CM, Staff RT, Wischik DJ, *et al.* Tau aggregation inhibitor therapy: an exploratory phase 2 study in mild or moderate Alzheimer's disease. J Alzheimers Dis 2015; 44(2): 705-20.
[http://dx.doi.org/10.3233/JAD-142874] [PMID: 25550228]

[353]   Alzforum.org [homepage on the internet]. accessed August 2020 . Available at: https://www.alzforum. org/therapeutics/lmtm

[354]   Wilcock GK, Gauthier S, Frisoni GB, *et al.* Potential of low dose leuco-methylthioninium bis(hydromethansulfonate) (LTMT) monotherapy for treatment of mild Alzheimer's disease: cohort

analysis as modified primary outcome in a phase III clinical trial. J Alzheimers Dis 2018; 61(1): 435-57.
[http://dx.doi.org/10.3233/JAD-170560] [PMID: 29154277]

[355] Gauthier S, Feldman HH, Schneider LS, *et al.* Efficacy and safety of tau-aggregation inhibitor therapy in patients with mild or moderate Alzheimer's disease: a randomised, controlled, double-blind, parallel-arm, phase 3 trial. Lancet 2016; 388(10062): 2873-84.
[http://dx.doi.org/10.1016/S0140-6736(16)31275-2] [PMID: 27863809]

[356] Shiells H, Schelter BO, Bentham P, *et al.* Concentration-dependent activity of hydromethylthionine on clinical decline and brain atrophy in a randomized controlled trial in behavioral variant frontotemporal dementia. J Alzheimers Dis 2020; 75(2): 501-19.
[http://dx.doi.org/10.3233/JAD-191173] [PMID: 32280089]

[357] Taniguchi S, Suzuki N, Masuda M, *et al.* Inhibition of heparin-induced tau filament formation by phenothiazines, polyphenols, and porphyrins. J Biol Chem 2005; 280(9): 7614-23.
[http://dx.doi.org/10.1074/jbc.M408714200] [PMID: 15611092]

[358] Crowe A, Huang W, Ballatore C, *et al.* Identification of aminothienopyridazine inhibitors of tau assembly by quantitative high-throughput screening. Biochemistry 2009; 48(32): 7732-45.
[http://dx.doi.org/10.1021/bi9006435] [PMID: 19580328]

[359] Prati F, Cavalli A, Bolognesi ML. Navigating the chemical space of multitargeted-directed ligands: from hybrids to fragments in Alzheimer's disease. Molecules 2016; 21(4): 466.
[http://dx.doi.org/10.3390/molecules21040466] [PMID: 27070562]

[360] Baum L, Lam CWK, Cheung SKK, *et al.* Six-month randomized, placebo-controlled, double-blind, pilot clinical trial of curcumin in patients with Alzheimer disease. J Clin Psychopharmacol 2008; 28(1): 110-3.
[http://dx.doi.org/10.1097/jcp.0b013e318160862c] [PMID: 18204357]

[361] Ringman JM, Frautschy SA, Teng E, *et al.* Oral curcumin for Alzheimer's disease: tolerability and efficacy in a 24-week randomized, double blind, placebo-controlled study. Alzheimers Res Ther 2012; 4(5): 43.
[http://dx.doi.org/10.1186/alzrt146] [PMID: 23107780]

[362] Frautschy SA, Cole GM. Why pleiotropic interventions are needed for Alzheimer's disease. Mol Neurobiol 2010; 41(2-3): 392-409.
[http://dx.doi.org/10.1007/s12035-010-8137-1] [PMID: 20437209]

[363] Okuda M, Hijikuro I, Fujita Y, *et al.* PE859, a novel tau aggregation inhibitor, reduces aggregated tau and prevents onset and progression of neural dysfunction *in vivo*. PLoS One 2015; 10(2): e0117511.
[http://dx.doi.org/10.1371/journal.pone.0117511] [PMID: 25659102]

[364] Okuda M, Fujita Y, Hijikuro I, *et al.* PE859, A Novel Curcumin Derivative, Inhibits Amyloid-β and Tau Aggregation, and Ameliorates Cognitive Dysfunction in Senescence-Accelerated Mouse Prone 8. J Alzheimers Dis 2017; 59(1): 313-28.
[http://dx.doi.org/10.3233/JAD-161017] [PMID: 28598836]

[365] Wu X, Kosaraju J, Tam KY. SLM, a novel carbazole-based fluorophore attenuates okadaic acid-induced tau hyperphosphorylation *via* down-regulating GSK-3β activity in SH-SY5Y cells. Eur J Pharm Sci 2017; 110: 101-8.
[http://dx.doi.org/10.1016/j.ejps.2017.03.037] [PMID: 28359686]

[366] Wu X, Kosaraju J, Zhou W, Tam KY. SLOH, a carbazole-based fluorophore, mitigates neuropathology and behavioral impairment in the triple-transgenic mouse model of Alzheimer's disease. Neuropharmacology 2018; 131: 351-63.
[http://dx.doi.org/10.1016/j.neuropharm.2018.01.003] [PMID: 29309769]

[367] Gandini A, Bartolini M, Tedesco D, *et al.* Tau-centric multitargeted approach for Alzheimer's disease: development of the first-in-class dual glycogen synthase kinase 3beta and tau aggregation inhibitors. J Med Chem 2018; 61(17): 7640-56.

[http://dx.doi.org/10.1021/acs.jmedchem.8b00610] [PMID: 30078314]

[368] Pérez-Areales FJ, Betari N, Viayna A, *et al.* Design, synthesis and multitarget biological profiling of second-generation anti-Alzheimer rhein-huprine hybrids. Future Med Chem 2017; 9(10): 965-81.
[http://dx.doi.org/10.4155/fmc-2017-0049] [PMID: 28632395]

[369] Tabrizi S, Leavitt B, Kordasiewicz H, *et al.* Effects of IONIS-HTTRx in patients with early Huntington's disease; results of the first HTT-lowering drug trial. Neurology 2018; 90(15 Suppl): CT.002.

[370] Evers MM, Toonen LJ, van Roon-Mom WM. Antisense oligonucleotides in therapy for neurodegenerative disorders. Adv Drug Deliv Rev 2015; 87: 90-103.
[http://dx.doi.org/10.1016/j.addr.2015.03.008] [PMID: 25797014]

[371] McClorey G, Banerjee S. Cell-penetrating peptides to enhance delivery of oligonucleotide-based therapeutics. Biomedicines 2018; 6(2): 6.
[http://dx.doi.org/10.3390/biomedicines6020051] [PMID: 29734750]

[372] DeVos SL, Goncharoff DK, Chen G, *et al.* Antisense reduction of tau in adult mice protects against seizures. J Neurosci 2013; 33(31): 12887-97.
[http://dx.doi.org/10.1523/JNEUROSCI.2107-13.2013] [PMID: 23904623]

[373] Mignon L, Kordasiewicz H, Lane R, *et al.* Design of the first-in-human study of IONIS-MAPTRx, a tau-lowering antisense oligonucleotide, in patients with Alzheimer's disease. Neurology 2018; 90: S2.006.

[374] Rodriguez-Martin T, Anthony K, Garcia-Blanco MA, Mansfield SG, Anderton BH, Gallo JM. Correction of tau mis-splicing caused by FTDP-17 MAPT mutations by spliceosome-mediated RNA trans-splicing. Hum Mol Genet 2009; 18(17): 3266-73.
[http://dx.doi.org/10.1093/hmg/ddp264] [PMID: 19498037]

[375] Richter W, Menniti FS, Zhang HT, Conti M. PDE4 as a target for cognition enhancement. Expert Opin Ther Targets 2013; 17(9): 1011-27.
[http://dx.doi.org/10.1517/14728222.2013.818656] [PMID: 23883342]

[376] Conti M, Beavo J. Biochemistry and physiology of cyclic nucleotide phosphodiesterases: essential components in cyclic nucleotide signaling. Annu Rev Biochem 2007; 76: 481-511.
[http://dx.doi.org/10.1146/annurev.biochem.76.060305.150444] [PMID: 17376027]

[377] Barad M, Bourtchouladze R, Winder DG, Golan H, Kandel E. Rolipram, a type IV-specific phosphodiesterase inhibitor, facilitates the establishment of long-lasting long-term potentiation and improves memory. Proc Natl Acad Sci USA 1998; 95(25): 15020-5.
[http://dx.doi.org/10.1073/pnas.95.25.15020] [PMID: 9844008]

[378] Gallant M, Aspiotis R, Day S, *et al.* Discovery of MK-0952, a selective PDE4 inhibitor for the treatment of long-term memory loss and mild cognitive impairment. Bioorg Med Chem Lett 2010; 20(22): 6387-93.
[http://dx.doi.org/10.1016/j.bmcl.2010.09.087] [PMID: 20933411]

[379] Vitolo OV, Sant'Angelo A, Costanzo V, Battaglia F, Arancio O, Shelanski M. Amyloid beta -peptide inhibition of the PKA/CREB pathway and long-term potentiation: reversibility by drugs that enhance cAMP signaling. Proc Natl Acad Sci USA 2002; 99(20): 13217-21.
[http://dx.doi.org/10.1073/pnas.172504199] [PMID: 12244210]

[380] Gong B, Vitolo OV, Trinchese F, Liu S, Shelanski M, Arancio O. Persistent improvement in synaptic and cognitive functions in an Alzheimer mouse model after rolipram treatment. J Clin Invest 2004; 114(11): 1624-34.
[http://dx.doi.org/10.1172/JCI22831] [PMID: 15578094]

[381] Fleischhacker WW, Hinterhuber H, Bauer H, *et al.* A multicenter double-blind study of three different doses of the new cAMP-phosphodiesterase inhibitor rolipram in patients with major depressive disorder. Neuropsychobiology 1992; 26(1-2): 59-64.

[http://dx.doi.org/10.1159/000118897] [PMID: 1475038]

[382] Gurney ME, Nugent RA, Mo X, *et al.* Design and synthesis of selective phosphodiesterase 4D (PDE4D) allosteric inhibitors for the treatment of fragile X syndrome and other brain disorders. J Med Chem 2019; 62(10): 4884-901.
[http://dx.doi.org/10.1021/acs.jmedchem.9b00193] [PMID: 31013090]

[383] Zhang C, Xu Y, Chowdhary A, *et al.* Memory enhancing effects of BPN14770, an allosteric inhibitor of phosphodiesterase-4D, in wild-type and humanized mice. Neuropsychopharmacology 2018; 43(11): 2299-309.
[http://dx.doi.org/10.1038/s41386-018-0178-6] [PMID: 30131563]

[384] Cui SY, Yang MX, Zhang YH, *et al.* Protection from Amyloid β peptide-induced memory, biochemical, and morphological deficits by a phosphodiesterase-4D allosteric inhibitor. J Pharmacol Exp Ther 2019; 371(2): 250-9.
[http://dx.doi.org/10.1124/jpet.119.259986] [PMID: 31488603]

[385] Alzforum.org [homepage on the internet]. accessed August 2020 Available at: https://www.alzforum.org/news/conference-coverage/running-trial-results-ctad-conference

[386] Press release: Notice Regarding Tetra becoming a wholly owned subsidiary of Shionogi. available at: https://www.shionogi.com/global/en/news/2020/05/20200526.html

[387] Zilka N, Kovacech B, Barath P, Kontsekova E, Novák M. The self-perpetuating tau truncation circle. Biochem Soc Trans 2012; 40(4): 681-6.
[http://dx.doi.org/10.1042/BST20120015] [PMID: 22817716]

[388] Fitzpatrick AWP, Falcon B, He S, *et al.* Cryo-EM structures of tau filaments from Alzheimer's disease. Nature 2017; 547(7662): 185-90.
[http://dx.doi.org/10.1038/nature23002] [PMID: 28678775]

[389] Sato C, Barthélemy NR, Mawuenyega KG, *et al.* Tau kinetics in neurons and the human central nervous system. Neuron 2018; 98(4): 861-4.
[http://dx.doi.org/10.1016/j.neuron.2018.04.035] [PMID: 29772204]

[390] Vidarsson G, Dekkers G, Rispens T. IgG subclasses and allotypes: from structure to effector functions. Front Immunol 2014; 5: 520.
[http://dx.doi.org/10.3389/fimmu.2014.00520] [PMID: 25368619]

[391] Lee SH, Le Pichon CE, Adolfsson O, *et al.* Antibody-mediated targeting of tau *in vivo* does not require effector function and microglial engagement. Cell Rep 2016; 16(6): 1690-700.
[http://dx.doi.org/10.1016/j.celrep.2016.06.099] [PMID: 27475227]

[392] Lambert JC, Ibrahim-Verbaas CA, Harold D, *et al.* European Alzheimer's Disease Initiative (EADI); Genetic and Environmental Risk in Alzheimer's Disease; Alzheimer's Disease Genetic Consortium; Cohorts for Heart and Aging Research in Genomic Epidemiology. Meta-analysis of 74,046 individuals identifies 11 new susceptibility loci for Alzheimer's disease. Nat Genet 2013; 45(12): 1452-8.
[http://dx.doi.org/10.1038/ng.2802] [PMID: 24162737]

[393] Baumgart M, Snyder HM, Carrillo MC, Fazio S, Kim H, Johns H. Summary of the evidence on modifiable risk factors for cognitive decline and dementia: A population-based perspective. Alzheimers Dement 2015; 11(6): 718-26.
[http://dx.doi.org/10.1016/j.jalz.2015.05.016] [PMID: 26045020]

[394] Rissman RA, Staup MA, Lee AR, *et al.* Corticotropin-releasing factor receptor-dependent effects of repeated stress on tau phosphorylation, solubility, and aggregation. Proc Natl Acad Sci USA 2012; 109(16): 6277-82.
[http://dx.doi.org/10.1073/pnas.1203140109] [PMID: 22451915]

[395] Mudher A, Colin M, Dujardin S, *et al.* What is the evidence that tau pathology spreads through prion-like propagation? Acta Neuropathol Commun 2017; 5(1): 99.
[http://dx.doi.org/10.1186/s40478-017-0488-7] [PMID: 29258615]

[396] Kontsekova E, Zilka N, Kovacech B, Skrabana R, Novak M. Identification of structural determinants on tau protein essential for its pathological function: novel therapeutic target for tau immunotherapy in Alzheimer's disease. Alzheimers Res Ther 2014; 6(4): 45.
[http://dx.doi.org/10.1186/alzrt277] [PMID: 25478018]

[397] Sarkar M, Kuret J, Lee G. Two motifs within the tau microtubule-binding domain mediate its association with the hsc70 molecular chaperone. J Neurosci Res 2008; 86(12): 2763-73.
[http://dx.doi.org/10.1002/jnr.21721] [PMID: 18500754]

[398] Bournazos S, DiLillo DJ, Ravetch JV. The role of Fc-FcγR interactions in IgG-mediated microbial neutralization. J Exp Med 2015; 212(9): 1361-9.
[http://dx.doi.org/10.1084/jem.20151267] [PMID: 26282878]

[399] Asuni AA, Boutajangout A, Quartermain D, Sigurdsson EM. Immunotherapy targeting pathological tau conformers in a tangle mouse model reduces brain pathology with associated functional improvements. J Neurosci 2007; 27(34): 9115-29.
[http://dx.doi.org/10.1523/JNEUROSCI.2361-07.2007] [PMID: 17715348]

[400] Rajamohamedsait H, Rasool S, Rajamohamedsait W, Lin Y, Sigurdsson EM. Prophylactic active tau immunization leads to sustained reduction in both tau and amyloid-beta pathologies in 3xTg mice. Sci Rep 2017; 7(1): 17034.
[http://dx.doi.org/10.1038/s41598-017-17313-1] [PMID: 29213096]

[401] Boutajangout A, Quartermain D, Sigurdsson EM. Immunotherapy targeting pathological tau prevents cognitive decline in a new tangle mouse model. J Neurosci 2010; 30(49): 16559-66.
[http://dx.doi.org/10.1523/JNEUROSCI.4363-10.2010] [PMID: 21147995]

[402] Troquier L, Caillierez R, Burnouf S, et al. Targeting phospho-Ser422 by active Tau Immunotherapy in the THYTau22 mouse model: a suitable therapeutic approach. Curr Alzheimer Res 2012; 9(4): 397-405.
[http://dx.doi.org/10.2174/156720512800492503] [PMID: 22272619]

[403] Novak P, Zilka N, Zilkova M, et al. AADvac1, an active immunotherapy for Alzheimer's disease and non-Alzheimer tauopathies: an overview of preclinical and clinical development. J Prev Alzheimers Dis 2019; 6(1): 63-9.
[PMID: 30569088]

[404] Kontsekova E, Zilka N, Kovacech B, Novak P, Novak M. First-in-man tau vaccine targeting structural determinants essential for pathological tau-tau interaction reduces tau oligomerisation and neurofibrillary degeneration in an Alzheimer's disease model. Alzheimers Res Ther 2014; 6(4): 44.
[http://dx.doi.org/10.1186/alzrt278] [PMID: 25478017]

[405] Novak P, Schmidt R, Kontsekova E, et al. Safety and immunogenicity of the tau vaccine AADvac1 in patients with Alzheimer's disease: a randomised, double-blind, placebo-controlled, phase 1 trial. Lancet Neurol 2017; 16(2): 123-34.
[http://dx.doi.org/10.1016/S1474-4422(16)30331-3] [PMID: 27955995]

[406] Novak P, Schmidt R, Kontsekova E, et al. FUNDAMANT: an interventional 72-week phase 1 follow-up study of AADvac1, an active immunotherapy against tau protein pathology in Alzheimer's disease. Alzheimers Res Ther 2018; 10(1): 108.
[http://dx.doi.org/10.1186/s13195-018-0436-1] [PMID: 30355322]

[407] Alzforum.org Alzforum.org [homepage on the internet]. accessed August 2020 available at https://www.alzforum.org/news/conference-coverage/active-tau-vaccine-hints-sl-wing-neurodegeneration

[408] Hickman DT, López-Deber MP, Ndao DM, et al. Sequence-independent control of peptide conformation in liposomal vaccines for targeting protein misfolding diseases. J Biol Chem 2011; 286(16): 13966-76.
[http://dx.doi.org/10.1074/jbc.M110.186338] [PMID: 21343310]

[409]   Theunis C, Crespo-Biel N, Gafner V, *et al*. Efficacy and safety of a liposome-based vaccine against protein Tau, assessed in tau.P301L mice that model tauopathy. PLoS One 2013; 8(8): e72301.
[http://dx.doi.org/10.1371/journal.pone.0072301] [PMID: 23977276]

[410]   ISRCTN registry,available at:
[http://dx.doi.org/10.1186/ISRCTN13033912]

[411]   Press release: AC Immune advances phospho-tau Alzheimer's vaccine in phase 1b/2a study. July 16, 2020, available at: https://ir.acimmune.com/news-releases/news-release-details/ac-immune-adv-nces-phospho-tau-alzheimers-vaccine-phase-1b2a

[412]   Boimel M, Grigoriadis N, Lourbopoulos A, Haber E, Abramsky O, Rosenmann H. Efficacy and safety of immunization with phosphorylated tau against neurofibrillary tangles in mice. Exp Neurol 2010; 224(2): 472-85.
[http://dx.doi.org/10.1016/j.expneurol.2010.05.010] [PMID: 20546729]

[413]   Bi M, Ittner A, Ke YD, Götz J, Ittner LM. Tau-targeted immunization impedes progression of neurofibrillary histopathology in aged P301L tau transgenic mice. PLoS One 2011; 6(12): e26860.
[http://dx.doi.org/10.1371/journal.pone.0026860] [PMID: 22174735]

[414]   Ando K, Kabova A, Stygelbout V, *et al*. Vaccination with Sarkosyl insoluble PHF-tau decrease neurofibrillary tangles formation in aged tau transgenic mouse model: a pilot study. J Alzheimers Dis 2014; 40 (Suppl. 1): S135-45.
[http://dx.doi.org/10.3233/JAD-132237] [PMID: 24614899]

[415]   Richter M, Mewes A, Fritsch M, Krügel U, Hoffmann R, Singer D. Doubly phosphorylated peptide vaccines to protect transgenic P301L mice against Alzheimer's disease like tau aggregation. Vaccines (Basel) 2014; 2(3): 601-23.
[http://dx.doi.org/10.3390/vaccines2030601] [PMID: 26344748]

[416]   Ji M, Xie XX, Liu DQ, *et al*. Hepatitis B core VLP-based mis-disordered tau vaccine elicits strong immune response and alleviates cognitive deficits and neuropathology progression in Tau.P301S mouse model of Alzheimer's disease and frontotemporal dementia. Alzheimers Res Ther 2018; 10(1): 55.
[http://dx.doi.org/10.1186/s13195-018-0378-7] [PMID: 29914543]

[417]   Yanamandra K, Kfoury N, Jiang H, *et al*. Anti-tau antibodies that block tau aggregate seeding *in vitro* markedly decrease pathology and improve cognition *in vivo*. Neuron 2013; 80(2): 402-14.
[http://dx.doi.org/10.1016/j.neuron.2013.07.046] [PMID: 24075978]

[418]   Combs B, Hamel C, Kanaan NM. Pathological conformations involving the amino terminus of tau occur early in Alzheimer's disease and are differentially detected by monoclonal antibodies. Neurobiol Dis 2016; 94: 18-31.
[http://dx.doi.org/10.1016/j.nbd.2016.05.016] [PMID: 27260838]

[419]   Johnson GV, Seubert P, Cox TM, Motter R, Brown JP, Galasko D. The tau protein in human cerebrospinal fluid in Alzheimer's disease consists of proteolytically derived fragments. J Neurochem 1997; 68(1): 430-3.
[http://dx.doi.org/10.1046/j.1471-4159.1997.68010430.x] [PMID: 8978756]

[420]   Amadoro G, Corsetti V, Florenzano F, *et al*. AD-linked, toxic NH2 human tau affects the quality control of mitochondria in neurons. Neurobiol Dis 2014; 62: 489-507.
[http://dx.doi.org/10.1016/j.nbd.2013.10.018] [PMID: 24411077]

[421]   Florenzano F, Veronica C, Ciasca G, *et al*. Extracellular truncated tau causes early presynaptic dysfunction associated with Alzheimer's disease and other tauopathies. Oncotarget 2017; 8(39): 64745-78.
[http://dx.doi.org/10.18632/oncotarget.17371] [PMID: 29029390]

[422]   Zhou Y, Shi J, Chu D, *et al*. Relevance of phosphorylation and truncation of tau to the ethiopathogenesis of Alzheimer's disease. Front Aging Neurosci 2018; 10: 27.

[http://dx.doi.org/10.3389/fnagi.2018.00027] [PMID: 29472853]

[423] Courade J-P, Angers R, Mairet-Coello G, *et al.* Epitope determines efficacy of therapeutic anti-Tau antibodies in a functional assay with human Alzheimer Tau. Acta Neuropathol 2018; 136(5): 729-45.
[http://dx.doi.org/10.1007/s00401-018-1911-2] [PMID: 30238240]

[424] Qureshi IA, Tirucherai G, Ahlijanian MK, Kolaitis G, Bechtold C, Grundman M. A randomized, single ascending dose study of intravenous BIIB092 in healthy participants. Alzheimers Dement (N Y) 2018; 4: 746-55.
[http://dx.doi.org/10.1016/j.trci.2018.10.007] [PMID: 30581980]

[425] Boxer AL, Qureshi I, Ahlijanian M, *et al.* Safety of the tau-directed monoclonal antibody BIIB092 in progressive supranuclear palsy: a randomised, placebo-controlled, multiple ascending dose phase 1b trial. Lancet Neurol 2019; 18(6): 549-58.
[http://dx.doi.org/10.1016/S1474-4422(19)30139-5] [PMID: 31122495]

[426] Alzforum.org [homepage on the internet], accessed August 2020,   Available at:https://www. alzforum.org/ news/research-news/gosuranemab-biogens-anti-tau-immunotherapy- does-not-fly-psp

[427] Alzforum.org, [homepage on the internet], accessed August 2020, . Available at:https://www. alzforum.org/ news/research-news/2019-year-hope-alzheimers-research

[428] Alzforum.org [homepage on the internet], accessed August 2020, available at: . https://www. alzforum.org/news/conference-coverage/treating-tau-finally-clinical-candidates-are-stepping-ring

[429] Alzforum.org [homepage on the internet], accesed August 2020, available at: . https://www. alzforum.org/news/conference-coverage/high-dose-av-and-tau-immunotherapies-complete-initial-safety-tests

[430] Braak H, Del Tredici K. Alzheimer's pathogenesis: is there neuron-to-neuron propagation? Acta Neuropathol 2011; 121(5): 589-95.
[http://dx.doi.org/10.1007/s00401-011-0825-z] [PMID: 21516512]

[431] Yanamandra K, Jiang H, Mahan TE, *et al.* Anti-tau antibody reduces insoluble tau and decreases brain atrophy. Ann Clin Transl Neurol 2015; 2(3): 278-88.
[http://dx.doi.org/10.1002/acn3.176] [PMID: 25815354]

[432] Alzforum.org [homepage on the internet], accessed August 2020, available at: . https://www. alzforum.org/news/research-news/abbvies-tau-antibody-flops-progressive-supranuclear-palsy

[433] Walls KC, Ager RR, Vasilevko V, Cheng D, Medeiros R, LaFerla FM. p-Tau immunotherapy reduces soluble and insoluble tau in aged 3xTg-AD mice. Neurosci Lett 2014; 575: 96-100.
[http://dx.doi.org/10.1016/j.neulet.2014.05.047] [PMID: 24887583]

[434] Alzforum.org   [homepage   on   the   internet],   accessed   August   2020:   . https://www.alzforum.org/news/conference-coverage/block-ta-s-proteopathic-spread-antibody-must-attack-its-mid-region

[435] International Congress of Parkinson's Disease and Movement Disorders ® 2019 Late-Breaking Abstract 3, Available at: 2019.https://www.mdscongress.org/Congress-Branded/Congress-20-9-Files/2019Late-BreakingAbstractsPublicationFile.pdf

[436] Press Release. UCB enters into collaboration with Roche to develop antibody treatment for people living   with   Alzheimer's   Disease   Available   at:https://www.ucb.com/stories-media/Pres--Releases/article/UCB-enters-into-collaboration-with-Roche-to-develop-antibody-tr-atment-for-people-living-with-Alzheimer-s-Disease

[437] Vander Zanden CM, Chi EY. Passive immunotherapies targeting amyloid beta and tau oligomers in Alzheimer's disease. J Pharm Sci 2020; 109(1): 68-73.
[http://dx.doi.org/10.1016/j.xphs.2019.10.024] [PMID: 31647950]

[438] Alam R, Driver D, Wu S, *et al.* Preclinical characterization of an antibody [LY3303560] targeting aggregated tau. Alzheimers Dement 2017; 13: 592-P593.

[http://dx.doi.org/10.1016/j.jalz.2017.07.227]

[439]   Lathuilière A, Valdés P, Papin S, *et al.* Motifs in the tau protein that control binding to microtubules and aggregation determine pathological effects. Sci Rep 2017; 7(1): 13556.
[http://dx.doi.org/10.1038/s41598-017-13786-2] [PMID: 29051562]

[440]   Chai X, Wu S, Murray TK, *et al.* Passive immunization with anti-Tau antibodies in two transgenic models: reduction of Tau pathology and delay of disease progression. J Biol Chem 2011; 286(39): 34457-67.
[http://dx.doi.org/10.1074/jbc.M111.229633] [PMID: 21841002]

[441]   Umeda T, Eguchi H, Kunori Y, *et al.* Passive immunotherapy of tauopathy targeting pSer413-tau: a pilot study in mice. Ann Clin Transl Neurol 2015; 2(3): 241-55.
[http://dx.doi.org/10.1002/acn3.171] [PMID: 25815351]

[442]   Hoskin JL, Sabbagh MN, Al-Hasan Y, Decourt B. Tau immunotherapies for Alzheimer's disease. Expert Opin Investig Drugs 2019; 28(6): 545-54.
[http://dx.doi.org/10.1080/13543784.2019.1619694] [PMID: 31094578]

[443]   Czerkowicz J, Chen W, Wang Q, *et al.* Pan-Tau Antibody BIIb076 exhibits promising safety and biomarker profile in Cynomolgus monkey toxicity study. Alzheimers Dement 2017; 13(7): 1271.
[http://dx.doi.org/10.1016/j.jalz.2017.06.1903]

[444]   Alzforum.org [homepage on the internet], last updated 31 Oct 2019, accessed August 2020, . Available at:https://www.alzforum.org/therapeutics/biib076

[445]   Buée L, Bussière T, Buée-Scherrer V, Delacourte A, Hof PR. Tau protein isoforms, phosphorylation and role in neurodegenerative disorders. Brain Res Brain Res Rev 2000; 33(1): 95-130.
[http://dx.doi.org/10.1016/S0165-0173(00)00019-9] [PMID: 10967355]

[446]   Alzforum.org, [homepage on the internet], updated 20 Nov 2015, accessed August 2020, available at: . https://www.alzforum.org/therapeutics/rg7345

[447]   Rosenqvist N, Asuni AA, Andersson CR, *et al.* Highly specific and selective anti-pS396-tau antibody C10.2 targets seeding-competent tau. Alzheimers Dement (N Y) 2018; 4: 521-34.
[http://dx.doi.org/10.1016/j.trci.2018.09.005] [PMID: 30386817]

[448]   Nakamura K, Greenwood A, Binder L, *et al.* Proline isomer-specific antibodies reveal the early pathogenic tau conformation in Alzheimer's disease. Cell 2012; 149(1): 232-44.
[http://dx.doi.org/10.1016/j.cell.2012.02.016] [PMID: 22464332]

[449]   Kondo A, Shahpasand K, Mannix R, *et al.* Antibody against early driver of neurodegeneration cis P-tau blocks brain injury and tauopathy. Nature 2015; 523(7561): 431-6.
[http://dx.doi.org/10.1038/nature14658] [PMID: 26176913]

[450]   Gilman S, Koller M, Black RS, *et al.* AN1792(QS-21)-201 Study Team. Clinical effects of Abeta immunization (AN1792) in patients with AD in an interrupted trial. Neurology 2005; 64(9): 1553-62.
[http://dx.doi.org/10.1212/01.WNL.0000159740.16984.3C] [PMID: 15883316]

[451]   Beyond amyloid: New approaches to Alzheimer's disease treatment. EBioMedicine 2020; 51: 102648.
[http://dx.doi.org/10.1016/j.ebiom.2020.102648] [PMID: 32007187]

[452]   Cure Alzheimer's Fund Biogen announces intention to file with FDA for approval for new Alzheimer's drug Aducanumab .Available at: https://curealz.org/news-and-events/aducanumab/

[453]   Sabbagh MN. Alzheimer's disease drug development pipeline 2020. J Prev Alzheimers Dis 2020; 7(2): 66-7.
[PMID: 32236392]

[454]   Cummings J, Lee G, Ritter A, Sabbagh M, Zhong K. Alzheimer's disease drug development pipeline: 2019. Alzheimers Dement (N Y) 2019; 5: 272-93.
[http://dx.doi.org/10.1016/j.trci.2019.05.008] [PMID: 31334330]

CHAPTER 3

Implication of Dehydroepiandrosterone on Dementia Related to Oxidative Stress

Adriane Bello Klein[1]**, Alexandre Castro**[1]**, Alex Sander da Rosa Araujo**[1]**, Daiane da Rocha Janner**[2] **and Maria Helena Vianna Metello Jacob**[3,*]

[1] *Federal University of Rio Grande do Sul (UFRGS), Institute of Basic Health Sciences, Physiology, R. Sarmento Leite, 500 – CEP 90035-190- Porto Alegre, RS,Brazil*

[2] *Federal University of Rio de Janeiro (UFRJ),Carlos Chagas Filho Institute of Biophysics – Neurogenesis Lab, Av. Carlos Chagas Filho, 373–CEP 21941-902, Rio de Janeiro, RJ,Brazil*

[3] *Montfort Hospital – Institut du Savoir, 745, Montreal Road, K1K 0T1 Ottawa, ON/Canada*

Abstract: The number of people living with dementia will increase worldwide over the coming decades as the population ages. The aging of the brain is associated with oxidative stress. Evidence of increased oxidative stress has been seen in Alzheimer's disease (the most common cause of dementia), contributing to the formation of amyloid plaques and neurofibrillary tangles. Dehydroepiandrosterone is a physiologically active steroid hormone that declines with aging and is associated with aging-related neurodegeneration. Exogenous dehydroepiandrosterone can exert an antioxidant or prooxidant effect, depending on the dose and tissue specificity. Dehydroepiandrosterone biosynthesis in the brains of rats, bovines, and humans can be mediated by prooxidant agents, such as Fe^{2+} and β-amyloid peptides. A-β can provoke an increase in oxygen free radicals in cells, and this rise in reactive oxygen species modulates dehydroepiandrosterone levels. Also, studies have demonstrated that dehydroepiandrosterone treatment may modulate Akt (a serine/threonine kinase implicated in neuronal survival), and its activation could be changed with aging. Despite the numerous studies, the mechanism of action of dehydroepiandrosterone and its relationship with dementia or improvement in behaviours associated with memory and motor activity should still be elucidated as relates to dosage, temporal treatment window, besides its acute and chronic effects. A better understanding of the physiological role of dehydroepiandrosterone in the aging process may be of benefit to the development of novel strategies in the treatment of dementia.

* **Corresponding author Maria Helena vianna Metello Jacob :** Montfort Hospital – Institut du Savoir, 745, Montreal Road, K1K 0T1 Ottawa, ON,Canada; Tel: (+1) 613-746-4621; E-mail: mhvmjacob@hotmail.com
# There is no first author, and all authors contributed similarly.

Dr. José Juan Antonio Ibarra Arias (Eds.)

Keywords: Aging, Akt, Dementia, Dehydroepiandrosterone, Dehydroepiandrosterone Sulphate, Oxidative Stress, Neuroprotection, Neuro- steroids, Reactive Oxygen Species, Steroids.

INTRODUCTION

The number of people living with dementia will increase worldwide over the coming decades as the population ages. Current estimates suggest that the number of adults living with dementia will increase slightly less than twofold in Europe, somewhat more than twofold in North America, threefold in Asia, and fourfold in Latin America and Africa from 2015 to 2050 [1]. Dementia is a global pandemic, and the majority of people living with dementia live in low- and middle-income countries where access to health services, support, care, and social protection is extremely limited.

A low level of education is a well-known risk factor for dementia, which is generally associated with low socioeconomic status and reduced access to health care, from the prenatal period to older ages. Therefore, a low level of education could increase the risk of dementia by limiting the adequate diagnosis and treatment of co-morbidities, especially diabetes mellitus and cardiovascular disease, as well as being often associated with insufficient nutritional status [2]. These associated factors may explain the rise in numbers; while 37% of people living with dementia live in high-income countries, 63% live in low-and-middle-income countries [1].

Aging, an inevitable and natural process, is related to increased oxidative stress (OS) and chronic inflammation [3, 4]. Similarly, aging of the brain is associated with OS and cumulative inflammation, which explains why older people predispose to developing neurodegenerative pathologies. Inflammation is a normal physiological process that is essential to maintain homeostasis. Under normal conditions, the inflammation-repair cycle works efficiently during youthful years, and then it is affected by aging, producing a relatively impaired ability to regenerate damaged tissues and thus contributing to a more pro-inflammatory phenotype [4]. Age-related neuroinflammation appears to be associated with the fact that microglia faces important functional and immunophenotypic changes with aging, being exposed to augmented pro-inflammatory responses [5]. In fact, aging can be defined as a loss of homeostasis due to chronic OS affecting especially the regulatory systems (endocrine, immune, and nervous systems) [6].

The presence of enhanced reactive oxygen species (ROS) in neurons leads to OS in the central nervous system (CNS), presenting a big threat to its integrity which can induce neurodegenerative disorders [7]. Dehydroepiandrosterone (DHEA), which is a physiologically active steroid hormone, declines with aging and is associated with aging-related neurodegeneration. Also seen as a neurosteroid, DHEA has also been found to protect rat and human hippocampal neuronal cells against OS-induced cellular damage [8]. Given the neuroprotective and cognitive-enhancing properties of DHEA, this neuroactive steroid [9, 10] may be of particular importance in the treatment of neurodegeneration diseases, including dementia. DHEA presents a dual effect linked to OS. Exogenous DHEA can exert an antioxidant [11 - 14] or prooxidant [15 - 18] effect depending on the dose and tissue specificity.

Much remains inconclusive about DHEA's mode of action in the human body or the rationale for its age-related decline. Given the commonly accepted phenomenon of accelerated aging in many chronic diseases, knowledge about the possible role of DHEA as a therapeutic or preventative benefit should be investigated. A better understanding of the physiological role of DHEA in the aging process in CNS may be of benefit to the development of novel strategies in the treatment of dementia. It is important to note that DHEA has been used as an anti-aging supplement, but there is a lack of information based on randomized controlled trials studies on DHEA supplementation and health outcomes. Although mechanisms of action have been pointed out in experimental models, further large clinical trials are necessary to better identify the clinical role of DHEA and to elucidate benefits *versus* potential risks. In this chapter, we contribute to a more comprehensive scenario of what is known about three important elements strongly implicated in dementia: aging, OS, and DHEA.

The Implication of Aging and Oxidative Stress in Dementia

Defining aging, as well as understanding by what mechanisms we age, is still a complicated task to gerontological science. There is a broad discussion of whether aging is a physiological process or a random set of damage events that result in the impairment of life conditions, disease development, and organismal system failure [19]. Moreover, the use of both aging and senescence terms as synonymous is also controversial; and even though senescence can exert a relevant role in the progression of aging, it also seems to be involved with beneficial effects, such as neoplasm control, cellular plasticity, and stress response [19].

In the biological spectrum, on the other hand, aging can represent all cellular alterations (adverse or not) that may happen during the whole life process [20]. Factors associated with aging progression can be represented by reduced telomere maintenance, genetic and epigenetic alterations, immunological decline, and mitochondrial dysfunction. In this sense, the reduction of homeostasis control capacity determines deleterious changes in the cell function and conduce to cellular death [21]. Although several age-correlated causal proposes have risen, aging may be consequent to accrual many changes in the intra and extracellular environments, among which free radicals accumulation can be highlighted [22]. Nonetheless, the redox homeostasis maintenance, which is established by a balanced oxidant/antioxidant ratio, represents the "push-button switch mechanism" that regulates the intracellular signaling associated with aging [21]. However, excessive free radical production may provoke protein oxidation and activate death pathways, contributing to the probabilistic augmentation of age-related diseases development, such as dementia [23].

Neurodegenerative diseases may involve, in their pathophysiology, inflammatory response, mitochondrial dysfunction, amyloid-β (A-β) peptide deposition, and hyperphosphorylation of the tau protein [24 - 26]. Most of this process implicates in the augmentation of free radicals levels which disrupt redox homeostases of nervous cells, such as neurons and glial cells [27]. Since redox balance is broken, the intracellular environment may have undergone relevant distress [28]. In this scenario, it is not easy to determine if OS is either cause or consequence of neurodegenerative events in which dementia establishment is inserted. Moreover, free radicals appear to be chemical mediators of intracellular signaling that modulate the phosphorylation of pathways involved in cell survival and death [29]. Consequently, important synapses are not established what may spoil not only the working memory but also the long-term memory (memory consolidation) [30]. As the nervous system metabolism is predominantly aerobic, free radicals' production becomes relevant to explore.

Further, the CNS may consume a range of 20% to 25% of the oxygen available in arterial blood [31]. Although the mass of the brain represents about 2% of the total body weight, the biochemical necessity of this organ is huge [31, 32]. In this context, aerobic glucose metabolism provides the major source of ATP to the energy demand of neurons [33]. Glucose is almost totally oxidized into CO_2 and H_2O in the mitochondria matrix through the tricarboxylic acid cycle, which produces $NADH+H^+$ and $FADH_2$. Such reducing agents supply the electron transport chain to maintain ATP production, a process known as oxidative phosphorylation [34]. Even though mitochondrial activity is critical to supply cellular ATP demand, this organelle is also involved in ROS production during aerobic metabolism [32]. Usually, oxygen receives four electrons, the trivalent reduction process; however, around 5% of the total oxygen may be reduced monovalently, producing ROS, such as the superoxide radical anion ($O_2^{\cdot-}$), hydrogen peroxide (H_2O_2), and hydroxyl radical ($OH^{\cdot}$). $O_2^{\cdot-}$ may abstract electrons from other molecules, initiating a chain reaction that produces other free radicals. In this sense, the manganese superoxide dismutase (SOD-2) is the enzyme that catalyzes the dismutation process in the mitochondria, leading to the synthesis of H_2O_2 from $O_2^{-\cdot}$. There is also an isoform present in cytosol (SOD-1) that also dismutates $O_2^{-\cdot}$ into H_2O_2 and is copper and zinc-dependent. H_2O_2, in turn, is not a free radical, but may react with either transition metals, such as Fe^{+2}, which is known as the Fenton reaction, or with $O_2^{-\cdot}$, the Haber-Weiss reaction. Both reaction mechanisms have $OH^{\cdot}$ as the major final product. This ROS demonstrates huge reactivity and can damage membrane lipids, proteins, and DNA [35, 36].

The CNS structure is enriched by lipids and, thus, is very susceptible to lipid peroxidation. The plasmatic membrane of neurons is plentiful of polyunsaturated acids (PUFA), which can be more sensitive to free radicals attacks [36]. The increased ROS levels and consequent lipid peroxidation seem to be associated with the modification of processing and expression of transmembrane amyloid-β protein precursor (AβPP), leading to A-β accumulation in Alzheimer's disease (AD) [24]. Moreover, cognitive impairment is also related not only to augmented carbonyl group levels but also to oxidized proteins (3-nitrotyrosine (3-NO2-Tyr), meaning important ROS-mediated protein oxidation and cellular death [24]. The oxidation of nucleic acids (both nuclear and mitochondrial) may be represented by augmented 8-hydroxy-2'-deoxyguanosine (8-OHdG) and 8-OH-guanosine (8-OHG) levels in the brain, blood, and cerebrospinal fluid [37]. In the context of DNA oxidation, 8-OHdG and 8-OHG are reliable markers that may be used to verify the redox homeostasis disruption of genetic material in neurons. Considering the CNS's low levels of reduced glutathione (GSH), a relevant antioxidant, other antioxidant systems will be essential in maintaining the redox homeostasis of the cell [38]. Antioxidant systems include substances that prevent ROS formation by inhibiting prooxidant enzymes, chelating transition metals, or being cofactors of antioxidant enzymes. This constitutes the first line of antioxidant defense and can be defined as preventive antioxidants [39]. The second line of defense is constituted by the chain-breaker antioxidants, that scavenge ROS or other derived radicals responsible for the reactions of oxidative damage propagation. The third line of antioxidant defense can also be emphasized: it is formed by the repair antioxidants, such as phospholipases, peroxidases, acryl transferases [38].

In this scenario, the enzymatic antioxidant system plays a highlighted role in the CNS detoxification of ROS. The group, including superoxide dismutase (SOD), catalase (CAT), glutathione peroxidase (GPx), and glutathione-S-transferase (GST) enzymes, can protect cells against ROS-mediated damage [35]. The presence of mutated SOD-1 and SOD-2 in neurodegenerative diseases was reported, and these mutations are involved in apoptosis activation, mitochondrial dysfunction, and abnormalities in cytoskeleton proteins [38]. On the other hand, the SOD-1 super expression protects neurons against A-β neurotoxic effects [40].

CAT is an enzyme that protects the CNS in case of more severe OS, acting when ROS levels are high. It was demonstrated that treatment with CAT can improve neuronal survival after a challenge with A-β in neuron's culture [41]. GPx activity is more relevant when OS is milder, and the GPx-1 is the predominant isoform in neurons and astrocytes [42]. GST is activated by electrophilic compounds, being important for xenobiotics detoxification, such as chemotherapeutic drugs, carcinogens, environmental pollutants [43]. GST catalyzes the conjugation of GSH with these electrophilic substrates to inactivate them, facilitating their elimination from the body. In this process, GSH is consumed, forming oxidized glutathione (GSSG). Glutathione reductase (GR) is one enzyme responsible for the recovery of GSH from GSSG by transferring electrons from nicotinamide and adenine dinucleotide phosphate (NADPH) to the flavin dinucleotide and adenine (FAD and NAD). Another antioxidant system that helps in the control of reduced cysteines isthioredoxin (Trx) [44]. The Trx system is composed of different Trx and thioredoxin reductases (TrxR) isoforms. TrxR catalyzes the transfer of electrons from NADPH to Trx, promoting the conversion of oxidized Trx into its reduced form [45]. In the rat brain, the most abundant isoforms are Trx-1 (cytosol) and Trx-2 (mitochondria) [46]. There is a report of increased TrxR activity and decreased Trx-1 content in rats with AD [47]. In addition to its important effect as an antioxidant, the Trx system is also crucial for the modulation of signaling pathways involved in the immune and inflammatory processes of many conditions, including neurodegenerative diseases [48].

The chain-breaker antioxidants include many natural compounds that have aromatic rings substituted with hydroxyl groups, such as vitamin C, vitamin E, coenzyme Q (CoQ), and polyphenols [49]. This class of antioxidants may scavenge radicals formed during the lipid peroxidation process, such as peroxyl and alkoxyl radicals, avoiding or reducing the propagation stage. Some of these chain-breakers are fat-soluble (vitamin E, CoQ, beta-carotene), while others are water-soluble (vitamin C). They act as direct antioxidants, but also help in the regeneration of other antioxidants. One example is the role of CoQ and vitamin C in the recycling of tocopherol from the tocopheryl radical [38]. Besides that, cells can also count on indirect antioxidants that induce the endogenous antioxidant reserve. These compounds (*e.g.,* sulforaphane, catechins, curcumin, resveratrol) can stimulate intracellular antioxidants and phase II detoxification enzymes *via* activation of the Nrf-2-Keap-1-ARE signaling pathway. The enzymes regulated by Nrf2 include heme-oxygenase, GST, TrxR, Trx-1, CAT, SOD, peroxiredoxin, among others. Many pre-clinical studies have demonstrated benefits from treatment with Nrf2 inducers for cardiocerebral vascular diseases, but there are just a few clinical trials using these compounds [50].

As explained before, oxidative DNA damage increases the risk of age-related diseases, such as neurodegenerative disease (*e.g.,* dementia), and some studies have demonstrated that the consumption of micronutrients reduces the DNA damage and/or improves the DNA repair efficiency [51], acting as repair antioxidants. One issue that is necessary to take into account is the lack of appetite that makes older people more susceptible to malnourishment, and as a consequence, to have micronutrient deficiencies. Even though antioxidant defense is pivotal to avoid redox homeostasis disruption, it seems to become inefficient during the aging process [32]. Moreover, the genetic polymorphism of antioxidant enzymes may predispose OS-dependent cerebral injury to higher risk [32, 43]. Therefore, supplementation with indirect antioxidants (Nrf2 inducers) would be more effective to the elderly population once it could produce hermetic responses to protect cells by promoting the activation of "vitagenes" that express proteins of stress resistance, such as heme oxygenase-1, Trx, TrxR [52]. In general, moderate stressor agents, such as H_2O_2 or other ROS, trigger these adaptive responses. Thus, for redox homeostasis, the presence of low H_2O_2 concentrations is important to signalize cellular effects, such as proliferation, recruitment of immune cells, cellular differentiation, metabolism regulation [29]. However, when ROS levels become cytotoxic, intracellular signaling is driven to cellular death through the activation of apoptosis, necrosis, and/or autophagy. In this context, when an imbalance in the ROS/antioxidant ratio is established, caused by either augmented ROS production or reduced antioxidant levels, it can be defined as a state of OS [28]. OS can activate apoptosis signaling and provoke loss of cerebral mass and dementia [29].

OS has been implicated in the progression of AD, as well as in the progression of other causes of dementia, such as Parkinson's disease and Huntington's disease, which are defined by the progressive loss of specific neuronal cells. Besides, these diseases are associated with the formation of protein aggregates [53]. In these diseases, disturbance in the function or loss of integrity of the endoplasmic reticulum due to OS can be caused, for example, by the accumulation and deposit of unfolded proteins and by changes in calcium homeostasis within this organelle [54]. In addition, evidence suggests that mitochondria play a central role in neurodegenerative diseases due to their role as a critical regulator of cell death [55]. Mutations in mitochondrial DNA and OS contribute to aging, being the biggest risk factors for neurodegenerative diseases [56, 57]. This organelle is the main source of ROS in cells, generating superoxide anions in the intermembrane space through the leakage of electrons that combine with molecular oxygen in complex III, in a process that is dependent on the membrane potential, and in the matrix through an undefined site of complex I [58]. In addition, mitochondria actively participate in calcium homeostasis [57] and are involved in several processes that lead to cell death, including the release of cytochrome c [59]. It is believed that all these processes are interconnected and that an imbalance in these functions may be involved in the pathophysiology of several causes of dementia [57].

AD is the most common cause of dementia, accounting for 50% to 60% of dementia syndromes [60]. In the United States (U.S.), nearly six million people are affected by Alzheimer's dementia. Due to its aged-related nature [61, 62], there are projections of between 15–20 million people to have this disease within the next 30 years. Usually, the first symptom is the loss of episodic memory for recent events. The memory deficit evolves insidiously and progresses slowly over time, also compromising semantic memory (vocabulary, for example) and working or operational memory (memorizing a phone number, for example). The decline in other cognitive functions, such as attention and language, can either arise concurrently with amnesia or later [63]. Alzheimer's diagnosis is obtained only after histopathological analysis of autopsy materials. Such analysis reveals neuronal loss in the pyramidal layers of the cerebral cortex and intense synaptic degenerations, both at the hippocampal and neocortical levels, in addition to identifying extracellular deposits of amyloid-β (A-β) peptide [53].The deposit of these peptides has neurotoxic effects, causing the breakdown of calcium homeostasis, induction of OS, mitochondrial and synaptic dysfunction [64]. A-β is derived by proteolytic cleavage induced by beta and gamma secretases. Under normal conditions, A-β is degraded by enzymes and removed from the brain through a balance between efflux and influx, mediated by the LDL receptor protein and receptors for advanced glycosylation end products. In AD there is an imbalance between the production and the clearance of these peptides, leading to the deposition of oligomers in the extracellular space, causing inhibition of the long-term hippocampal potential and synaptic plasticity [60]. In mitochondria, A-β reduces the potential of the mitochondrial membrane, inhibits the electron transport chain, decreases the respiratory rate and induces the release of ROS, favoring the occurrence of OS and apoptosis. These peptides can also cause neurotoxicity through the direct production of these reactive species, through the interaction of A-β with Cu^{2+}, Fe^{2+} and Zn^{2+} [65]. In the presence of these transition metals, it is observed that H_2O_2 is catalytically converted into the highly toxic $OH^{\cdot}$ [66]. It is also observed that A-β is strongly related to oxidative damage of lipid membranes, making them more hydrophilic and changing the function of transporters, enzymes, and cellular receptors located on these membranes. In patients with Alzheimer's dementia, there was an increase in the production of 4-hydroxy-2-trans-nonenal (4HNE), one of the main products of lipid peroxidation, causing oxidative modifications of the GLT1 glutamate transporter and, thus, contributing to neurodegeneration [66, 67]. In addition, it is known that 4HNE binds to the GST antioxidant enzyme, causing a reduction in its activity [66]. In fact, previous studies have shown that lipid peroxidation promoted important alterations in the plasmatic neuron membrane of AD patients brains [67].

Dehydroepiandrosterone, Hormonal Derangement, and Dementia

DHEA, What is it?

In humans, DHEA is the most abundantly produced steroid by adrenal glands, and its sulfated form called dehydroepiandrosterone sulfate (DHEAS) is 20-fold that of any other circulating steroid hormone [68]. Steroid hormones derive from cholesterol and are chiefly synthesized in the adrenals, the gonads, and the placenta. The synthesis of steroid hormones requires several sequential enzymatic reactions to convert cholesterol into glucocorticoids, mineralocorticoids, or sexual hormones. Circulating DHEAS works as a DHEA reserve, the conversion isperformed by sulfotransferases found in several tissues [69]. DHEA molecular formula is $C_{19}H_{28}O_2$ and its molecular weight, 288.43 (g/mol). In blood, DHEA's half-life is of approximately 1-3 hours, while that of DHEAS is of 1 - 20 hours [70]. DHEA can be converted into many other metabolites besides DHEAS, depending on the target tissue and the cell type.

Huge amounts of DHEA are synthesized during fetal life as a precursor for the placental estrogen. Production is interrupted after birth and resumed between 5 and 7 years of age [71]. DHEA and DHEAS levels reach their peak between the second and third life decades and then start to decline around 2% every year [72]. In humans, plasmatic DHEA concentrations are of approximately 1-4 ng/mL (0.003-0.015 μmol/L), while the circulating DHEAS concentrations are fairly higher (3 -10 μmol/L) [70].

DHEA as a Neuro-Steroid

DHEA and DHEAS were the first steroids identified in rat brains even after gonadectomy and adrenalectomy [73]. Therefore, these steroids can be synthesized *de novo* (from cholesterol) independently from the peripheral secretion [9, 10]. The expression "neuro-steroid" was then adopted to refer to steroids synthesized both in the CNS and the peripheral nervous system (PNS) by several cell types, including astrocytes and neurons. DHEA is a major neuro-steroid identified in the brain. In the human brain, DHEA concentrations have been proven to be higher than those in circulation, while DHEAS concentrations are lower, which not only reinforces the theory of local synthesis, but also shows the importance of this hormone centrally as a major neuro-steroid.Then, the main DHEA producing cells in the CNS are neurons and astrocytes [74 - 76]. Based on the discoveries of the function of these steroids on the CNS, investigations were developed for the elucidation of the mechanisms and identification of receptors that mediate their biological effects, as well as the participation of these receptors in neurodegenerative, age-related illnesses.

DHEA and DHEAS, regardless of their local synthesis, are neuroactive steroids that modulate the brain functions through a multiplicity of mechanisms. At different regions of the encephalon, DHEA concentrations vary by behavioral and environmental circumstances such as aggressiveness and stress [77]. However, the mechanisms by which DHEA and DHEAS work are not fully understood. Those can operate either by interaction with receptors related to various neurotransmitters systems [10, 78 - 81], by modulating serotonin, dopamine, and norepinephrine release [82, 83], or due to their conversion to other steroid sexual hormones [84].

DHEA's Receptor

Until now, no nuclear receptor specific for DHEA or DHEAS has been unveiled, and no genomic direct effect was reported. However, some effects can be exerted by DHEA when converted into testosterone or dihydrotestosterone, by linking itself to the androgenic receptors, or by turning into estradiol and its connection to the estrogen receptors. In this way, they work in agreement with the classical model of the steroid hormones by linking to cytosolic cognate receptors. This is traditionally called genomic action and answers to this kind of signaling can take several hours [85 - 87]. However, actions deployed both by DHEA and DHEAS are qualified as non-genomic, which do not depend on gene transcription or protein synthesis and imply in the modulation of cytoplasmatic regulatory proteins or on proteins linked to the plasmatic membrane [79, 81, 88]. DHEA and DHEAS modulate different kinds of membrane receptors, such as type A gamma-aminobutyric acid receptor (GABAA), ionotropic glutamatergic receptors N-methyl-D-aspartate (NMDA), sigma-1 (σ1) receptor, type I metabotropic glutamatergic (mGluR5) as well as serotoninergic (5-HT) receptors [79, 81, 88 - 91]. The metabolism of neurotransmitters is an important consideration in the pathology of all neurological diseases, and neurotransmitter augmentation strategies could decrease the behavioral symptoms of dementia, and these medications are in current use in patients [92, 93]. The activation of membrane receptors provides evidence for DHEA and DHEAS quick and non-genomic modulatory actions, reinforcing their beneficial effects for a spectrum of neurological troubles associated with the activation or suppression of these receptors [79, 87, 88].

Hormonal Derangement and Dementia

Dementia is a progressive and terminal illness characterized by impairments in memory, reasoning, thinking, and communication. For people with dementia, the ability to take care of themselves, make decisions and plan for the future on a daily basis degenerates as the disease advances [94]. There are different subtypes of dementia, and AD is the most common one, corresponding to 50-70% of dementia cases in the elderly. Environmental and genetic factors make people susceptible to dementia, and the interaction between the multiple risk factors (*i.e.,* advanced age, female sex, cardiovascular disease, diabetes mellitus type 2, low educational level) still needs elucidation [95].

It is becoming increasingly clear that stress is highly implicated in the development and progression of diseases. Neurodegenerative diseases comprise the detrimental loss of motor and cognitive functions which is stressful and can also disturb the neural circuits that temporize stress responses. The mechanisms that associate stress to changes in Hypothalamic-Pituitary-Adrenal (HPA) axis tone and increased risk for dementia are not fully understood [96]. However, many experimental animal models show that stress increases AD-related pathogenesis. Exposure to stress increases the generation of A-β peptide expression and the expression of Amyloid Precursor Protein (APP) in mice and rats, and both aspects are considered fundamental to AD etiology. Stress not only increases A-β peptide production but also elevates its discharge in amyloid plaques [97 - 100] and precipitates diminution of cognitive performance in animal models of AD [101, 102].

Brown *et al*. (2003) measured the levels of DHEA in AD and age-matched control brains. They have found that DHEA levels are significantly higher in the brain of AD patients(and maximal in the hippocampus of AD patients) [103]. This may reflect increased OS in the AD brain, potentially due to the actions of A-β. DHEA-S has been shown to enhance the *in vitro* release of hippocampal acetylcholine [104], a neurotransmitter likely involved in memory processes and impairment in AD [105].

Concerning the etiology of stress-related illnesses (*i.e.,* depression and anxiety), sex and sex hormones are relevant factors to take into consideration. It is well known that men and women differ in the predisposition of an extensive range of stress-related illnesses such as neurodegenerative, psychiatric, and neurological disorders. However, men have a higher prevalence of ADHD and autism (neurodevelopmental disorders) [106, 107], while women experience a higher prevalence of depression and anxiety (stress-related disorders) in adulthood [108 - 110]. In fact, the relationship between CNS and sex steroids in women and men is a complex and intricate process, which depends mainly on the type and number of steroid receptors and the target cell types [111]. Neurodegenerative diseases (*e.g.,* dementia) implicate the devastating loss of motor and cognitive functions, which is particularly stressful in itself, but it also brings disruption of neural circuits producing atypical emotional　and aggressive behaviors. Indeed, the feed-forward relationship between diseases and stress is a well-known cycle of modern societies [96].

Serious hormonal changes along the female lifespan (puberty, pregnancy, menopause) are accompanied by an increased prevalence of stress-related illnesses [112]. Cortisol is chronically augmented in patients with stress-related illnesses [113, 114], and also cortisol reactivity seems to be modified. The capacity of the HPA-axis to respond to stressors has been named as cortisol awakening response (CAR) in humans, and it represents an adaptive ability to react to stress. Depression, chronic fatigue syndrome, or PSTD can be the result of CAR loss [115]. Normally, studies have shown that CAR is tougher in women than in men due to a slower decline of cortisol [116, 117]. Indeed, depression plays an important role in sex differences related to the physiological stress response.

Apart from sex differences, people who experienced late-life depression showed a two-fold risk of a dementia diagnosis, including AD [118]. Interestingly, the presence of depressive symptoms has been related to low levels of DHEA [119, 120]. Higher cortisol responses were detected in men with depressive symptoms under stressful circumstances, while reduced cortisol responses in women (also related to depressive symptoms) were detected [115, 121]. Corroborating the idea that sex hormones contribute to the stress response difference between women and men, studies indicate that before puberty, girls show an increased cortisol response when compared to boys [117], and puberty seems to be the period when girls have a reduction in cortisol elevations [122, 123]. Additionally, women's menstrual cycle phases appear to temper the sex difference in the cortisol response [124, 125].

DHEA and DHEAS have been shown to have specific functional "anti-glucocorticoid" repercussions (antagonize the actions of glucocorticoids on the CNS) that may equalize the harmful effects of increased cortisol exposure and also a multifaceted neuroprotective role [87, 126]. Animal *in vivo* studies have shown that low DHEA concentrations can exert protective properties against CNS damage related to OS, glutamate, and glucocorticoids [127 - 130]. Using high-resolution magnetic resonance imaging, Jin *et al.* [131] investigated the relationship between hippocampal volume and peripheral measures of cortisol and DHEA in individuals with major depressive disorder. Studies have shown controversial associations between peripheral levels of cortisol and hippocampal volume in neuropsychiatric diagnoses, such as depression [132 - 135]. This cross-sectional study shows that hippocampal volume was significantly negatively correlated with the cortisol/DHEA ratio, suggesting that serum DHEA levels may be more relevant to hippocampal volume in individuals with the major depressive disorder when compared to healthy ones. The molecular mechanisms by which DHEA probably decreases some effects of cortisol are not completely understood. There has been a great deal of interest in the lay and medical press concerning the potential therapeutic uses of DHEA. Notwithstanding, it is paramount to highlight that no well-designed clinical trials have clearly demonstrated the utility and safety of long-term DHEA supplementation yet.

Aging, Oxidative Stress, Dementia, and Dehydroepiandrosterone

Two centuries ago, the average life was around 24 years caused by the high child mortality, precarious hygiene, and the inexistence of efficient treatments for infectious diseases. However, with the development of hygiene principles, the huge advancement of medicine, and abundant food production, the average life expectancy in many developed countries reached 80 years or more [136]. Across the globe, people are living longer, and population aging is increasing dramatically. The population aged 60 years and older is expected to total 2 billion worldwide by 2050. The population aged 80 years or older is estimated to be 125 million today, and there will be almost 434 million people in this age group by 2050 [137]. Concomitantly with the increase in life expectancy, the causes of death were also changed, infectious diseases being progressively outweighed by chronic-neurodegenerative illnesses, like AD and Parkinson's [138].

In the aim of slowing the functional loss associated with aging during an individual's life, several functional approaches are being developed as potential therapies. These include the modulation of metabolic ways of signaling using small molecules, rejuvenation with stem cells, and the elimination of senescent cells accumulated during aging. Among the discussed and proposed strategies aiming at healthy aging are a nutritionally balanced diet and regular physical exercise [139, 140]. Caloric restriction, defined as reduced ingestion of calories without lack of nutrition, can fade aging effects and prolong useful life in various species reducing oxidative damage in distinct types of tissues - particularly of the nervous and cardiovascular systems- increasing life expectancy [141]. At present, a series of new medicines, as well as stem cells treatments and cell reprogramming strategies, are being evaluated for their potential anti-aging effects [142, 143].

One of the pillars of anti-aging clinical practices is the hormonal replacement. Several studies have pointed out aging hormonal dysfunctions as reasons for skin aging, low bone mineral density, lean mass loss, sexual dysfunction, cognitive and motor damages, as well as depression and anxiety symptoms observed in the aged population [144, 145]. This fact stimulated the large-scale use of "hormonal" supplements aiming at reversing aging effects and improving the quality of life of the elderly. However, the use of hormonal therapies is also present in the lives of youngsters and adults to improve physical performance and well-being. Among the hormones associated with benefits for the overall health of the individual is DHEA [87, 145, 146]. DHEA is regulated as a controlled substance in virtually every developed country in the world, with the singular exception of the U.S. Until 1994, likewise other androgens, DHEA was regulated as a controlled substance in the U.S. This was changed by the passage into law of the 'Dietary Supplement Health and Education Act of 1994," whereby DHEA, based on a 'presumption' of safety (upon its apparent lack of toxicity in rats and mice), was exempted from clinical trials. Nevertheless, DHEA is widely available online, as well as from supermarkets and pharmacies in the U.S. DHEA's status as a 'dietary supplement,' particularly used by men, is linked to the belief that its ingestion will offset the decline in DHEAS that naturally occurs with aging. The highly spread self-administration of this hormone is highly questioned because of commercial viability out of the pharmaceutical network, besides the lack of proper and strong scientific evidence in humans [144, 146, 147].

DHEA and Oxidative Stress Mechanisms

ROS generation has been implicated in all types of neurodegenerative diseases [148]. The interaction of DHEA and DHEAS with membrane receptors promotes the activation of intracellular pathways such as mitogen-activated kinases (MAPK), protein kinase A (PKA) and protein kinase B (PKB/Akt), and activation of the transcription factor of the AMPc (CREB) - responsive linkage protein in a culture of hippocampal rat cells and cells of cell lineage derived from the rat suprarenal marrow (PC12) deprived of serum, promoting anti-apoptotic effect [149, 150].

Akt, a serine/threonine kinase, stimulates growth pathways and also inhibits apoptotic cycles. It acts as a survival kinase by phosphorylating many apoptosis-regulatory molecules [151]. The Akt pathways can be activated by moderate levels of H_2O_2, indicating that OS can cause changes in proteins, modulating the regulation of signal transduction [152]. Akt is implicated in neuronal survival, and its activation can be modified in aging according to alterations in signalizing molecules, reduced cellular responses, or a functional decrease in hormonal synthesis [153]. *In vivo* studies (rats) have been demonstrating that DHEA treatment may modulate Akt, and these effects are dose and time-dependent as well as age and tissue-dependent [15, 154 - 156].

Besides, a further mechanism associated with the DHEA effects is the maintenance of the redox balance in hippocampal cells of diabetic rats, promoting inhibition of the chronic activation of the kappa β nuclear transcription factor, reducing its nuclear translocation, and consequently protecting this structure from the deleterious effects of hyperglycemia and the ROS [157]. Fig. (**1**) illustrates the effects of the mechanisms of actions by which DHEA and DHEAS provide their effects.

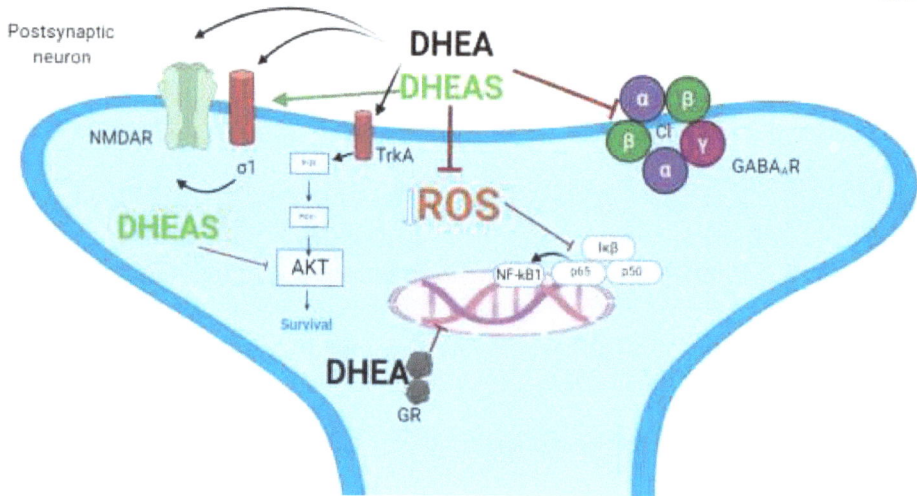

Fig. (1). Mechanisms of action of DHEA and DHEAS in neurons. This image summarizes many of the actions of both hormones described in detail in the text.DHEA and DHEAS have inhibitory effects (red blocking arrow) at the GABAA receptor. DHEA and DHEAS act as agonists (black arrow) at the σ1 receptor, which subsequently may activate the NMDA receptor, and DHEA influences embryonic neurite growth through stimulation (black arrow) of the NMDA receptor. DHEA increases (black arrow) kinase activity of Akt and decreases apoptosis, while DHEAS decreases (red blocking arrow) Akt and increases apoptosis. DHEA and DHEAS inhibit (red blocking arrow) reactive oxygen species (ROS) activation of transcription mediated by NF-κB, and DHEA inhibits (red blocking arrow) nuclear translocation of the glucocorticoid receptor (GR).

Evidence of augmented OS has been seen in the AD brain [158 - 160], contributing to the formation of amyloid plaques and neurofibrillary tangles [161, 162]. The A-β peptide is a possible source of OS (causing increases in ROS *via* distinct mechanisms) in the AD brain and also a component of AD plaques [163]. A-β can activate the microglia to produce free radicals [164, 165] or acquire a free radical state on its own [166].

Over the years, it has been reported that, unlike other neurosteroids, DHEA biosynthesis in rats, bovines, and the human brain is mediated by an OS-mediated mechanism, independent of the cytochrome P450 17α-hydroxylase/17,20-lyase (CYP17A1) enzyme activity found in the periphery. This alternative pathway is induced by pro-oxidant agents, such as Fe^{2+} and A-β peptide [167]. A-β can provoke an increase in oxygen free radicals in cells, and this rise in ROS can modify DHEA levels. Adding $FeSO_4$ to glial cells resulted in a 5-to10-fold increase in DHEA [168]. In human serum, the presence of the DHEA precursor using a Fe^{2+}-based reaction determined the amounts of DHEA formed in 86 subjects [167]. Furthermore, the *in vitro* oxidation of sera from age-matched control individuals led to an augmentation of more than 50% in DHEA levels when compared to respective baseline levels. Such augmentation was significantly higher than that verified in AD patients (14% and 3% increases in the mild and severe AD groups, respectively) measured using the mini-mental state examination. Interestingly, the correlation between the percent DHEA increase in the oxidative pathway and the cognitive impairment seemed to be stronger in women than in men.

DHEA, Aging and Dementia

Contrary to the synthesis of cortisol, which is not affected by aging, the main age-related change in the human supra-renal cortex is the reduction in the DHEA and DHEAS biosynthesis [169,170]. Since in humans DHEA plasmatic concentrations diminish with age, experimental and pre-clinic studies on their biological effects would point out that restoration to levels close to those observed in youth could increase the feeling of well-being, reduce symptoms of depression, protect the brain against age-related damages and increase life expectancy. Therefore, exogenous supplementation with DHEA may bring beneficial effects to the aged population [87, 144, 145, 171, 172].

In rodents, treatment with DHEA and DHEAS is associated with the improvement of memory in old animals. Markowski *et al.* (2001) described the improvement in retention memory and work memory of 18-20-month mice which received DHEAS (1.5 mg/day for 5 days) before the aquatic maze escape test. Animals treated with DHEAS exhibited lower latency to find the aquatic platform [173]. Chen *et al.* (2014) observed that for old male rats of 18-20 months compared with young animals, the plasma testosterone level was reduced by around 75-80%; also, they featured less dendritic arborization and loss of cortical and hippocampal neurons. Besides, when submitted to chronic light stress, aged animals exhibited lower performance for memory and spatial learning. However, treatment with DHEAS (*i.p.* 4 mg/kg for 5 days) reversed the behavioural deficits and partially restored the loss of cortical neurons in these animals. However, no effect on the dendritic shrinking of the affected neurons could be observed [11]. The effect of the chronic treatment with DHEA (8 mg) was also studied for middle-aged male rats (14 months) submitted to chronic light stress for 6 weeks. Aging reduced the number of neuronal cells of the hippocampal dentate gyrus, and the treatment aided in the reduction in the damage. Further, DHEA treatment eased dendritic maturation of new immature hippocampal dentate gyrus neurons, improving neural plasticity even under the reported stressful conditions when submitted to chronic light stress [174]. In another study, *in vitro* treatment with DHEA for 15 minutes on 12-months rats, cortical synaptosomes reduced the basal glutamate release (15% and 20% for DHEA 100nM e 1mM, respectively). The authors suggest that the inhibitory DHEA effect could be linked to its protective role against excitotoxicity caused by overstimulation of the glutamatergic system and aging [175].

Yin *et al.* (2015) observed that the chronic treatment with DHEA (8 mg/kg/day for 8 weeks by gavage) improved the antioxidant profile of various tissues, such as the liver, heart, and brain, and that the treatment mitigated the physiological atrophy of these tissues when compared with tissues of old, non-treated animals. For treated old animals, SOD activity reduction was successfully reversed. These results indicate that DHEA chronic administration, by improving the antioxidant profile of aged rats, could soothe the systemic oxidative effects associated with the aging process [176]. Neuroprotective effects associated with aging and chronic DHEA treatment could also be due to signaling pathways modulation related to cell survival (*e.g.,* Akt pathway). Old animals (24 months) who received a subcutaneous injection of DHEA (10mg/kg once a week for 5 weeks) featured increased Akt hippocampal phosphorylation, as compared with similar, non-treated animals. The treatment effect could be observed for young (3 months of age) animals and was specific to the hippocampus since, in the hypothalamus, DHEA did not intervene in the Akt activation for both ages [154].

For humans, the relationship between the DHEA and DHEAS blood levels and cognitive, anxiolytic, and depressive behavioural aspects has been widely investigated in adult and aged populations [147, 171, 177, 178]. A significant decrease in serum DHEA and DHEAS levels in AD patients, when compared to control patients, is clearly demonstrated in the literature [179 - 185]. For nearly 20 years, reduced DHEA and DHEAS blood levels have also been related to a triggering potential poor cognitive impairment and tendency to neurodegenerative illnesses, including the AD in the elderly, by providing neuroprotective actions against the toxicity of the amyloid protein and reducing lipoperoxidation in human brain tissue [179, 186]. In 2003, Wolkowitz *et al.* observed improved cognition (assessed by the dopamine Evaluation Scale) after supplementation with DHEA (50 mg/day orally/3 months). However, for these individuals' improvement was no longer significant after 6 months of treatment [126]. On the other hand, Yamada *et al.* (2010) demonstrated that DHEA treatment (25 mg/day, for 6 months) in aged women with light to moderate cognitive impairment provided both more autonomy for basic daily tasks and improved cognitive function (verbal fluency) [187].

Hildreth *et al.* (2013), upon evaluating three cognitive domains (operational memory, executive function, and text processing speed) for a group of men (n=49) and women (n=54), of ages 60-88 years having low DHEAS blood concentrations (<3,8 µmol/L, 140 µg/dL), demonstrated a positive association between DHEAS and working memory [177]. However, in 2005, a study containing a huge cohort of old men called the *Massachusetts Male Aging Study* did not identify any significant association between endogenous DHEAS levels and working memory, speed, attention, and spatial memory [188]. In the DHEA and Well-Ness (DAWN) study, performed with more than 200 middle-aged men and women and healthy elderly with no signs of cognitive impairment, oral supplementation for one year with DHEA (50 mg/day) did not confer any beneficial effect on the cognitive function [189].

Maggio *et al.* (2015) screened studies in humans that examined the relationship between DHEA and DHEAS and the cognitive function for both sexes [171]. In spite of the existence of transversal evidence of a positive association between DHEA and DHEAS with cognitive function, these authors reported that longitudinal studies with the oral use of DHEA (50 mg/day) in normal or insane adult individuals led to conflicting reactions and inconsistent results, with no evidence on the utility of the DHEA treatment for cognitive improvement in older individuals.

De Menezes *et al.* (2016), through a clinical research systematic review study, evidenced a positive correlation between DHEAS blood levels with overall cognitive effects (better perception, learning, memory, and attention), both in men and women. However, studies aiming at evaluating the effect of DHEA treatment did not demonstrate improvement in cognitive functions nor worsened their performance. These authors point out that just one study using DHEA (300 mg) in young men featured cognitive improvement [147]. Therefore, before supplementing DHEA for improvement of cognition in clinical practice, more studies are required to assure the safety of the procedure.

The approaches by Maggio *et al.* [171] and De Menezes *et al.* [147] based on systematic clinical studies do not confirm the possibility of adopting DHEA treatment as a supplement for improving cognitive function and well-being of non-insane individuals, being them either middle-aged or older adults. In conclusion, available data lead to the rejection of the hypothesis of such supplementation for improving or maintaining memory or other cognitive functions in older individuals. Despite the recommendation of daily doses of around 10 to 50 mg/day, DHEA is also sold as 100 mg, 200 mg, and even 400 mg/day, dosages of even 1600 mg/day [87, 190] being prescribed. However, Safiulina and colleagues (2006) demonstrated that high DHEA concentrations and their prolonged use can lead to neurotoxic effects. In the primary culture of the cerebellum and cortical granular cells of neonatal rats, DHEA treatment impaired the energetic metabolism, promoting cell death through the mitochondrial respiratory chain I complex inhibition [191]. In this same study, chronic DHEA supplementation in mice promoted neuronal loss of the brain and hippocampal cortex, inducing motor and cognitive deficits. Despite the numerous studies, many questions related to the mechanism of action of these steroids and their relationship with psychiatric troubles or improvement in behaviors associated with memory and motor activity should still be elucidated as relates to dosage, temporal treatment window, and its acute and chronic effects.

CONCLUSION

In aging, successful interventions are synonymous with health preservation and maintenance of useful life, enabling preservation of function, autonomy, and productivity, lessen morbidity, incapacity, and functional deficits throughout life. In spite of uncertainties as for DHEA effects, it has been approved by the USA FDA in 2016 for post-menopausal intravaginal treatment and is also sold in various countries without prescription as an "anti-aging" supplement [144, 146, 147]. DHEA presents a huge variety of pharmacological applications due to potential anti-cancer, anti-allergic, antidiabetic, cardiovascular properties, and also having potential for obesity treatment. It may be beneficial for the treatment of autoimmune disorders like immune modulation, lupus erythematosus, hormonal problems, and muscle building. DHEA is known as an anti-aging hormone in osteoporosis and dementia [192]. However, DHEA as a hormonal treatment requires more studies related to its acute effects on different motor and cognitive aspects of everyday life linked to aging and dementia.

There is no doubt that the literature reviewed here provides evidence that DHEA has significant effects on the CNS. The modulation of cognitive performance is one of the potential effects of DHEA. Unfortunately, clinical research shows a paradox between the measurable, significant results obtained in animals and the vague results obtained in humans. Limitations of translating the results of studies with cells or animals should be recognized. Additionally, there is limited evidence from controlled trials that DHEA ameliorates cognitive function in normal middle-aged or elderly people as well as in individuals with dementia. It is a fact that the theoretical approach presenting beneficial effects of DHEA on cognitive function in elderly people or people suffering from dementia is persuasive. However, the point is that a final clarification of the potential of DHEA in age-related cognitive decline and dementia is necessary, and more studies with demented individuals are fundamental. Extremely few studies are extensive enough to conclude the effects of DHEA on age-related diseases and aging [145].

More than that, it will be important to evaluate deeper acceptable routes, effective dosages, and duration of administration, factors that remain unclear. DHEA is freely sold in health food stores and online worldwide. Future research should be focused on pharmacokinetics and also on the pharmaceutical preparation of DHEA formulations [193].

Worldwide dementia cases will rapidly increase over the next decades due to the increasing life expectancy and the aging of populations. Accordingly, more comprehensive knowledge of dementia should be spread among laypeople using health care services to promote early diagnosis and better support after it. More importantly, the media should share information and strategies to reduce the risk of dementia, encouraging behaviors to reduce lifestyle risk factors. Ideally, a supplement that could be taken regularly without any major side effects capable of reducing oxidative stress and increasing the energy levels of neurons would be a great solution for dementia [194]. DHEA has been reported to have neuroprotective and antioxidant effects, showing promising and potential benefits in prevention and treatments. However, it warrants further investigation to elucidate the specific effects of DHEA (and DHEAS) as a novel therapeutic agent(s) before application in clinical practice regarding neurodegenerative disorders, including AD and other dementias [195].

CONSENT FOR PUBLICATION

Not applicable.

CONFLICT OF INTEREST

The authors declare no conflict of interest, financial or otherwise.

ACKNOWLEDGEMENTS

Declared none.

REFERENCES

[1]   The epidemiology and impact of dementia: current state and future trends. World Health Organization 2019.

[2]   Nitrini R, Barbosa MT, Dozzi Brucki SM, Yassuda MS, Caramelli P. Current trends and challenges on dementia management and research in Latin America. J Glob Health 2020; 10(1): 010362.
[http://dx.doi.org/10.7189/jogh.10.010362] [PMID: 32566153]

[3]   Straub RHML, Miller LE, Schölmerich J, Zietz B. Cytokines and hormones as possible links between endocrinosenescence and immunosenescence. J Neuroimmunol 2000; 109(1): 10-5.
[http://dx.doi.org/10.1016/S0165-5728(00)00296-4] [PMID: 10969175]

[4]   Petersen KSSC. Ageing-associated oxidative stress and inflammation are alleviated by products from grapes. Oxidative Med Cell Longev 2016; pp. (3): 1-12.
[http://dx.doi.org/10.1155/2016/6236309]

[5]   Barrientos RMFM, Frank MG, Watkins LR, Maier SF. Aging-related changes in neuroimmune-endocrine function: implications for hippocampal-dependent cognition. Horm Behav 2012; 62(3): 219-27.
[http://dx.doi.org/10.1016/j.yhbeh.2012.02.010] [PMID: 22370243]

[6]   Liguori I, Russo G, Aran L, *et al.* Sarcopenia: assessment of disease burden and strategies to improve outcomes. Clin Interv Aging 2018; 13: 913-27.
[http://dx.doi.org/10.2147/CIA.S149232] [PMID: 29785098]

[7]   Singh E, Devasahayam G. Neurodegeneration by oxidative stress: a review on prospective use of small molecules for neuroprotection. Mol Biol Rep 2020; 47(4): 3133-40.
[http://dx.doi.org/10.1007/s11033-020-05354-1] [PMID: 32162127]

[8]   Bastianetto S, Ramassamy C, Poirier J, Quirion R. Dehydroepiandrosterone (DHEA) protects hippocampal cells from oxidative stress-induced damage. Brain Res Mol Brain Res 1999; 66(1-2): 35-41.
[http://dx.doi.org/10.1016/S0169-328X(99)00002-9] [PMID: 10095075]

[9]   Baulieu EE. Neurosteroids: of the nervous system, by the nervous system, for the nervous system. Recent Prog Horm Res 1997; 52: 1-32.
[PMID: 9238846]

[10]   Baulieu EE. Neuroactive neurosteroids: dehydroepiandrosterone (DHEA) and DHEA sulphate. Acta Paediatr Suppl 1999; 88(433): 78-80.
[http://dx.doi.org/10.1111/j.1651-2227.1999.tb14408.x] [PMID: 10626550]

[11]   Chen JR, Tseng GF, Wang YJ, Wang TJ. Exogenous dehydroisoandrosterone sulfate reverses the dendritic changes of the central neurons in aging male rats. Exp Gerontol 2014; 57: 191-202.
[http://dx.doi.org/10.1016/j.exger.2014.06.010] [PMID: 24929010]

[12]   Ding X, Wang D, Li L, Ma H. Dehydroepiandrosterone ameliorates H_2O_2-induced Leydig cells oxidation damage and apoptosis through inhibition of ROS production and activation of PI3K/Akt pathways. Int J Biochem Cell Biol 2016; 70: 126-39.
[http://dx.doi.org/10.1016/j.biocel.2015.11.018] [PMID: 26643608]

[13]   Li L, Yao Y, Jiang Z, Zhao J, Cao J, Ma H. Dehydroepiandrosterone prevents H_2O_2-Induced BRL-3A cell oxidative damage through activation of PI3K/Akt pathways rather than MAPK pathways. Oxid Med Cell Longev 2019; 2019: 2985956.

[14]   Abdelazeim SA, Shehata NI, Aly HF, Shams SGE. Amelioration of oxidative stress-mediated apoptosis in copper oxide nanoparticles-induced liver injury in rats by potent antioxidants. Sci Rep 2020; 10(1): 10812.
[http://dx.doi.org/10.1038/s41598-020-67784-y] [PMID: 32616881]

[15]   Jacob MHVM, Janner D da R, MP Jahn, LC Kucharski, Klein A Belló, MFM Ribeiro. Age related effects of DHEA on peripheral markers of oxidative stress. Cell Biochemistry and Function: Cellular biochemistry and its modulation by active agents or disease. 2010; 28(1): 52-7.

[16]   Jacob MHVM, Fernandes RO, Bonetto JHP, *et al.* DHEA treatment effects on redox environment in skeletal muscle of young and aged healthy rats. Curr Aging Sci 2018; 11(2): 126-32.
[http://dx.doi.org/10.2174/1874609811666180803125723] [PMID: 30073935]

[17]   Goldfarb AH, McIntosh MK, Boyer BT. Vitamin E attenuates myocardial oxidative stress induced by DHEA in rested and exercised rats. J Appl Physiol 1996; 80(2): 486-90.
[http://dx.doi.org/10.1152/jappl.1996.80.2.486] [PMID: 8929588]

[18]   Mastrocola R, Aragno M, Betteto S, *et al.* Pro-oxidant effect of dehydroepiandrosterone in rats is mediated by PPAR activation. Life Sci 2003; 73(3): 289-99.
[http://dx.doi.org/10.1016/S0024-3205(03)00287-X] [PMID: 12757836]

[19]   Schmeer C, Kretz A, Wengerodt D, Stojiljkovic M, Witte OW. Dissecting aging and senescence-current concepts and open lessons. Cells 2019; 8(11): 1446.

[http://dx.doi.org/10.3390/cells8111446] [PMID: 31731770]

[20]   da Costa JP, Vitorino R, Silva GM, Vogel C, Duarte AC, Rocha-Santos T. A synopsis on aging-Theories, mechanisms and future prospects. Ageing Res Rev 2016; 29: 90-112.
[http://dx.doi.org/10.1016/j.arr.2016.06.005] [PMID: 27353257]

[21]   Sohal RS, Orr WC. The redox stress hypothesis of aging. Free Radic Biol Med 2012; 52(3): 539-55.
[http://dx.doi.org/10.1016/j.freeradbiomed.2011.10.445] [PMID: 22080087]

[22]   Zhang H, Davies KJA, Forman HJ. Oxidative stress response and Nrf2 signaling in aging. Free Radic Biol Med 2015; 88(Pt B): 314-36.
[http://dx.doi.org/10.1016/j.freeradbiomed.2015.05.036] [PMID: 26066302]

[23]   Mecocci P, Boccardi V, Cecchetti R, *et al.* A long journey into aging, brain aging, and Alzheimer's disease following the oxidative stress tracks. J Alzheimers Dis 2018; 62(3): 1319-35.
[http://dx.doi.org/10.3233/JAD-170732] [PMID: 29562533]

[24]   O'Brien RJ, Wong PC. Amyloid precursor protein processing and Alzheimer's disease. Annu Rev Neurosci 2011; 34: 185-204.
[http://dx.doi.org/10.1146/annurev-neuro-061010-113613] [PMID: 21456963]

[25]   Wyss-Coray T. Ageing, neurodegeneration and brain rejuvenation. Nature 2016; 539(7628): 180-6.
[http://dx.doi.org/10.1038/nature20411] [PMID: 27830812]

[26]   Laurent C, Buée L, Blum D. Tau and neuroinflammation: What impact for Alzheimer's disease and tauopathies? Biomed J 2018; 41(1): 21-33.
[http://dx.doi.org/10.1016/j.bj.2018.01.003] [PMID: 29673549]

[27]   Tönnies E, Trushina E. Oxidative stress, synaptic dysfunction, and Alzheimer's disease. J Alzheimers Dis 2017; 57(4): 1105-21.
[http://dx.doi.org/10.3233/JAD-161088] [PMID: 28059794]

[28]   Sies H. Oxidative stress. Academic Press 1985.

[29]   Sies H. Hydrogen peroxide as a central redox signaling molecule in physiological oxidative stress: Oxidative eustress. Redox Biol 2017; 11: 613-9.
[http://dx.doi.org/10.1016/j.redox.2016.12.035] [PMID: 28110218]

[30]   Scheff SW, Ansari MA, Mufson EJ. Oxidative stress and hippocampal synaptic protein levels in elderly cognitively intact individuals with Alzheimer's disease pathology. Neurobiol Aging 2016; 42: 1-12.
[http://dx.doi.org/10.1016/j.neurobiolaging.2016.02.030] [PMID: 27143416]

[31]   Aoyama K, Watabe M, Nakaki T. Regulation of neuronal glutathione synthesis. J Pharmacol Sci 2008; 108(3): 227-38.
[http://dx.doi.org/10.1254/jphs.08R01CR] [PMID: 19008644]

[32]   Salminen LE, Paul RH. Oxidative stress and genetic markers of suboptimal antioxidant defense in the aging brain: a theoretical review. Rev Neurosci 2014; 25(6): 805-19.
[http://dx.doi.org/10.1515/revneuro-2014-0046] [PMID: 25153586]

[33]   Rooijackers HM, Wiegers EC, Tack CJ, van der Graaf M, de Galan BE. Brain glucose metabolism during hypoglycemia in type 1 diabetes: insights from functional and metabolic neuroimaging studies. Cell Mol Life Sci 2016; 73(4): 705-22.
[http://dx.doi.org/10.1007/s00018-015-2079-8] [PMID: 26521082]

[34]   Lieberman M, Peet A. Marks' basic medical biochemistry: a clinical approach: Wolters Kluwer 2018.

[35]   Dröge W. Free radicals in the physiological control of cell function. Physiol Rev 2002; 82(1): 47-95.
[http://dx.doi.org/10.1152/physrev.00018.2001] [PMID: 11773609]

[36]   Singh A, Kukreti R, Saso L, Kukreti S. Oxidative stress: a key modulator in neurodegenerative diseases. Molecules 2019; 24(8): 1583.
[http://dx.doi.org/10.3390/molecules24081583] [PMID: 31013638]

[37] Dizdaroglu M. Chemical determination of free radical-induced damage to DNA. Free Radic Biol Med 1991; 10(3-4): 225-42.
[http://dx.doi.org/10.1016/0891-5849(91)90080-M] [PMID: 1650738]

[38] Lee KH, Cha M, Lee BH. Neuroprotective Effect of Antioxidants in the Brain. Int J Mol Sci 2020; 21(19): 7152.
[http://dx.doi.org/10.3390/ijms21197152] [PMID: 32998277]

[39] Pisoschi AM, Pop A. The role of antioxidants in the chemistry of oxidative stress: A review. Eur J Med Chem 2015; 97: 55-74.
[http://dx.doi.org/10.1016/j.ejmech.2015.04.040] [PMID: 25942353]

[40] Iadecola C, Zhang F, Niwa K, *et al.* SOD1 rescues cerebral endothelial dysfunction in mice overexpressing amyloid precursor protein. Nat Neurosci 1999; 2(2): 157-61.
[http://dx.doi.org/10.1038/5715] [PMID: 10195200]

[41] Nell HJ, Au JL, Giordano CR, *et al.* Targeted antioxidant, catalase-SKL, reduces beta-amyloid toxicity in the rat brain. Brain Pathol 2017; 27(1): 86-94.
[http://dx.doi.org/10.1111/bpa.12368] [PMID: 26919450]

[42] Taylor JM, Ali U, Iannello RC, Hertzog P, Crack PJ. Diminished Akt phosphorylation in neurons lacking glutathione peroxidase-1 (Gpx1) leads to increased susceptibility to oxidative stress-induced cell death. J Neurochem 2005; 92(2): 283-93.
[http://dx.doi.org/10.1111/j.1471-4159.2004.02863.x] [PMID: 15663476]

[43] Dasari S, Gonuguntla S, Ganjayi MS, Bukke S, Sreenivasulu B, Meriga B. Genetic polymorphism of glutathione S-transferases: Relevance to neurological disorders. Pathophysiology 2018; 25(4): 285-92.
[http://dx.doi.org/10.1016/j.pathophys.2018.06.001] [PMID: 29908890]

[44] Lillig CH, Berndt C, Holmgren A. Glutaredoxin systems. Biochimica et Biophysica Acta (BBA)-. General Subjects 2008; 1780(11): 1304-17.
[http://dx.doi.org/10.1016/j.bbagen.2008.06.003]

[45] Lee S, Kim SM, Lee RT. Thioredoxin and thioredoxin target proteins: from molecular mechanisms to functional significance. Antioxid Redox Signal 2013; 18(10): 1165-207.
[http://dx.doi.org/10.1089/ars.2011.4322] [PMID: 22607099]

[46] Silva-Adaya D, Gonsebatt ME, Guevara J. Thioredoxin system regulation in the central nervous system: experimental models and clinical evidence. Oxid Med Cell Longev 2014 2014; 590808.
[http://dx.doi.org/10.1155/2014/590808]

[47] Lovell MA, Xie C, Gabbita SP, Markesbery WR. Decreased thioredoxin and increased thioredoxin reductase levels in Alzheimer's disease brain. Free Radic Biol Med 2000; 28(3): 418-27.
[http://dx.doi.org/10.1016/S0891-5849(99)00258-0] [PMID: 10699754]

[48] Ren X, Zou L, Zhang X, *et al.* Redox signaling mediated by thioredoxin and glutathione systems in the central nervous system. Antioxid Redox Signal 2017; 27(13): 989-1010.
[http://dx.doi.org/10.1089/ars.2016.6925] [PMID: 28443683]

[49] Patel M. Targeting oxidative stress in central nervous system disorders. Trends Pharmacol Sci 2016; 37(9): 768-78.
[http://dx.doi.org/10.1016/j.tips.2016.06.007] [PMID: 27491897]

[50] Cheng L, Zhang H, Wu F, Liu Z, Cheng Y, Wang C. Role of Nrf2 and its activators in cardiocerebral vascular disease. Oxid Med Cell Longev 2014; 2014. 590808

[51] Kaźmierczak-Barańska J, Boguszewska K, Karwowski BT. Nutrition Can Help DNA Repair in the Case of Aging. Nutrients 2020; 12(11): 3364.
[http://dx.doi.org/10.3390/nu12113364] [PMID: 33139613]

[52] Concetta Scuto M, Mancuso C, Tomasello B, *et al.* Curcumin, hormesis and the nervous system. Nutrients 2019; 11(10): 2417.

[http://dx.doi.org/10.3390/nu11102417] [PMID: 31658697]

[53] Li S, Selkoe DJ. A mechanistic hypothesis for the impairment of synaptic plasticity by soluble Aβ oligomers from Alzheimer's brain. J Neurochem 2020; 154(6): 583-97.
[http://dx.doi.org/10.1111/jnc.15007] [PMID: 32180217]

[54] Paschen W, Frandsen A. Endoplasmic reticulum dysfunction--a common denominator for cell injury in acute and degenerative diseases of the brain? J Neurochem 2001; 79(4): 719-25.
[http://dx.doi.org/10.1046/j.1471-4159.2001.00623.x] [PMID: 11723164]

[55] Wang X, Wang W, Li L, Perry G, Lee HG, Zhu X. Oxidative stress and mitochondrial dysfunction in Alzheimer's disease. Biochim Biophys Acta 2014; 1842(8): 1240-7.
[http://dx.doi.org/10.1016/j.bbadis.2013.10.015] [PMID: 24189435]

[56] Mattson MP. Neuronal life-and-death signaling, apoptosis, and neurodegenerative disorders. Antioxid Redox Signal 2006; 8(11-12): 1997-2006.
[http://dx.doi.org/10.1089/ars.2006.8.1997] [PMID: 17034345]

[57] Butterfield DA, Boyd-Kimball D. Mitochondrial oxidative and nitrosative stress and alzheimer disease. Antioxidants 2020; 9(9): 818.
[http://dx.doi.org/10.3390/antiox9090818] [PMID: 32887505]

[58] van der Bliek AM, Sedensky MM, Morgan PG. Cell biology of the mitochondrion. Genetics 2017; 207(3): 843-71.
[http://dx.doi.org/10.1534/genetics.117.300262] [PMID: 29097398]

[59] Liu X, Kim CN, Yang J, Jemmerson R, Wang X. Induction of apoptotic program in cell-free extracts: requirement for dATP and cytochrome c. Cell 1996; 86(1): 147-57.
[http://dx.doi.org/10.1016/S0092-8674(00)80085-9] [PMID: 8689682]

[60] Blennow K, de Leon MJ, Zetterberg H. Alzheimer's disease. Lancet 2006; 368(9533): 387-403.
[http://dx.doi.org/10.1016/S0140-6736(06)69113-7] [PMID: 16876668]

[61] Katzman R, Saitoh T. Advances in Alzheimer's disease. FASEB J 1991; 5(3): 278-86.
[http://dx.doi.org/10.1096/fasebj.5.3.2001787] [PMID: 2001787]

[62] Nelson PT, Braak H, Markesbery WR. Neuropathology and cognitive impairment in Alzheimer disease: a complex but coherent relationship. J Neuropathol Exp Neurol 2009; 68(1): 1-14.
[http://dx.doi.org/10.1097/NEN.0b013e3181919a48] [PMID: 19104448]

[63] Mega MS. Differential diagnosis of dementia: clinical examination and laboratory assessment. Clin Cornerstone 2002; 4(6): 53-65.
[http://dx.doi.org/10.1016/S1098-3597(02)90036-0] [PMID: 12739331]

[64] Panza F, Lozupone M, Logroscino G, Imbimbo BP. A critical appraisal of amyloid-β-targeting therapies for Alzheimer disease. Nat Rev Neurol 2019; 15(2): 73-88.
[http://dx.doi.org/10.1038/s41582-018-0116-6] [PMID: 30610216]

[65] Finefrock AE, Bush AI, Doraiswamy PM. Current status of metals as therapeutic targets in Alzheimer's disease. J Am Geriatr Soc 2003; 51(8): 1143-8.
[http://dx.doi.org/10.1046/j.1532-5415.2003.51368.x] [PMID: 12890080]

[66] Pérez MA, Arancibia SR. Estrés oxidativo y neurodegeneración:¿ causa o consecuencia. Arch Neurocien (Mex) 2007; 12(1): 45-54.

[67] Reed T, Perluigi M, Sultana R, *et al.* Redox proteomic identification of 4-hydroxy-2-nonenal-modified brain proteins in amnestic mild cognitive impairment: insight into the role of lipid peroxidation in the progression and pathogenesis of Alzheimer's disease. Neurobiol Dis 2008; 30(1): 107-20.
[http://dx.doi.org/10.1016/j.nbd.2007.12.007] [PMID: 18325775]

[68] Nakamura S, Yoshimura M, Nakayama M, *et al.* Possible association of heart failure status with synthetic balance between aldosterone and dehydroepiandrosterone in human heart. Circulation 2004; 110(13): 1787-93.

[http://dx.doi.org/10.1161/01.CIR.0000143072.36782.51] [PMID: 15364798]

[69] Komesaroff PA. Unravelling the enigma of dehydroepiandrosterone: moving forward step by step. Endocrinology 2008; 149(3): 886-8.
[http://dx.doi.org/10.1210/en.2007-1787] [PMID: 18292199]

[70] Webb SJ, Geoghegan TE, Prough RA, Michael Miller KK. The biological actions of dehydroepiandrosterone involves multiple receptors. Drug Metab Rev 2006; 38(1-2): 89-116.
[http://dx.doi.org/10.1080/03602530600569877] [PMID: 16684650]

[71] Nippoldt TB, Nair KS. Is there a case for DHEA replacement? Baillieres Clin Endocrinol Metab 1998; 12(3): 507-20.
[http://dx.doi.org/10.1016/S0950-351X(98)80286-3] [PMID: 10332570]

[72] Genazzani AD, Lanzoni C, Genazzani AR. Might DHEA be considered a beneficial replacement therapy in the elderly? Drugs Aging 2007; 24(3): 173-85.
[http://dx.doi.org/10.2165/00002512-200724030-00001] [PMID: 17362047]

[73] Corpéchot C, Robel P, Axelson M, Sjövall J, Baulieu E-E. Characterization and measurement of dehydroepiandrosterone sulfate in rat brain. Proc Natl Acad Sci USA 1981; 78(8): 4704-7.
[http://dx.doi.org/10.1073/pnas.78.8.4704] [PMID: 6458035]

[74] Plassart-Schiess E, Baulieu E-E. Neurosteroids: recent findings. Brain Res Brain Res Rev 2001; 37(1-3): 133-40.
[http://dx.doi.org/10.1016/S0165-0173(01)00113-8] [PMID: 11744081]

[75] Schumacher M, Akwa Y, Guennoun R, *et al.* Steroid synthesis and metabolism in the nervous system: trophic and protective effects. J Neurocytol 2000; 29(5-6): 307-26.
[http://dx.doi.org/10.1023/A:1007152904926] [PMID: 11424948]

[76] Zwain IH, Yen SS. Neurosteroidogenesis in astrocytes, oligodendrocytes, and neurons of cerebral cortex of rat brain. Endocrinology 1999; 140(8): 3843-52.
[http://dx.doi.org/10.1210/endo.140.8.6907] [PMID: 10433246]

[77] Baulieu EE, Robel P, Schumacher M. Neurosteroids: beginning of the story. Int Rev Neurobiol 2001; 46(46): 1-32.
[PMID: 11599297]

[78] Majewska MD, Demirgören S, Spivak CE, London ED. The neurosteroid dehydroepiandrosterone sulfate is an allosteric antagonist of the GABAA receptor. Brain Res 1990; 526(1): 143-6.
[http://dx.doi.org/10.1016/0006-8993(90)90261-9] [PMID: 1964106]

[79] Hill M, Dušková M, Stárka L. Dehydroepiandrosterone, its metabolites and ion channels. J Steroid Biochem Mol Biol 2015; 145: 293-314.
[http://dx.doi.org/10.1016/j.jsbmb.2014.05.006] [PMID: 24846830]

[80] Dong Y, Zheng P. Dehydroepiandrosterone sulphate: action and mechanism in the brain. J Neuroendocrinol 2012; 24(1): 215-24.
[http://dx.doi.org/10.1111/j.1365-2826.2011.02256.x] [PMID: 22145821]

[81] Prough RA, Clark BJ, Klinge CM. Novel mechanisms for DHEA action. J Mol Endocrinol 2016; 56(3): R139-55.
[http://dx.doi.org/10.1530/JME-16-0013] [PMID: 26908835]

[82] Yu Q, Hao S, Wang H, Song X, Shen Q, Kang J. Depression-like behavior in a dehydroepiandrosterone-induced mouse model of polycystic ovary syndrome. Biology of reproduction 2016; 95(4): 79-10.
[http://dx.doi.org/10.1095/biolreprod.116.142117]

[83] Zajda ME, Krzascik P, Hill M, Majewska MD. Psychomotor and rewarding properties of the neurosteroids dehydroepiandrosterone sulphate and androsterone: effects on monoamine and steroid metabolism. Acta Neurobiol Exp (Warsz) 2012; 72(1): 65-79.
[PMID: 22508085]

[84]   Arlt W, Callies F, Allolio B. DHEA replacement in women with adrenal insufficiency--pharmacokinetics, bioconversion and clinical effects on well-being, sexuality and cognition. Endocr Res 2000; 26(4): 505-11.
[http://dx.doi.org/10.3109/07435800009048561] [PMID: 11196420]

[85]   Mellon SH, Griffin LD, Compagnone NA. Biosynthesis and action of neurosteroids. Brain Res Brain Res Rev 2001; 37(1-3): 3-12.
[http://dx.doi.org/10.1016/S0165-0173(01)00109-6] [PMID: 11744070]

[86]   DHEA, important source of sex steroids in men and even more in women. Labrie F. Progress in brain research 2010; 182: pp. 97-148.
[http://dx.doi.org/10.1016/S0079-6123(10)82004-7]

[87]   Maninger N, Wolkowitz OM, Reus VI, Epel ES, Mellon SH. Neurobiological and neuropsychiatric effects of dehydroepiandrosterone (DHEA) and DHEA sulfate (DHEAS). Front Neuroendocrinol 2009; 30(1): 65-91.
[http://dx.doi.org/10.1016/j.yfrne.2008.11.002] [PMID: 19063914]

[88]   Stárka L, Dušková M, Hill M. Dehydroepiandrosterone: a neuroactive steroid. J Steroid Biochem Mol Biol 2015; 145: 254-60.
[http://dx.doi.org/10.1016/j.jsbmb.2014.03.008] [PMID: 24704258]

[89]   Dubrovsky BO. Steroids, neuroactive steroids and neurosteroids in psychopathology. Prog Neuropsychopharmacol Biol Psychiatry 2005; 29(2): 169-92.
[http://dx.doi.org/10.1016/j.pnpbp.2004.11.001] [PMID: 15694225]

[90]   Xu Y, Tanaka M, Chen L, Sokabe M. DHEAS induces short-term potentiation *via* the activation of a metabotropic glutamate receptor in the rat hippocampus. Hippocampus 2012; 22(4): 707-22.
[http://dx.doi.org/10.1002/hipo.20932] [PMID: 21484933]

[91]   Moriguchi S, Shinoda Y, Yamamoto Y, *et al.* Stimulation of the sigma-1 receptor by DHEA enhances synaptic efficacy and neurogenesis in the hippocampal dentate gyrus of olfactory bulbectomized mice. PLoS One 2013; 8(4): e60863.
[http://dx.doi.org/10.1371/journal.pone.0060863] [PMID: 23593332]

[92]   Huey ED, Putnam KT, Grafman J. A systematic review of neurotransmitter deficits and treatments in frontotemporal dementia. Neurology 2006; 66(1): 17-22.
[http://dx.doi.org/10.1212/01.wnl.0000191304.55196.4d] [PMID: 16401839]

[93]   Snowden SG, Ebshiana AA, Hye A, *et al.* Neurotransmitter imbalance in the brain and alzheimer's disease pathology. J Alzheimers Dis 2019; 72(1): 35-43.
[http://dx.doi.org/10.3233/JAD-190577] [PMID: 31561368]

[94]   Downs M, Bowers B. Excellence in dementia care: Research into practice: McGraw-Hill Education (UK). 2014.

[95]   Lewczuk P, Riederer P, O'Bryant SE, *et al.* Cerebrospinal fluid and blood biomarkers for neurodegenerative dementias: An update of the Consensus of the Task Force on Biological Markers in Psychiatry of the World Federation of Societies of Biological Psychiatry. World J Biol Psychiatry 2018; 19(4): 244-328.
[http://dx.doi.org/10.1080/15622975.2017.1375556] [PMID: 29076399]

[96]   Justice NJ. The relationship between stress and Alzheimer's disease. Neurobiol Stress 2018; 8: 127-33.
[http://dx.doi.org/10.1016/j.ynstr.2018.04.002] [PMID: 29888308]

[97]   Rothman SM, Herdener N, Camandola S, Texel SJ, Mughal MR. 3xTgAD mice exhibit altered behavior and elevated Aβ after chronic mild social stress. Neurobiology of aging 2012; 33(4): 830.

[98]   Baglietto-Vargas D, Chen Y, Suh D, *et al.* Short-term modern life-like stress exacerbates Aβ-pathology and synapse loss in 3xTg-AD mice. J Neurochem 2015; 134(5): 915-26.
[http://dx.doi.org/10.1111/jnc.13195] [PMID: 26077803]

[99] Justice NJ, Huang L, Tian J-B, *et al.* Posttraumatic stress disorder-like induction elevates β-amyloid levels, which directly activates corticotropin-releasing factor neurons to exacerbate stress responses. J Neurosci 2015; 35(6): 2612-23.
[http://dx.doi.org/10.1523/JNEUROSCI.3333-14.2015] [PMID: 25673853]

[100] Lesuis SL, Maurin H, Borghgraef P, Lucassen PJ, Van Leuven F, Krugers HJ. Positive and negative early life experiences differentially modulate long term survival and amyloid protein levels in a mouse model of Alzheimer's disease. Oncotarget 2016; 7(26): 39118-35.
[http://dx.doi.org/10.18632/oncotarget.9776] [PMID: 27259247]

[101] Han B, Yu L, Geng Y, *et al.* Chronic stress aggravates cognitive impairment and suppresses insulin associated signaling pathway in APP/PS1 mice. J Alzheimers Dis 2016; 53(4): 1539-52.
[http://dx.doi.org/10.3233/JAD-160189] [PMID: 27392857]

[102] Han B, Wang J-H, Geng Y, *et al.* Chronic stress contributes to cognitive dysfunction and hippocampal metabolic abnormalities in APP/PS1 mice. Cell Physiol Biochem 2017; 41(5): 1766-76.
[http://dx.doi.org/10.1159/000471869] [PMID: 28365686]

[103] Brown RC, Han Z, Cascio C, Papadopoulos V. Oxidative stress-mediated DHEA formation in Alzheimer's disease pathology. Neurobiol Aging 2003; 24(1): 57-65.
[http://dx.doi.org/10.1016/S0197-4580(02)00048-9] [PMID: 12493551]

[104] Rhodes ME, Li P-K, Burke AM, Johnson DA. Enhanced plasma DHEAS, brain acetylcholine and memory mediated by steroid sulfatase inhibition. Brain Res 1997; 773(1-2): 28-32.
[http://dx.doi.org/10.1016/S0006-8993(97)00867-6] [PMID: 9409701]

[105] Kása P, Rakonczay Z, Gulya K. The cholinergic system in Alzheimer's disease. Prog Neurobiol 1997; 52(6): 511-35.
[http://dx.doi.org/10.1016/S0301-0082(97)00028-2] [PMID: 9316159]

[106] Faraone SV. Attention deficit hyperactivity disorder and premature death. Lancet 2015; 385(9983): 2132-3.
[http://dx.doi.org/10.1016/S0140-6736(14)61822-5]

[107] Schuck RK, Flores RE, Fung LK. Brief report: Sex/gender differences in symptomology and camouflaging in adults with autism spectrum disorder. J Autism Dev Disord 2019; 49(6): 2597-604.
[http://dx.doi.org/10.1007/s10803-019-03998-y] [PMID: 30945091]

[108] Irvine K, Laws KR, Gale TM, Kondel TK. Greater cognitive deterioration in women than men with Alzheimer's disease: a meta analysis. J Clin Exp Neuropsychol 2012; 34(9): 989-98.
[http://dx.doi.org/10.1080/13803395.2012.712676] [PMID: 22913619]

[109] Albert PR. Why is depression more prevalent in women? J Psychiatry Neurosci 2015; 40(4): 219-21.
[http://dx.doi.org/10.1503/jpn.150205] [PMID: 26107348]

[110] Olff M. Sex and gender differences in post-traumatic stress disorder: an update. Eur J Psychotraumatol 2017; 8(sup4): 1351204.
[http://dx.doi.org/10.1080/20008198.2017.1351204]

[111] Vegeto E, Villa A, Della Torre S, *et al.* The role of sex and sex Hormones in Neurodegenerative Diseases. Endocr Rev 2020; 41(2): 273-319.
[http://dx.doi.org/10.1210/endrev/bnz005] [PMID: 31544208]

[112] Slavich GM, Sacher J. Stress, sex hormones, inflammation, and major depressive disorder: Extending Social Signal Transduction Theory of Depression to account for sex differences in mood disorders. Psychopharmacology (Berl) 2019; 236(10): 3063-79.
[http://dx.doi.org/10.1007/s00213-019-05326-9] [PMID: 31359117]

[113] Keller J, Gomez R, Williams G, *et al.* HPA axis in major depression: cortisol, clinical symptomatology and genetic variation predict cognition. Mol Psychiatry 2017; 22(4): 527-36.
[http://dx.doi.org/10.1038/mp.2016.120] [PMID: 27528460]

[114] Szeszko PR, Lehrner A, Yehuda R. Glucocorticoids and hippocampal structure and function in PTSD. Harv Rev Psychiatry 2018; 26(3): 142-57.
[http://dx.doi.org/10.1097/HRP.0000000000000188] [PMID: 29734228]

[115] Zorn JV, Schür RR, Boks MP, Kahn RS, Joëls M, Vinkers CH. Cortisol stress reactivity across psychiatric disorders: A systematic review and meta-analysis. Psychoneuroendocrinology 2017; 77: 25-36.
[http://dx.doi.org/10.1016/j.psyneuen.2016.11.036] [PMID: 28012291]

[116] Wüst S, Wolf J, Hellhammer DH, Federenko I, Schommer N, Kirschbaum C. The cortisol awakening response - normal values and confounds. Noise Health 2000; 2(7): 79-88.
[PMID: 12689474]

[117] Hollanders JJ, van der Voorn B, Rotteveel J, Finken MJJ. Is HPA axis reactivity in childhood gender-specific? A systematic review. Biol Sex Differ 2017; 8(1): 23.
[http://dx.doi.org/10.1186/s13293-017-0144-8] [PMID: 28693541]

[118] Barnes DE, Yaffe K, Byers AL, McCormick M, Schaefer C, Whitmer RA. Midlife *vs* late-life depressive symptoms and risk of dementia: differential effects for Alzheimer disease and vascular dementia. Arch Gen Psychiatry 2012; 69(5): 493-8.
[http://dx.doi.org/10.1001/archgenpsychiatry.2011.1481] [PMID: 22566581]

[119] Barrett-Connor E, von Mühlen D, Laughlin GA, Kripke A. Endogenous levels of dehydroepiandrosterone sulfate, but not other sex hormones, are associated with depressed mood in older women: the Rancho Bernardo Study. J Am Geriatr Soc 1999; 47(6): 685-91.
[http://dx.doi.org/10.1111/j.1532-5415.1999.tb01590.x] [PMID: 10366167]

[120] Michael A, Jenaway A, Paykel ES, Herbert J. Altered salivary dehydroepiandrosterone levels in major depression in adults. Biol Psychiatry 2000; 48(10): 989-95.
[http://dx.doi.org/10.1016/S0006-3223(00)00955-0] [PMID: 11082473]

[121] Powers SI, Laurent HK, Gunlicks-Stoessel M, Balaban S, Bent E. Depression and anxiety predict sex-specific cortisol responses to interpersonal stress. Psychoneuroendocrinology 2016; 69: 172-9.
[http://dx.doi.org/10.1016/j.psyneuen.2016.04.007] [PMID: 27107208]

[122] Cameron CA, McKay S, Susman EJ, Wynne-Edwards K, Wright JM, Weinberg J. Cortisol stress response variability in early adolescence: Attachment, affect and sex. J Youth Adolesc 2017; 46(1): 104-20.
[http://dx.doi.org/10.1007/s10964-016-0548-5] [PMID: 27468997]

[123] Stroud LR, Papandonatos GD, D'Angelo CM, Brush B, Lloyd-Richardson EE. Sex differences in biological response to peer rejection and performance challenge across development: A pilot study. Physiol Behav 2017; 169: 224-33.
[http://dx.doi.org/10.1016/j.physbeh.2016.12.005] [PMID: 27939363]

[124] Stephens MAC, Mahon PB, McCaul ME, Wand GS. Hypothalamic-pituitary-adrenal axis response to acute psychosocial stress: Effects of biological sex and circulating sex hormones. Psychoneuroendocrinology 2016; 66: 47-55.
[http://dx.doi.org/10.1016/j.psyneuen.2015.12.021] [PMID: 26773400]

[125] Montero-López E, Santos-Ruiz A, García-Ríos MC, Rodríguez-Blázquez M, Rogers HL, Peralta-Ramírez MI. The relationship between the menstrual cycle and cortisol secretion: Daily and stress-invoked cortisol patterns. Int J Psychophysiol 2018; 131: 67-72.
[http://dx.doi.org/10.1016/j.ijpsycho.2018.03.021] [PMID: 29605399]

[126] Wolkowitz OM, Kramer JH, Reus VI, *et al.* DHEA treatment of Alzheimer's disease: a randomized, double-blind, placebo-controlled study. Neurology 2003; 60(7): 1071-6.
[http://dx.doi.org/10.1212/01.WNL.0000052994.54660.58] [PMID: 12682308]

[127] Li H, Klein G, Sun P, Buchan AM. Dehydroepiandrosterone (DHEA) reduces neuronal injury in a rat model of global cerebral ischemia. Brain Res 2001; 888(2): 263-6.

[http://dx.doi.org/10.1016/S0006-8993(00)03077-8] [PMID: 11150483]

[128] Li Z, Cui S, Zhang Z, *et al.* DHEA-neuroprotection and -neurotoxicity after transient cerebral ischemia in rats. J Cereb Blood Flow Metab 2009; 29(2): 287-96.
[http://dx.doi.org/10.1038/jcbfm.2008.118] [PMID: 18854841]

[129] Karishma KK, Herbert J. Dehydroepiandrosterone (DHEA) stimulates neurogenesis in the hippocampus of the rat, promotes survival of newly formed neurons and prevents corticosterone-induced suppression. Eur J Neurosci 2002; 16(3): 445-53.
[http://dx.doi.org/10.1046/j.1460-9568.2002.02099.x] [PMID: 12193187]

[130] Newman AE, MacDougall-Shackleton SA, An YS, Kriengwatana B, Soma KK. Corticosterone and dehydroepiandrosterone have opposing effects on adult neuroplasticity in the avian song control system. J Comp Neurol 2010; 518(18): 3662-78.
[http://dx.doi.org/10.1002/cne.22395] [PMID: 20653028]

[131] Jin RO, Mason S, Mellon SH, *et al.* Cortisol/DHEA ratio and hippocampal volume: A pilot study in major depression and healthy controls. Psychoneuroendocrinology 2016; 72: 139-46.
[http://dx.doi.org/10.1016/j.psyneuen.2016.06.017] [PMID: 27428086]

[132] Sapolsky RM. Glucocorticoids and hippocampal atrophy in neuropsychiatric disorders. Arch Gen Psychiatry 2000; 57(10): 925-35.
[http://dx.doi.org/10.1001/archpsyc.57.10.925] [PMID: 11015810]

[133] O'Brien JT, Lloyd A, McKeith I, Gholkar A, Ferrier N. A longitudinal study of hippocampal volume, cortisol levels, and cognition in older depressed subjects. Am J Psychiatry 2004; 161(11): 2081-90.
[http://dx.doi.org/10.1176/appi.ajp.161.11.2081] [PMID: 15514410]

[134] Videbech P, Ravnkilde B. Hippocampal volume and depression: a meta-analysis of MRI studies. Am J Psychiatry 2004; 161(11): 1957-66.
[http://dx.doi.org/10.1176/appi.ajp.161.11.1957] [PMID: 15514393]

[135] Colla M, Kronenberg G, Deuschle M, *et al.* Hippocampal volume reduction and HPA-system activity in major depression. J Psychiatr Res 2007; 41(7): 553-60.
[http://dx.doi.org/10.1016/j.jpsychires.2006.06.011] [PMID: 17023001]

[136] Waldron H. Trends in mortality differentials and life expectancy for male social security-covered workers, by socioeconomic status. Soc Secur Bull 2007; 67(3): 1-28.
[PMID: 18605216]

[137] Ageing and Health Key Facts. World Health Organization 2018.

[138] Simões CCS. Breve histórico do processo demográfico. Brasil: uma visão geográfica e ambiental no início do século XXI Rio de Janeiro: IBGE 2016.
[http://dx.doi.org/10.21579/isbn.9788524043864_cap.2]

[139] Barnes JN. Exercise, cognitive function, and aging. Adv Physiol Educ 2015; 39(2): 55-62.
[http://dx.doi.org/10.1152/advan.00101.2014] [PMID: 26031719]

[140] Kandola A, Hendrikse J, Lucassen PJ, Yücel M. Aerobic exercise as a tool to improve hippocampal plasticity and function in humans: practical implications for mental health treatment. Front Hum Neurosci 2016; 10: 373.
[http://dx.doi.org/10.3389/fnhum.2016.00373] [PMID: 27524962]

[141] Barja G. Endogenous oxidative stress: relationship to aging, longevity and caloric restriction. Ageing Res Rev 2002; 1(3): 397-411.
[http://dx.doi.org/10.1016/S1568-1637(02)00008-9] [PMID: 12067594]

[142] López-Otín C, Blasco MA, Partridge L, Serrano M, Kroemer G. The hallmarks of aging. Cell 2013; 153(6): 1194-217.
[http://dx.doi.org/10.1016/j.cell.2013.05.039] [PMID: 23746838]

[143] Ocampo A, Reddy P, Belmonte JCI. Anti-aging strategies based on cellular reprogramming. Trends

Mol Med 2016; 22(8): 725-38.
[http://dx.doi.org/10.1016/j.molmed.2016.06.005] [PMID: 27426043]

[144] Maggio M, Lauretani F, De Vita F, *et al.* Multiple hormonal dysregulation as determinant of low physical performance and mobility in older persons. Curr Pharm Des 2014; 20(19): 3119-48.
[http://dx.doi.org/10.2174/13816128113196660062] [PMID: 24050169]

[145] Samaras N, Samaras D, Frangos E, Forster A, Philippe J. A review of age-related dehydroepiandrosterone decline and its association with well-known geriatric syndromes: is treatment beneficial? Rejuvenation Res 2013; 16(4): 285-94.
[http://dx.doi.org/10.1089/rej.2013.1425] [PMID: 23647054]

[146] Allolio B, Arlt W. DHEA treatment: myth or reality? Trends Endocrinol Metab 2002; 13(7): 288-94.
[http://dx.doi.org/10.1016/S1043-2760(02)00617-3] [PMID: 12163230]

[147] de Menezes KJ, Peixoto C, Nardi AE, Carta MG, Machado S, Veras AB. Dehydroepiandrosterone, its sulfate and cognitive functions. Clin Pract Epidemiol Ment Health 2016; 12: 24-37.
[http://dx.doi.org/10.2174/1745017901612010024] [PMID: 27346998]

[148] Coyle JT, Puttfarcken P. Oxidative stress, glutamate, and neurodegenerative disorders. Science 1993; 262(5134): 689-95.
[http://dx.doi.org/10.1126/science.7901908] [PMID: 7901908]

[149] Charalampopoulos I, Alexaki VI, Tsatsanis C, *et al.* Neurosteroids as endogenous inhibitors of neuronal cell apoptosis in aging. Ann N Y Acad Sci 2006; 1088(1): 139-52.
[http://dx.doi.org/10.1196/annals.1366.003] [PMID: 17192562]

[150] Charalampopoulos I, Remboutsika E, Margioris AN, Gravanis A. Neurosteroids as modulators of neurogenesis and neuronal survival. Trends Endocrinol Metab 2008; 19(8): 300-7.
[http://dx.doi.org/10.1016/j.tem.2008.07.004] [PMID: 18771935]

[151] Mullonkal CJ, Toledo-Pereyra LH. Akt in ischemia and reperfusion. J Invest Surg 2007; 20(3): 195-203.
[http://dx.doi.org/10.1080/08941930701366471] [PMID: 17613695]

[152] Tanaka T, Nakamura H, Yodoi J, Bloom ET. Redox regulation of the signaling pathways leading to eNOS phosphorylation. Free Radic Biol Med 2005; 38(9): 1231-42.
[http://dx.doi.org/10.1016/j.freeradbiomed.2005.01.002] [PMID: 15808421]

[153] Cole GM, Frautschy SA. The role of insulin and neurotrophic factor signaling in brain aging and Alzheimer's Disease. Exp Gerontol 2007; 42(1-2): 10-21.
[http://dx.doi.org/10.1016/j.exger.2006.08.009] [PMID: 17049785]

[154] Janner DdaR, Jacob MH, Jahn MP, Kucharski LCR, Ribeiro MFM. Dehydroepiandrosterone effects on Akt signaling modulation in central nervous system of young and aged healthy rats. J Steroid Biochem Mol Biol 2010; 122(4): 142-8.
[http://dx.doi.org/10.1016/j.jsbmb.2010.07.006] [PMID: 20691781]

[155] Jacob MH, Janner DdaR, Belló-Klein A, Llesuy SF, Ribeiro MF. Dehydroepiandrosterone modulates antioxidant enzymes and Akt signaling in healthy Wistar rat hearts. J Steroid Biochem Mol Biol 2008; 112(1-3): 138-44.
[http://dx.doi.org/10.1016/j.jsbmb.2008.09.008] [PMID: 18848627]

[156] Jacob MHVM, Janner DdaR, Jahn MP, Kucharski LCR, Belló-Klein A, Ribeiro MFM. DHEA effects on myocardial Akt signaling modulation and oxidative stress changes in aged rats. Steroids 2009; 74(13-14): 1045-50.
[http://dx.doi.org/10.1016/j.steroids.2009.08.005] [PMID: 19699218]

[157] Aragno M, Mastrocola R, Brignardello E, *et al.* Dehydroepiandrosterone modulates nuclear factor-kappaB activation in hippocampus of diabetic rats. Endocrinology 2002; 143(9): 3250-8.
[http://dx.doi.org/10.1210/en.2002-220182] [PMID: 12193536]

[158] Subbarao KV, Richardson JS. Iron-dependent peroxidation of rat brain: a regional study. J Neurosci

Res 1990; 26(2): 224-32.
[http://dx.doi.org/10.1002/jnr.490260212] [PMID: 2366265]

[159] Mecocci P, MacGarvey U, Beal MF. Oxidative damage to mitochondrial DNA is increased in Alzheimer's disease. Ann Neurol 1994; 36(5): 747-51.
[http://dx.doi.org/10.1002/ana.410360510] [PMID: 7979220]

[160] Smith MA, Rottkamp CA, Nunomura A, Raina AK, Perry G. Oxidative stress in Alzheimer's disease. Biochimica et Biophysica Acta (BBA)-. Molecular Basis of Disease 2000; 1502(1): 139-44.
[http://dx.doi.org/10.1016/S0925-4439(00)00040-5]

[161] Smith MA, Kutty RK, Richey PL, *et al.* Heme oxygenase-1 is associated with the neurofibrillary pathology of Alzheimer's disease. Am J Pathol 1994; 145(1): 42-7.
[PMID: 8030754]

[162] Dyrks T, Dyrks E, Hartmann T, Masters C, Beyreuther K. Amyloidogenicity of beta A4 and beta A4-bearing amyloid protein precursor fragments by metal-catalyzed oxidation. J Biol Chem 1992; 267(25): 18210-7.
[http://dx.doi.org/10.1016/S0021-9258(19)37174-1] [PMID: 1517249]

[163] Joachim CL, Selkoe DJ. The seminal role of beta-amyloid in the pathogenesis of Alzheimer disease. Alzheimer Dis Assoc Disord 1992; 6(1): 7-34.
[http://dx.doi.org/10.1097/00002093-199205000-00003] [PMID: 1605946]

[164] Klegeris A, McGeer PL. β-amyloid protein enhances macrophage production of oxygen free radicals and glutamate. J Neurosci Res 1997; 49(2): 229-35.
[http://dx.doi.org/10.1002/(SICI)1097-4547(19970715)49:2<229::AID-JNR11>3.0.CO;2-W] [PMID: 9272645]

[165] Klegeris A, Walker DG, McGeer PL. Activation of macrophages by Alzheimer β amyloid peptide. Biochem Biophys Res Commun 1994; 199(2): 984-91.
[http://dx.doi.org/10.1006/bbrc.1994.1326] [PMID: 7510964]

[166] Hensley K, Carney JM, Mattson MP, *et al.* A model for beta-amyloid aggregation and neurotoxicity based on free radical generation by the peptide: relevance to Alzheimer disease. Proc Natl Acad Sci USA 1994; 91(8): 3270-4.
[http://dx.doi.org/10.1073/pnas.91.8.3270] [PMID: 8159737]

[167] Rammouz G, Lecanu L, Papadopoulos V. Oxidative stress-mediated brain dehydroepiandrosterone (DHEA) formation in Alzheimer's disease diagnosis. Front Endocrinol (Lausanne) 2011; 2: 69.
[http://dx.doi.org/10.3389/fendo.2011.00069] [PMID: 22654823]

[168] Cascio C, Brown RC, Liu Y, Han Z, Hales DB, Papadopoulos V. Pathways of dehydroepiandrosterone formation in rat brain glia. J Steroid Biochem Mol Biol 2000; 75(2-3): 177-86.
[http://dx.doi.org/10.1016/S0960-0760(00)00163-1] [PMID: 11226834]

[169] Orentreich N, Brind JL, Vogelman JH, Andres R, Baldwin H. Long-term longitudinal measurements of plasma dehydroepiandrosterone sulfate in normal men. J Clin Endocrinol Metab 1992; 75(4): 1002-4.
[PMID: 1400863]

[170] Labrie F, Bélanger A, Cusan L, Gomez J-L, Candas B. Marked decline in serum concentrations of adrenal C19 sex steroid precursors and conjugated androgen metabolites during aging. J Clin Endocrinol Metab 1997; 82(8): 2396-402.
[http://dx.doi.org/10.1210/jcem.82.8.4160] [PMID: 9253307]

[171] Maggio M, De Vita F, Fisichella A, *et al.* DHEA and cognitive function in the elderly. J Steroid Biochem Mol Biol 2015; 145: 281-92.
[http://dx.doi.org/10.1016/j.jsbmb.2014.03.014] [PMID: 24794824]

[172] Souza-Teodoro LH, de Oliveira C, Walters K, Carvalho LA. Higher serum dehydroepiandrosterone sulfate protects against the onset of depression in the elderly: Findings from the English Longitudinal

Study of Aging (ELSA). Psychoneuroendocrinology 2016; 64: 40-6.
[http://dx.doi.org/10.1016/j.psyneuen.2015.11.005] [PMID: 26600009]

[173] Markowski M, Ungeheuer M, Bitran D, Locurto C. Memory-enhancing effects of DHEAS in aged mice on a win-shift water escape task. Physiol Behav 2001; 72(4): 521-5.
[http://dx.doi.org/10.1016/S0031-9384(00)00446-7] [PMID: 11282135]

[174] Herrera-Pérez JJ, Martínez-Mota L, Jiménez-Rubio G, *et al.* Dehydroepiandrosterone increases the number and dendrite maturation of doublecortin cells in the dentate gyrus of middle age male Wistar rats exposed to chronic mild stress. Behav Brain Res 2017; 321: 137-47.
[http://dx.doi.org/10.1016/j.bbr.2017.01.007] [PMID: 28062256]

[175] Lhullier FL, Riera NG, Nicolaidis R, *et al.* Effect of DHEA glutamate release from synaptosomes of rats at different ages. Neurochem Res 2004; 29(2): 335-9.
[http://dx.doi.org/10.1023/B:NERE.0000013735.50736.0a] [PMID: 15002728]

[176] Yin FJ, Kang J, Han NN, Ma HT. Effect of dehydroepiandrosterone treatment on hormone levels and antioxidant parameters in aged rats. Genet Mol Res 2015; 14(3): 11300-11.
[http://dx.doi.org/10.4238/2015.September.22.24] [PMID: 26400361]

[177] Hildreth KL, Gozansky WS, Jankowski CM, Grigsby J, Wolfe P, Kohrt WM. Association of serum dehydroepiandrosterone sulfate and cognition in older adults: sex steroid, inflammatory, and metabolic mechanisms. Neuropsychology 2013; 27(3): 356-63.
[http://dx.doi.org/10.1037/a0032230] [PMID: 23688217]

[178] Peixoto C, Grande AJ, Mallmann MB, Nardi AE, Cardoso A, Veras AB. Dehydroepiandrosterone (DHEA) for depression: a systematic review and meta-analysis. CNS & Neurological Disorders-Drug Targets (Formerly Current Drug Targets-CNS & Neurological Disorders) 2018; 17(9): 706-11.

[179] Sunderland T, Merril CR, Harrington MG, *et al.* Reduced plasma dehydroepiandrosterone concentrations in Alzheimer's disease. Lancet 1989; 2(8662): 570.
[http://dx.doi.org/10.1016/S0140-6736(89)90700-9] [PMID: 2570275]

[180] Yanase T, Fukahori M, Taniguchi S, *et al.* Serum dehydroepiandrosterone (DHEA) and DHEA-sulfate (DHEA-S) in Alzheimer's disease and in cerebrovascular dementia. Endocr J 1996; 43(1): 119-23.
[http://dx.doi.org/10.1507/endocrj.43.119] [PMID: 8732462]

[181] Bernardi F, Lanzone A, Cento RM, *et al.* Allopregnanolone and dehydroepiandrosterone response to corticotropin-releasing factor in patients suffering from Alzheimer's disease and vascular dementia. Eur J Endocrinol 2000; 142(5): 466-71.
[http://dx.doi.org/10.1530/eje.0.1420466] [PMID: 10802523]

[182] Hillen T, Lun A, Reischies FM, Borchelt M, Steinhagen-Thiessen E, Schaub RT. DHEA-S plasma levels and incidence of Alzheimer's disease. Biol Psychiatry 2000; 47(2): 161-3.
[http://dx.doi.org/10.1016/S0006-3223(99)00217-6] [PMID: 10664834]

[183] Murialdo G, Barreca A, Nobili F, *et al.* Relationships between cortisol, dehydroepiandrosterone sulphate and insulin-like growth factor-I system in dementia. J Endocrinol Invest 2001; 24(3): 139-46.
[http://dx.doi.org/10.1007/BF03343833] [PMID: 11314741]

[184] Genedani S, Rasio G, Cortelli P, *et al.* Studies on homocysteine and dehydroepiandrosterone sulphate plasma levels in Alzheimer's disease patients and in Parkinson's disease patients. Neurotox Res 2004; 6(4): 327-32.
[http://dx.doi.org/10.1007/BF03033443] [PMID: 15545016]

[185] Aldred S, Mecocci P. Decreased dehydroepiandrosterone (DHEA) and dehydroepiandrosterone sulfate (DHEAS) concentrations in plasma of Alzheimer's disease (AD) patients. Arch Gerontol Geriatr 2010; 51(1): e16-8.
[http://dx.doi.org/10.1016/j.archger.2009.07.001] [PMID: 19665809]

[186] Cardounel A, Regelson W, Kalimi M. Dehydroepiandrosterone protects hippocampal neurons against neurotoxin-induced cell death: mechanism of action. Proc Soc Exp Biol Med 1999; 222(2): 145-9.

[http://dx.doi.org/10.1046/j.1525-1373.1999.d01-124.x] [PMID: 10564538]

[187] Yamada S, Akishita M, Fukai S, *et al.* Effects of dehydroepiandrosterone supplementation on cognitive function and activities of daily living in older women with mild to moderate cognitive impairment. Geriatr Gerontol Int 2010; 10(4): 280-7.
[http://dx.doi.org/10.1111/j.1447-0594.2010.00625.x] [PMID: 20497239]

[188] Fonda SJ, Bertrand R, O'Donnell A, Longcope C, McKinlay JB. Age, hormones, and cognitive functioning among middle-aged and elderly men: cross-sectional evidence from the Massachusetts Male Aging Study. J Gerontol A Biol Sci Med Sci 2005; 60(3): 385-90.
[http://dx.doi.org/10.1093/gerona/60.3.385] [PMID: 15860479]

[189] Kritz-Silverstein D, von Mühlen D, Laughlin GA, Bettencourt R. Effects of dehydroepiandrosterone supplementation on cognitive function and quality of life: the DHEA and Well-Ness (DAWN) Trial. J Am Geriatr Soc 2008; 56(7): 1292-8.
[http://dx.doi.org/10.1111/j.1532-5415.2008.01768.x] [PMID: 18482290]

[190] Nestler JE, Barlascini CO, Clore JN, Blackard WG. Dehydroepiandrosterone reduces serum low density lipoprotein levels and body fat but does not alter insulin sensitivity in normal men. J Clin Endocrinol Metab 1988; 66(1): 57-61.
[http://dx.doi.org/10.1210/jcem-66-1-57] [PMID: 2961787]

[191] Safiulina D, Peet N, Seppet E, Zharkovsky A, Kaasik A. Dehydroepiandrosterone inhibits complex I of the mitochondrial respiratory chain and is neurotoxic *in vitro* and *in vivo* at high concentrations. Toxicol Sci 2006; 93(2): 348-56.
[http://dx.doi.org/10.1093/toxsci/kfl064] [PMID: 16849397]

[192] Sahu P, Gidwani B, Dhongade HJ. Pharmacological activities of dehydroepiandrosterone: A review. Steroids 2020; 153: 108507.
[http://dx.doi.org/10.1016/j.steroids.2019.108507] [PMID: 31586606]

[193] Racchi M, Balduzzi C, Corsini E. Dehydroepiandrosterone (DHEA) and the aging brain: flipping a coin in the "fountain of youth". CNS Drug Rev 2003; 9(1): 21-40.
[http://dx.doi.org/10.1111/j.1527-3458.2003.tb00242.x] [PMID: 12595910]

[194] Vegh C, Stokes K, Ma D, *et al.* A bird's-eye view of the multiple biochemical mechanisms that propel pathology of Alzheimer's disease: recent advances and mechanistic perspectives on how to halt the disease progression targeting multiple pathways. J Alzheimers Dis 2019; 69(3): 631-49.
[http://dx.doi.org/10.3233/JAD-181230] [PMID: 31127770]

[195] Strac DS, Konjevod M, Perkovic MN, Tudor L, Erjavec GN, Pivac N. Dehydroepiandrosterone (DHEA) and its Sulphate (DHEAS) in Alzheimer's Disease. Curr Alzheimer Res 2020; 17(2): 141-57.
[http://dx.doi.org/10.2174/1567205017666200317092310] [PMID: 32183671]

CHAPTER 4

Emerging Nanotherapeutic Strategies in Alzheimer's Disease

Soheila Montazersaheb¹, Elham Ahmadian², Solmaz Maleki Dizaj³, Yalda Jahanbani⁴, Soodabeh Davaran⁴, Irada Huseynova⁵, Cumali Keskin⁶, Renad I. Zhdanov⁷,⁸, Rovshan Khalilov⁷,⁹,¹⁰ and Aziz Eftekhari⁷,¹¹,*

¹ Molecular Medicine Research Center, Tabriz University of Medical Sciences, Tabriz, Iran

² Kidney Research Center, Tabriz University of Medical Sciences, Tabriz, Iran

³ Dental and Periodontal Research Center, Tabriz University of Medical Sciences, Tabriz, Iran

⁴ Department of Medicinal Chemistry, School of Pharmacy, Tabriz University of Medical Sciences, Tabriz, Iran

⁵ Institute of Molecular Biology & Biotechnologies, Azerbaijan National Academy of Sciences, 11 Izzat Nabiyev, Baku AZ 1073, Azerbaijan

⁶ Medical Laboratory Techniques, Vocational Higher School of Healthcare Studies, Mardin Artuklu University, Mardin, Turkey

⁷ Russian Institute for Advanced Study, Moscow State Pedagogical University, 1/1, Malaya Pirogovskaya St, Moscow, 119991, Russian Federation

⁸ Kazan Federal University, 420008, Kremlevskaya 18, Kazan, Russia

⁹ Department of Biophysics and Biochemistry, Baku State University, Baku, Azerbaijan

¹⁰ Institute of Radiation Problems, Azerbaijan National Academy of Science, Baku, Azerbaijan

¹¹ Pharmacology and Toxicology Department, Maragheh University of Medical Sciences, Maragheh, Iran

Abstract: Recent nanotechnological advancements have opened new windows of hope for the treatment of neurodegenerative diseases and all of these new methods are rapidly evolving. Alzheimer's disease (AD) treatment based on neuroprotective and neurogenerative techniques has been advancing rapidly in recent decades, and the use of nanotechnology developments such as polymers, emulsions, lipo-carriers, solid lipid carriers, carbon nanotubes, and metal-based carriers has been very effective in both diagnosis and treatment methods. Targeted drug delivery is one of the most important concerns in the AD treatment approaches because the 'blood-brain barrier' or the 'blood-cerebrospinal fluid barrier is a serious obstacle for delivering a therapeutic agent to the desired location, but the use of nanocarriers have provided acceptable results in this area. It seems that the use of nanotechnology in approving a successful treatment

---

* **Corresponding author Aziz Eftekhari:** Maragheh University of Medical Sciences, Maragheh, Iran; Tel: +989364655619; E-mail:ftekhari@ymail.com

method for Alzheimer's is inevitable and this review has collected a complete copy of the recent nanotechnology-based approaches for the treatment of AD.

Keywords: Alzheimer's Disease, Amyloid β-protein, Dementia, Drug Delivery, Magnetic Nanoparticles, Nanomedicine, Nanotechnology, Neurodegenerative Diseases, Polymeric Nanoparticles, Targeted Drug Delivery.

INTRODUCTION

Alzheimer's disease (AD) is a progressive degenerative disease of the central nervous system in the elderly population, accounting for an estimated 60–80% of dementia cases worldwide. This devastating disease is characterized by the irreversible loss of neurons and synapses in the brain, resulting in cognitive and intellectual deficits as well as alterations in behavioral traits such as anhedonia- and anxiety-like symptoms [1]. The progressive and brain decay is attributed to the formation of senile plaques and neurofibrillary tangles (NFTs) in the hippocampus, which are the most prominent pathological hallmarks of AD. Extracellular senile plaques are formed by amyloid-β42 peptide (Aβ42) and intracellular NFTs composed of hyperphosphorylated TAU protein [2]. It is also important to note that the formation of these plaques can start approximately 20 years before appearance of the clinical symptoms in patients with AD [3]. The histopathologic features of AD are strongly associated with hippocampal neuronal degeneration, synaptic dysfunction and loss, and aneuploidy. In addition, neuronal inflammation, mitochondrial dysfunction, and impaired lymphatic system have a major role in the course of AD. The pathological mechanisms in AD can enhance the production of ROS, resulting in oxidative stress-mediated signaling cascades. In this context, increasing cell stress, microglial dysfunctions, and upregulation of inflammatory cytokines lead to neuronal cell death in AD. Apart from these, various physiological parameters and lifestyle factors are also involved in the onset of AD [4, 5]. Collectively, the complex pathophysiology of AD results in cholinergic neurotransmission deficits in the brain of patients with AD. With this background, there is an urgent need to deal with this chronic disease. Multiple efforts have been made to slow the degenerative process; however, little success has been achieved [6]. In general, two types of treatments are available for patients with AD; symptomatic treatments and targeting approaches [7]. Nanotechnology-based medicine is a novel therapeutic approach that offers opportunities for prevention, diagnosis, and AD therapy. Promising anti-AD strategies are based on targeting the cholinergic system, Aβ protein, and other AD-related factors [8].

In general, AD is diagnosed with memory, cognition, and behavioral impairment (dementia), all of these complications occur as a result of progressive neurodegeneration. AD in the final stages may eventually lead to mood fluctuations and fatal delirium [1]. Due to the progress of different communities towards aging, AD has become a global concern. According to the statistics provided, nearly 36 million inhabitants have been reported with dementia-AD since 2010; the number of reported clinical cases is expected to reach 65.7 million by 2030 [9]. Numerous clinical data suggested that during AD, severe impairment in cholinergic–neurotransmitter systems occurs; it seems the cause of this defect is suppression of acetylcholine (responsible for neural synapse) by Acetyl-cholinesterase (AChE) activity [10]. In addition to the AChE activity, the activation of the glutamatergic system also plays an important role in AD pathology [11]. Based on two processes mentioned that occur during AD, few drug candidates such as tacrine, donepezil, rivastigmine, galantamine (AChE inhibitors), and memantine (NMDA inhibitor) were developed by researchers [12, 13]. Despite all the investments of pharmaceutical bodies such as Merck & Co., Lily, Pfizer, *etc.* and the efforts of researchers, these drugs in terms of pharmacokinetic (half-life) profiles of the drugs in the biological system and existence of side effects did not provide the desired results [14, 15]. These drugs have shown successful results in pre-clinical studies but their efficacy in human trials was not satisfactory; of course, patient withdrawal in clinical trials and not retention to a full study term, has been a big problem to access comprehensive information [16]. But in addition to these, therapeutic failures can be caused by factors such as poor pharmacokinetics or low bioavailability, chemical nature (absorption in biological – blood brain barrier system), volatility (oxidation, hydrolysis) of the drugs in question [16 - 18].

Biodistribution, targeting and BBB permeability are the main challenges for drug delivery into AD. One of the main important areas to solve these challenges is the use of nanotechnology approaches in drug delivery systems to achieve improved bioavailability and kinetic profile of different types of drugs in biological systems [19]. Functionalized NPs resulted in a greater biodistribution of the formulation in the brain than the intravenous administration of the NPS [20]. Nanotechnology made possible targeting and safe delivery of drugs to different parts of the body [21]. The controlled release profile of drugs improves with using of sustained release of nano-drug delivery systems [22]. Accordingly, the nanotechnology-based delivery systems can influence the BBB permeability *via* tethering AD drugs onto the surfaces of the nanoparticles [23]. It is important that the use of nanotechnology approaches in drug delivery systems improved the bioavailability and kinetic profile of different types of drugs in biological systems significantly [19 , 22].

NEURON TARGETING IN AD

At present, conventional medical treatments have yielded modest efficacy for neurodegenerative disorders such as AD. This may be due to the difficulties of targeting specific neuronal populations. On the whole, unspecific delivery of therapeutic agents to tissues and cells out of the target site (healthy cells) causes various side effects and hamper clinical efficacy. Achieving improved clinical outcomes relies on the delivery of the therapeutics at proper doses to diseased cells and tissues [7, 24].

Multiple factors can affect the entry of drugs into the brain region, and among those, the blood-brain barrier (BBB) is the most prominent one that acts as a protective layer around the brain. In other words, the unique structure and low permeability of BBB is the main obstacle for systemically administered drugs. In addition to this, the chemical instability of developed drugs is another hurdle in neuronal diseases which limits their utilization [25]. In general, the brain vasculature has a crucial role in CNS homeostasis. The BBB consists of endothelial cells with specific traits that limit the free movement of a variety of molecules from the blood to the CNS. Besides, these specialized endothelial cells can facilitate the transfer of harmful products from the CNS to the blood. Both functions are critical for the proper function of neuronal cells and the protection of the CNS from harmful stimuli. This leaves the door open for developing specific carriers to encapsulate the drugs to overcome the barriers in passage across BBB.

Nanotechnology-based technology offers the best prospects for the delivery of drugs with improved therapeutic efficacy. This strategy is increasingly being applied in AD. It is well documented that nanomaterials improve therapeutics efficacy through enhancing the dosing efficacy of delivered agents, controlling cargo release, selective transport of drugs across the BBB after systemic administration, and more importantly, being functionalized for the specific target of interest in the brain [26]. Desired physicochemical properties and multifunctionality of these carriers gain much popularity for developing therapeutics for site-specific delivery of nanomedicine in AD. Brain accessibility to nanomedicines depends on systemic or intranasal administration. The intranasal delivery approach is a noninvasive strategy that can bypass the BBB [27].

In the case of non-specific formulation, the drug release into neurons and other brain cells in the vicinity, is known as passive neuronal targeting. However, the most desirable approach is active neuronal targeting, *via* decoration of nanoparticles by neuronal-specific ligands to reach only diseased neurons, instead of other brain cell types [27]. In this regard, developing drug-delivery strategies to target disease-related neurons in AD is of crucial importance to minimize off-target effects.

Understanding the characteristics of neurons and nanoparticles can help to achieve site-specific delivery [24]. For instance, neurons of the cerebral cortex and hippocampal region overexpress M1 and M2 muscarinic acetylcholine receptors, thereby using appropriate ligand can enable specific targeting toward impaired neurons in AD [28]. It is also important to note that type, surface charge, shape, and concentration of nano-carrier influence the delivery of drugs to the brain [29].

Despite the superiority of nanomedicine-based approaches, the formation of protein corona can affect drug release profiles and determine the pharmacological fate of these carriers. Upon contact with biological fluids like plasma and cerebrospinal fluid, functionalized nanomedicine can immediately absorb various kinds of biomolecules such as proteins, forming protein corona. This can lead to non-targeted interaction of nanomedicine and deposit of the nono-carrier in biological environments, affecting the functionality of nonocarriers [30]. Along this line, protein-coated nanomedicines can alter physicochemical characteristics of the nanoparticles in terms of size and surface features, leading to the aggregation of nanoparticles and subsequent events [31]. Relying on this fact, nanomedicines can lose their targeting specificity in biological systems due to blocking by the protein corona. Several modifications are needed to prevent protein corona formation as well as achieve enough therapeutic level in the target site of the brain. Using zwitterionic nanoparticles can prevent protein absorption to nanoparticles. Another approach that can overcome this issue is using engineered nanoparticles to interact with their binding molecules on the target cells.

It has also been reported that effective transportation through BBB can be achieved by the formation of the protein corona. A study has shown that coating surface of nanoparticles by Tween-80 causes interaction with apolipoprotein E in the bloodstream and the formation of the protein corona. These modified nanocomposites showed superior biocompatibility and uptake capacity in the brain [7]. In addition to this, protein corona can modify the surface chemistry of nanocarriers and improves drug delivery by enhancing the functionalization of nanoparticles. Collectively, protein corona formation on the surface of nanoparticles is a critical determinant of the fate of nanocarrier.

Therapeutical opportunities are restricted chiefly to the incapability of drugs to cross the blood-brain barrier (BBB) [8] or their poor oral bioavailability [32]. Several approaches have been established to overcome the BBB, including drug delivery systems, polymeric and solid lipid NPs (SLNs), microemulsions (MEs), hydrogels, liposomes, solid lipid carriers, and liquid crystals (LCs) [8, 32, 33]. The physicochemical properties of some therapeutical agents, such as their hydrophilicity, ionization, poor bioavailability, high molecular weight, extensive metabolization, and unwanted effects, can lead to weak pharmacotherapeutic impacts [8, 34]. Intranasal administrations could offer promising treatment choices to overcome these limitations and enhance the transportation of the drugs into the brain [35]. Other strategies have also been utilized to target NPs across the BBB, such as the conjugation of drugs/monoclonal antibodies against endogenous receptors in the BBB [20, 36 - 42] or using coated nano-devices [38, 41]. Anti-transferrin receptor antibody-functionalized nanomaterials can cross the BBB. The latter increases the uptake, and permeability of drugs [38]. Cell-penetrating peptides could also be functionalized with nanomaterials [42]. The immobilization of anti-transferrin monoclonal antibody against transferrin receptor in BBB has been used to develop mono- and dual-decorated liposomes. In this context, a peptide analog of apolipoprotein (PAA) has successfully targeted low-density lipoprotein receptors in BBB through vesicle transcytosis. Accordingly, dual ligands and monoclonal antibodies surged brain targeting to a greater extent in comparison with non-targeted liposomes. However, different results were obtained *in vivo*. This was postulated to be concerning the presence of serum proteins in the cell culture medium, which exhibits a crucial role in the function of nanoformulations [39]. The formation of gelatin by nasal fluid increases the residence time of nasal-administrated AD drugs and thus their targeting to the brain [37].

Detection of amyloid peptides (Aβ) as an important biomarker of AD has been addressed using different nanomaterials. Thioflavin-T entrapped in polymeric NPs have been developed as probes of Aβ [43]. Thioflavin-T-delivered nanosphere could also be used in targeted therapy of AD [43].

Polysorbate 80-coated poly (*n*-butyl cyanoacrylate) NPs have been designed to deliver rivastigmine to the brain tissue in vivo. It was reported that the interaction between microvessels of the endothelial cells in the brain and polysorbate 80 could efficiently increase drug entrance [44]. Targeting peptides such as TGN (targeting ligand present in the BBB) and QSH (targeting Aβ) have been coated in PEGylated poly(lactic acid) polymer which meaningfully surged the targeted delivery of the peptides into the brain [45]. The functionalization of NPs enhances their biodistribution. *Solanum tuberosum* lectin (STL)-modified fibroblast growth factor (bFGF) – entrapped PEG and PLGA NPs (STL-bFGF-NPs) have shown a greater distribution in intranasal administration. Also, PEG inhibits the aggregation of NPs in the nasal mucosa [46]. Cell-penetrating peptide (CPP)-modified liposomes have been developed to increase the distribution of rivastigmine, elevate its pharmacodynamics and hamper the adverse effects of the drug [47]. Polysorbate 80 coated tacrine-loaded NPs have shown a better biodistribution pattern [48].

NANOTECHNOLOGY IN AD THERAPY

Pathogenesis of AD

Mounting evidence suggests the new mitochondrial cascade hypothesis as the crucial mechanism of AD [49]. In these pathways, the generation of ROs and free radicals lead to neuronal damages [50, 51]. Thus, the neurons become more susceptible to apoptosis and excitotoxicity. Mitochondria have been considered as a target in treating AD. Latrepidine as a mitochondrial targeting moiety is being tested in stage II clinical trials in treating mild to moderate AD [52].

Another demonstrated pathogenesis for AD is the excitotoxicity hypothesis. In this context, Aβ can regulate the activity of the N-methyl-D-aspartate (NMDA) receptor, which is associated with synaptic flexibility and memory function of the neurons [53, 54]. When the NMDA receptors are over-excited, Ca^{2+} over-burden occurs which finally leads to cellular apoptosis [54].

Previous reports have suggested the pivotal role of acetylcholine (Ach) in memory and learning which terminated in the introduction of the cholinergic hypothesis as another pathogenesis of AD. Memory and cognition impairments in AD patients have been linked with reduced liberation of Ach and uptake of choline [53, 55]. Rivastigmine, donepezil, and galantamine as acetylcholinesterase inhibitors have been approved by FDA in the treatment of AD [54].

On the other hand, the tau protein hypothesis points towards the function of tua protein, which is connected with axonal transport-mediated microtubules. Accordingly, neurofibrillary tangles (NFTs) are produced from paired helical fragments. Nonetheless, the multifactorial pathogenesis of AD involving protein accumulation, lipid distribution, and inflammation [56].

Crossing the blood-brain barrier (BBB) and drug delivery to the CNS is the most challenging issue; thereby the development of novel strategies to overcome this obstacle is in demand. Revolutionary progress regarding nanotechnology-based approaches opens the way toward AD treatment. Nanotechnology-based strategies have fetched a lot of attention due to the efficacy and sustained release of the entrapped drugs. Some examples are given as   follows: polymer-based nanoparticles (NPs), lipid/lipoprotein-based NPs, liposomal carriers, optical imaging systems, metal/magnetic-based NPs, antibody-tethered NPs, nanoemulsion systems, dendrimers, nanocomposites, curcumin-encapsulated NPs, and other vehicles and strategies such as intranasally administered routes.

Nanotechnology-based Treatments for Alzheimer's Disease

Various physiological factors that affect the effectiveness of AD therapy are the BBB, the blood-cerebrospinal fluid (CSF) barrier, and p-glycoprotein-mediated drug efflux. However, potential advancements in nanotechnology may offer an innovative strategy to deal with these barriers *via* tethering AD drugs onto the surfaces of the nanoparticles. Accordingly, the nanotechnology-based delivery systems can overcome the above-mentioned problems (Fig. **1**) [23].

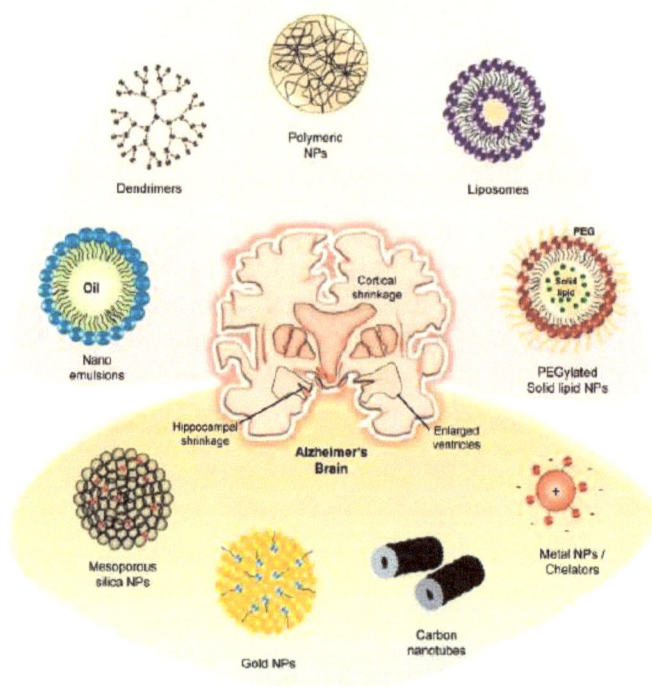

Fig. (1). Examples of some nano-carriers for the treatment of AD. Adopted from [23].

Polymeric Nanoparticles

Polymeric nanoparticles are effective carriers developed for targeted/controlled drug delivery. In this system, the drug bioavailability is improved by protection from degradation and environmental modulation [57].

In a paper reported by Kulkarni *et al.* (2010), an efficient polymeric n-butyl-2-cyanoacrylate (PBCA) was prepared to image the amyloid plaques for early detection of AD. The radio-iodinated 5-chloro-7-iodo-8-hydroxyquinoline (CQ) was incorporated within the developed structure. In consequence, compared to the 125I-CQ control, encapsulation of the 125I-CQ in PBCA showed great efficiency in a mouse model of AD [58].

In another study, Wilson *et al.* (2011) employed chitosan NPs containing rivastigmine for brain targeted therapy in AD. In detail, the prepared chitosan NPs had an average size of 47 ± 4 nm combined with rivastigmine through the emulsification method. In addition, zeta potential analysis revealed that coating the constructed particle with polysorbate 80 reduced its surface charge it. Besides, the biodistribution analysis disclosed that coated NP could alter the potential of uptake by various organs. This method offers a biphasic and Fickian drug release pattern [59].

Jaruszewski *et al.* (2012) provided evidence of amyloid β-protein (Aβ) aggregation in AD. They developed a chitosan-based drug delivery approach to overcome the BBB. The designed Immuno-nano vehicle is based on chitosan-coated poly lactic-co-glycolic acid (PLGA) NPs which are conjugated with an anti-Aβ antibody. *in vitro* studies revealed that the loaded nanocarriers could efficiently cross the BBB and target Aβ proteins. Moreover, chitosan-based technology improved the stability and aqueous dispersibility of the lyophilization process. Taking into consideration these features, it is speculated that these vehicles can be a good platform for therapeutic and diagnostic applications in AD and Parkinson's disease [60].

Zhang *et al.* (2013), on the other hand, used dual-functional NPs for drug delivery in the brain of AD mice. This system was based on surface conjugation of PEGylated polylactic acid (PLA) polymer with two targeting peptides, including TGN and QSH. TGN can specifically target ligands at the BBB, while QSH has a good affinity for Aβ 1-42, the main constituent of amyloid plaques. Noteworthy, the density of TGN and QSH on the surface of nanoparticles was very crucial for obtaining the enhanced and precise dual-targeting effects. In this case, the optimal targeting effect was attained with the molar ratio of 1:3 for TGN/maleimide and QSH/maleimide. The constructed NPs could precisely and efficiently target AD lesions in the brains of the AD model without cytotoxic impacts [45].

Over recent decades, various studies have demonstrated the diverse biological activity of curcumin as well as its remarkable neuroprotective effects. Despite the extraordinary pharmacological effects of curcumin, it has poor bioavailability and instability. Indeed, this natural compound is prone to be oxidized, hydrolyzed and decomposed, and not reach the biological system. With this background, this setback can be bypassed by the incorporation of curcumin into nanocapsules for easy delivery and crossing the BBB [61].

Moreover, an engineered long-circulating nano-carrier was constructed by Brambilla *et al.* (2012). This class of high-molecular-weight glycolated polyethylene could interact with the Aβ1-42 peptide both in serum and solution. The authors suggest that the engineered NPs have a value in capturing toxic forms of the Aβ peptide from the systemic circulation, thereby improve AD status through nanosink effects [62].

In another similar study, Mathew *et al.* (2012) applied a curcumin loaded-PLGA NPs conjugated with Tet-1 peptide [63]. Taken together, curcumin has the potential to be used as a therapeutic agent for AD therapy due to its pleiotropic effects [64, 65].

Growing evidence has shown that curcumin can act as a contrast substance in MRI for the detection of Aβ plaque in AD [66]. Despite the significant properties of curcumin, substantial hurdles remain for *in vivo* application. To combat these obstacles, Mathew and colleagues coated curcumin with water-soluble PLGA NPs conjugated with Tet-1 peptide. These particles were found to have an affinity for neurons and confirmed retrograde transport of NPs. This approach showed the potential of curcumin for destroying Aβ aggregates. It is worth noting that the developed NPs exhibited anti-oxidative effects without meaningful cytotoxicity [66].

In another work conducted by Elnaggar and his co-workers (2015), piperine-loaded chitosan NPs were utilized for intranasal delivery in a rat model of AD. The developed NPs could significantly improve cognitive functions, and interestingly, their efficacy is as good as observed with the standard drugs (donepezil). In addition, these carriers alleviated the piperine-induced nasal irritation in the brain, without any adverse effects [67]. Noteworthy, this was the first report regarding the anti-inflammatory and anti-apoptotic effects of piperine (PIP) in AD. Collectively, the developed mucoadhesive chitosan particles are a non-invasiveness strategy for safe PIP delivery. In addition, these vehicles are as efficient as oral administration, but with a lower dosage (about 20-folds) [67].

Dendrimer, as a new class of polymer of the 21ˢᵗ century, has the ability to solubilize sparingly soluble ingredients in aqueous media, thus acting as a potential drug carrier for AD [68]. Dendrimers are polymeric nanostructures, consisting of regular branched architecture. These particles are composed of three main components, including central core, interior dendritic structure, and exterior functionalized surface. In other words, in a stepwise synthesis procedure, the structure of dendrimers can be tethered. Based on the unique properties of dendrimers, these particles have gained considerable research interest in recent years [68].

Another study by Patel *et al*. showed that sialic-acid-conjugated dendrimers could mimic cell surface sialic acids, thereby attenuated Aβ-induced neurotoxic effects. Based on the results, the authors suggested that such agents might be helpful for the prevention of Aβ toxicity in AD [69].

It has been shown that polyamidoamine (PAMAM) dendrimers can act as a carrier for the delivery of carbamazepine(CBZ), a potent autophagy enhancer in the AD model. However, due to poor solubility and irregular gastrointestinal absorption, CBZ displays incomplete drug bioavailability. Accordingly, there is a need to use a higher dose to achieve desired therapeutic effects. Considering the above-mentioned points, increasing the solubility of CBZ improved the absorption rate and oral bioavailability and reduced the needed dose of CBZ to attain therapeutic concentrations. Indeed, the PAMAM dendrimers inhibit ß amyloid fibril formation and destroy the existing aggregation of ß amyloid. In this way, Igartúa *et al*. developed G4 PAMAM and G4.5 PAMAM dendrimers encapsulated 20 molecules of CBZ. This delivery system offers a controlled-release profile with less toxicity as compared to free CBZ [70].

In recent research, Aso *et al*. (2019) employed a novel type of dendrimers with a core of poly (propylene imine) and a shell of maltose-histidine (G4HisMal). This nanoparticle could significantly improve biocompatibility and the ability to cross BBB in the transgenic mouse model. Noteworthy, G4HisMal displayed synapse and memory protection in AD, as a result, improved the spatial memory deficits [71].

Lipid Nanoparticles

After the discovery of liposomes in the 1960s, special attention has been paid to these nano-medical devices due to their unique properties, including being low-toxic, non-immunogenic, biodegradable, biocompatible, and flexible format [72, 73].

Mourtas *et al*. (2014) provided multifunctional nano-liposomes for brain drug delivery. This structure incorporated the curcumin derivative with the surface decoration of the anti-transferrin antibody (TrF) as a BBB transport mediator [74]. In the analysis of post-mortem brains of AD, a high tendency for amyloid deposits was revealed. The authors of this study verified that binding of the TrF on the curcumin-liposome surface did not affect staining of Aβ deposit or prevent Aβ aggregation. On the other hand, the curcumin-PEG-lipid conjugate had no effect on brain-targeting potential, inferring that these new multifunctional NLs can be used for diagnostic and therapeutic intervention in AD patients [74].

Lipid nanoparticles (LNPs) such as solid lipid nanoparticles (SLNs) and nanostructured lipid carriers (NLCs) are one of the remarkable drug delivery systems as well as a good alternative carrier to liposomes, emulsions, and polymeric nanoparticles. Owing to various advantages such as physical stability, good drug release profile, and targeted delivery, SLNs have the potential to circumvent some of the limitations of colloidal carriers. NLCs as modified SLNs, are capable of improving stability and loading capacity. These systems have the potential to be applied in several fields, such as targeted drug delivery, resulting in an improvement in the physical stability and drug release profile [73].

In another study using an animal model of AD, Bernardi *et al.* used indomethacin-loaded lipid-core nanocapsules (IndOH-LNCs) to evaluate the possible effects of construction on amyloid beta (Aβ)1-42-induced cell damage and neuro-inflammation. They found that in organotypic culture, the developed carrier could attenuate Aβ-induced cell damage and neuroinflammatory responses. Moreover, IndOH-LNC increased interleukin-10 secretion accompanied by reducing the activation of glial and phosphorylation of c-jun N-terminal kinase [75]. It is also important to note that lipoprotein-based nanoparticles are a good candidate for diagnostic and therapeutic implications. Besides, the high binding affinity of these NPs to Aβ leads to the facile degradation of them [57].

Song *et al.* (2014) prepared lipoprotein-based nanoparticles to rescue the memory deficit in mice with AD. In brief, an Apo-lipoprotein E3-reconstituted high-density lipoprotein (ApoE3-rHDL) was constructed to accelerate the clearance of Aβ.42. After one hour of IV administration, around 0.4% ID/g of ApoE3-rHDL provided brain access. Four-week daily treatment with ApoE3-rHDL reduced Aβ deposition, diminished microgliosis, improved neurologic alterations, and importantly, rescued memory loss in an animal model. These observations suggest the potential of ApoE3–rHDL in therapeutic applications for patients suffering from AD. However, its toxic effects are not fully known [76].

In another study by Muntimadugu *et al.* (2016), intranasal delivery of nanoparticles encapsulated with tarenflurbil (TFB) was employed to gain brain access. For loading TBF, two nano-carriers were used, including poly lactide-c--glycolide (PLGA) nanoparticles (TFB-NPs) and solid lipid nanoparticles (TFB-SLNs). Both of them showed desirable brain bio-distribution, suggesting the potential use of this kind of administration in AD [77].

TFB is a selective lowering agent of Aβ42 as well as an enantiomer form of flurbiprofen.

γ-secretase performs the final cleavage of amyloid β protein precursor APP and releases of Aβ peptide. Accordingly, inhibition of γ-secretase has positive effects in AD through reducing the level of Aβ [78, 79].

Compared with a pure drug, *in vitro* release tests proved the sustained release performance of TFB from both nanocarriers TFB-NPs and TFB-SLNs, indicating long residence times at the target site. Pharmacokinetic evaluations also showed better circulation behavior of these nanoparticles. The absolute bioavailability was found as follows: TFB-NPs (i.n.) > TFB-SLNs (i.n.) > TFB solution (i.n.) > TFB suspension (oral). These encouraging findings demonstrated that TFB could be transported to the brain through the olfactory pathway when polymeric and lipidic nanoparticles administered intranasally [77].

Loureiro *et al.* (2017) provided evidence that grape extracts (skin and seed) had inhibitory effects on Aβ aggregation. Resveratrol is a natural compound found in the seeds and skins of grapes, exhibiting neuroprotective characteristics [80]. The authors developed a solid lipid nanoparticle (SLN) functionalized with the anti-transferrin receptor monoclonal antibody (OX26 mAb), as a possible carrier for improving cellular uptake in an *in vitro* model of the human BBB.

Decoration of SLN with OX-26 antibody led to increased uptake of the SLN by the brain when compared with non-functionalized SLNs or even with non-specific antibodies such as LB 509. Collectively, this kind of functionalization can be applied for targeting the BBB and may be used as a possible carrier for the delivery of other compounds to the brain [80].

Nanoemulsion-based systems are colloidal dispersions composing of essential components such as oil, water, and an emulsifier as a surfactant, at an appropriate ratio. Indeed, the surfactant is added to this system for stabilizing and making a metastable isotropic vehicle. Based on the diameter, these thermodynamically stable carriers are transparent or translucent. Using nanoemulsion-based delivery systems could affect the stability and solubility of the incorporated agents; thereby these structures can be applied as a promising delivery strategy in various biomedical applications [81].

Sood *et al.* designed curcumin-donepezil nanoemulsions for brain delivery *via* intranasal administration. These structures were stabilized by adding surfactants and a co-surfactant. The prepared vehicle significantly improved the solubility of curcumin for brain targeting in AD [82]. Also, Ferreira *et al.* prepared pomegranate seed oil nanoemulsions that contain ketoprofen. Pullulan was added to this carrier as a stabilizer agent. *in vitro* release profile revealed that nanoemulsions could release 95.0% of ketoprofen in 5 h. These nanoemulsions exhibited adequate physicochemical properties with a concomitant improvement of bioavailability that ultimately enhance brain permeation [83].

In another study by Hussein *et al.*, they synthesized carvacrol-based nanoemulsion to treat neurodegenerative disorders in a rat model of diabetes. The results revealed that carvacrol nanoemulsion could effectively attenuate neuronal injury and inflammation. Besides the level of homocysteine was decreased following treatment of this nanoemulsion. The observed findings are mediated by inhibition of TNFα-induced NF-κB signaling pathways [84].

Naringenin is the predominant flavonoid that possesses potent neuroprotective effects against free radicals and inflammatory activities. However, its bioavailability is limited by crossing biological membranes.

In this regard, Shadab *et al.* tested the neuroprotective effects of naringenin-based nanoemulsion. They observed the naringenin-loaded nanoemulsion could significantly alleviate the neurotoxic effects of β-amyloid plaque on human neuroblastoma cell lines (SH-SY5Y). The observed effects are probably mediated by down-regulation of APP and tau phosphorylation, indicating the potential therapeutic application of this strategy in AD therapy [85].

Other Nanoparticles

In a recent study, Haimov *et al.* (2019) developed metal-based nanoparticles as a drug carrier for mTHPC drug, a photosensitizer molecule. They used nanoparticles that were composed of either gold or magnetic carriers such as Ce-doped or Yb-doped -γ-Fe2O3 maghemite. This magnetic metal-based approach enhanced the penetration and accumulation of the drug in the tumor tissues [86].

In another similar research, promising results have been reported about applying cerium (Ce) oxide nanoparticles (CNPs), as an interesting antioxidant in biological systems. These carriers exhibited anti-inflammatory, anti-apoptotic, and radical scavenging activities. The various redox states of Ce provide radical scavenging properties, creating an antioxidant niche to repair brain cell damage. A substantial amount of work has also been performed on other metals with different oxidation states and toxicities. This nano metallic formulation could be a good candidate for the targeted delivery of CNS- drugs. Also, growing research has been done on multiple metals with different oxidation states [87].

DNA-nanoparticles conjugates can detect protein biomarkers at a low concentration of about 10–18 moles/liter. This detection system is also called a bio-barcode assay, using gold nanoparticles conjugated with an antibody for targeting specific proteins [61].

And in the following studies, Zhang *et al.* (2014) designed dual-functional delivery vehicles loaded with H102 (HKQLPFFEED) peptide, as a β-sheet breaker. TGN and QSH peptides were conjugated to the surfaces of NPs, with the aim of brain-targeting and accumulation in the brain for targeting Aβ42 peptides, respectively [11]. This dual-function delivery system offers a precise strategy for delivery of the β-sheet breaker to the brain lesions in the AD mice model. The results demonstrated that TQNP is a promising carrier for peptide or protein drugs, such as H102 to enter into the CNS and localization in the brain AD lesions. Taken together, these methods provide a specific therapeutic approach toward AD [88].

Cubosomes are cubic liquid crystalline nanoparticles composed of certain amphiphilic lipids with an appropriate ratio. Cubosomes are formed from curved bicontinuous lipid bilayers organized in a three-dimensional honeycombed structure, separating into two internal aqueous channels. These lipid-based nanoparticles can encapsulate amphiphilic, hydrophilic, and hydrophobic substances, offering as biocompatible carriers for a controlled release drug delivery for various routes of administration such as peptides and proteins and bioactive ingredients [89].

In another study, Elnaggar *et al.* (2015) used a novel Tween-integrated monoolein cubosomes (T-cubs), containing PIP for brain targeting in an animal model of AD. PIP as a phytopharmaceutical drug has the neuroprotective potential in AD therapy that is mediated by memory improvement. *in-vivo* studies revealed that T-cubs could dramatically enhance the effect of PIP on cognitive function, and restore to the normal value. In other words, PIP-loaded cubs have beneficial impacts in halting AD progression and improving the cognitive deficit which may be attributed to anti-apoptotic and anti-inflammatory effects. Interestingly, no brain toxicity was observed with this formulation on the kidneys and livers of experimental Wistar rats [90].

Magnetic nanoparticles (MNPs) are an attractive class of functional material for diagnostic and therapeutic applications [61]. Due to the inherent superparamagnetic features of these NPs, Do *et al.* (2016) designed a novel nanotechnology-based method to deliver therapeutic compounds to the brain. In this regard, an electromagnetic actuator was used to guide magnetic-containing agents. The developed carrier could cross the normal BBB when exposed to external electromagnetic interference of 28 mT and 0.43 T/m, and/or 79.8 mT and 1.39 T/m. The rate of MNP uptake and crossing the BBB was significantly increased *via* using a pulsed magnetic field. Moreover, brain localization of NPs was established by fluorescent MNPs. The application of 770-nm fluorescent carboxyl magnetic NPs showed the suitability of this approach for drug delivery [91].

Quantum dots, as the new generation of NPs, have gained much attention in recent years. Inorganic materials such as SiO2 is being able to cross the BBB. Accordingly, some inorganic compounds like SiO2 have been investigated for their ability to penetrate the BBB. Due to the high biocompatibility of silicon, silicon quantum dots are the promising carriers for diagnostic and therapeutic implications [92].

As discussed in the text, conjugation of moieties to the NPs can widen their application. In this context, the conjugation of antibodies with nanoparticles combines the specificity and selectivity of the antibodies with the unique properties of NPs. Antibodies are nanosized biological molecules that can bind and neutralize antigens and complement systems activation. The antibody-conjugated nanoparticles eliminate immune stimulations, thereby offering novel diagnostic applications for AD therapy [93].

Carradori *et al.* (2018) designed antibody-functionalized NPs that lead to memory recovery in a transgenic mouse model of AD [94]. To do so, they synthesized PEGylated biodegradable NPs which were surface-functionalized with an Aβ1-42 monoclonal antibody. Treatment of the experimental model of AD with antibody-decorated NPs resulted in the following events: full correction of memory deficit, a dramatic reduction in the level of Aβ soluble peptides in the brain, and significant increment in plasma level of the Aβ peptide. Further investigations are needed to develop antibody-decorated NPs with more potency and less toxicity [94].

Tamba *et al.* (2018) developed a novel tailor-made NPs in which the fluorescent [Ru(bpy)3]Cl2 dye was entrapped in silica NPs to monitor crossing across the BBB. They synthesized three kinds of SNPs derivatives for in vivo evaluation, including Batch fluorescently decorated mono-shell silica NPs (Ru@SNPs), the glucose-coated silica NPs (Ru@Glu-SNP), and the PEGylated glucose-coated silica NPs (Ru@Glu-PEG-SNP).

The findings pointed out the therapeutic application of engineered silica NPs derivatives for drug delivery across the BBB. The mechanism by which these vehicles cross the BBB is mediated *via* specific and non-specific interactions [95].

Another class of nanotechnology-based strategy is nanocomposite. Nanocomposites are solid multiphase materials, consisting of nanoscale structures with multiple dimensions. This kind of strategy offers a wide range of therapeutic applications in drug delivery for AD therapy, cancer therapy, stem cell-based therapy, biosensing, antimicrobial activity, and enzyme immobilization [96].

In recent years, growing studies have indicated that the tau pathway is strongly associated with clinical symptoms of AD. Hence, targeting Tau proteins can be a promising approach toward AD therapy. Chen *et al.* (2018) constructed a methylene blue as an aggregation inhibitor of Tau protein, loading by multifunctional nanocomposite (CeNC/IONC/MSN-T807). The established structure had a high affinity to bind to hyperphosphorylated tau and prevent tau-associated signaling pathways involved in the pathogenesis of AD. Both *in vitro* and in vivo experiments demonstrated that these nanocomposites could alleviate the AD symptoms *via* reducing mitochondrial-related oxidative stress, repression of tau hyperphosphorylation, and preventing neuronal cell death (Fig. **2**) [97].

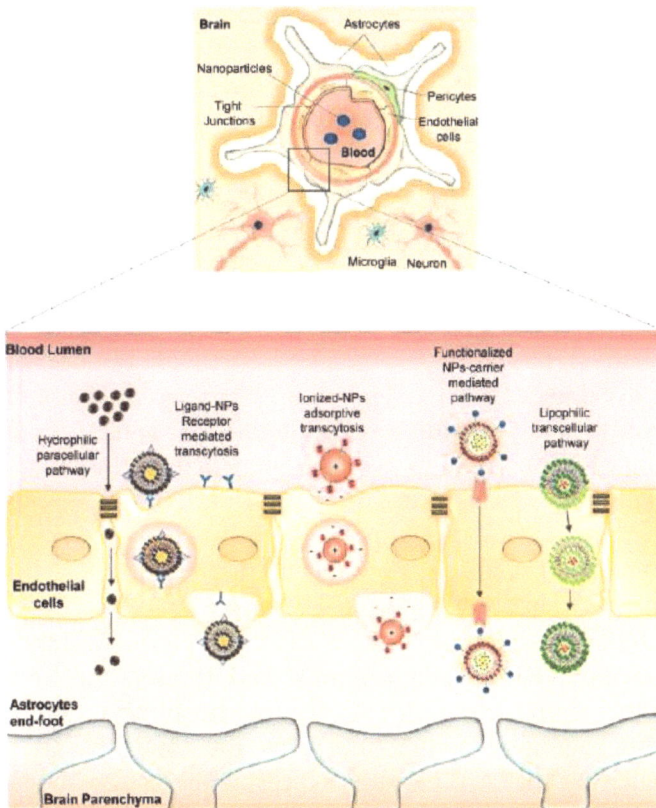

Fig. (2). The potential pathways for nanoparticle-based drug delivery to the brain in AD [23].

NANOTECHNOLOGY IN AD DIAGNOSIS

Biomarkers in AD Diagnosis

Since the symptoms of AD (*e.g.* cognitive dysfunction) appear in the later phases of the disease; thus, the identification of the pathological process of AD concerning the clinical phenotype is difficult. Therefore. Biomarkers are valuable tolls in early AD diagnosis in particular the prodromal AD [98]. As mentioned before Aβ (Aβ1-42 and Aβ1-40), P-tau181, and T-tau (6 isoforms) are three known biomarkers of the disease [99]. In addition, APOE4 has been considered as another important biomarker [99]. The sensitivity and specificity of these biomarkers are considerably high in the early-onset AD [100]. The ideal biosensor for each of these biomarkers could be chosen based on their structure. However, inadequate information regarding the exact pathogenesis of the disease and its heterogeneity might restrict their clinical application.

Accordingly, there is a need to have affordable, selective, and ultrasensitive molecular-based detection approaches. Collectively, the application of nanotechnology-based detection methods fulfills the early diagnosis of AD. In the era of molecular molecular-based diagnosis, detection can be performed either within the body or derived samples from the body. Indeed, nanotechnology provides early diagnosis of AD through signal transduction, a fundamental process in that cells transform biological signals and amplify enough to be recorded.

In other words, the potential of nanotechnology-based detection relied on the features of NPs such as physical parameters (electrical, optical, or magnetic), chemical and biological aspects [5, 101]. It has been shown, for example, the Fe^{3+} /Cu^{2+} interaction with Aβ peptide resulted in oxidative stress, as shown in Fig. (3), Oligomerization of Aβ peptide in the lipid bilayer of cell membranes leads to the formation of calcium channels in the membranes. The channels induce an imbalance in the homeostasis of calcium, thereby drive to oxidative stress. Moreover, membrane integrated Aβ peptide interacts with membrane lipids *via* methionine amino acid, leading to lipid peroxidation and production of 4-hydroxy-2-nonenal (4HNE). These events disrupt cell membranes leading to reactive oxygen species (ROS) accumulation and finally, the development of oxidative stress in the brain tissue. Moreover, the Phosphorylation of tau protein and subsequent aggregation is mediated by the 4HNE and other ROS. Intracellular aggregation of Aβ peptide is responsible for the induction of mitochondrial oxidative stress as well as calcium hemostasis imbalance. Interestingly, the resultant deficiency of mitochondrial electron transport chain causes superoxide anion overproduction, forming H2O2 or peroxynitrite ONOO (generating from nitric oxide). Besides, generation of the hydroxyl radical (OH*) is initiated by the interaction of Fe^{2+} and Cu^{2+} with H_2O_2, promoting membrane-related oxidative stress [102].

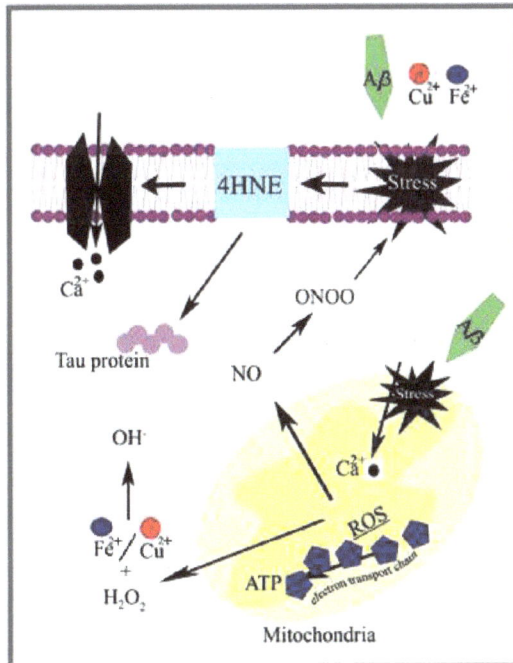

Fig. (3). Interaction of Fe^{3+} and Cu^{2+} with Aβ and the production of oxidative stress. Adopted from [103].

DNA-Nanoparticle Conjugates

Atomic-scale detection of biomarkers can be useful in many cases and is made possible by DNA-Nanoparticle conjugates [104, 105]. For example, the bio-barcode assay is a known method for the detection of ultra-low concentrations of biomarkers, the method of detection owing to carrier gold nanoparticles should match the specific antibody of the target biomarker with hundreds of DNA barcodes. Therefore, a single molecule of the biomarker may be traced by hundreds of DNA barcodes (a biological signal transformation), which could be additionally amplified by the polymerase chain reaction (PCR) technique [104, 105]. The reported results of researches show that bio-barcode essay can be a very useful and sensitive method for amyloid β-derived diffusible ligand (ADDL) detection in cerebrospinal fluid (CSF) samples of AD patients [105], ADDL concentrations in AD patients, and age-matched healthy control group varies significantly.

Another method for the detection of molecular biomarkers that has received a lot of attention recently is the use of localized surface plasmon resonance (LSPR) nanosensor. This nanosensor is ultra-sensitive, inexpensive, and based on singular optical properties of triangular silver nanoparticles. Any change in the external environment of nanoparticles such as target biomolecule concentration (the target biomolecule can be ADDL) cause a change in the refractive index of the surrounding magnetic field and different wavelength shifts for different concentrations are detectable.

Three major applications for LSPR nanosensors have been identified, as follows 1) investigation of Aβ oligomerization *in vivo* condition, the concentration of Aβ: is ultra-low at conditions *in vivo*, 2) AD patients screening, and 3) investigation of pharmaceutics and target molecules in drug discovery and development [102].

Protein biomarkers detection with attomolar sensitivity is possible by using a method known as bio- barcode assay, DNA-Nanoparticle conjugates are capable of trapping a single molecule of biomarker by hundreds of DNA barcodes. Biomarkers in ultra-low concentration can be detected by a biological signal transformation. In addition, the target biomarker that is trapped could be amplified by the polymerase chain reaction (PCR) technique [104, 105]. Haes AJ *et al.* [105] reported detection of amyloid β-derived diffusible ligand (ADDL) in cerebrospinal fluid (CSF) samples of AD patients in ultra-low concentration with atomic sensitivity that had been done by bio-barcode assay. According to the results of this study, there is a significant difference between concentrations of ADDL in AD diagnosed patients and age-matched healthy control groups. The difference in ADDL concentration between the two groups is reported as 1.7 fM and ca.200 aM; therefore the relationship between increased ADDL concentrations and the incidence of Alzheimer's disease becomes clear.

Early detection of AD biomarkers has become very important due to its essential role in early and effective treatment. Recently localized surface plasmon resonance (LSPR) nanosensors for the detection of AD biomarkers have been considered. This method is ultra-sensitive, inexpensive, and is based on singular optical properties of triangular silver nanoparticles (AgNPs), such a way that, changes in ADDL concentration in cerebrospinal fluid samples directly change the refractive index of the surrounding magnetic field. Eventually, different wavelength shifts for various concentrations of samples are detectable. In general, there are three main applications for (LSPR) nanosensors: 1) Investigation and study of the Aβ oligomerization at *in vivo* conditions, because ultra-low concentrations of Aβ, 2) Screening patients at risk for AD, 3) Studying pharmaceutics and target biomolecules interactions on the way to drug discovery

[102]. Considering all the above, it seems that the use of this technology can be very useful in the field of diagnosis and treatment.

Magnetic Nanoparticles

Iron oxide nanoparticles as a contrast agent in magnetic resonance imaging (MRI) in recent years have received a lot of attention [106, 107]. According to the published reports, iron oxide nanoparticles in the form of monocrystalline iron oxide nanoparticles (MION) and ultra-small superparamagnetic iron oxide (USPIO) nanoparticles as MRI contrast agents have been used for detection of amyloid peptide plaques in the brain of mouse models; samples were transgenic mouse models of AD [108]. In these studies, nanoparticles conjugated with Aβ were detection factors for amyloid plaques [107]. The use of this technique supply could be minimally invasive, especially when intravenous injection of MRI agent (nanoparticles) is used instead of intra-arterially injection [108]. Despite all the content mentioned, diagnosis of AD in early stages by these techniques is not possible, because detection agent (amyloid plaques) appears in later stages of the disease. Therefore the application of these methods for AD patients limit to monitoring effect of candidate drugs that reduce amyloid plaques level.

NIAD-4

Optical imaging through special near-infrared (NIR) fluorescent contrast agents are another attractive approach for in vivo condition molecular biomarkers detection [109]. Scattered light from these contrast agents because of their long wavelength, could penetrate through biological tissues. The ability to cross the BBB, detection of AD-related biomarkers such as Aβ, an appropriate absorption, and emission wavelength interval (600-800 nm), and a strong rigidification are essential factors for molecular AD diagnostic probes. Fluorescent molecule binding with a molecular target undergoes a significant conformational restriction is called origination. The substantial imaging contrast between bounded and unbounded fluorescent markers is the basis of detection. During the studies, NIAD-4 with the chemical formula [[5'-(4-Hydroxyphenyl) [2, 2'-bithiophen]-5-yl] methylene]-propanedinitrile was introduced as a NIR Alzheimer's dye. According to the recorded results, NIR Alzheimer's dye because of its specific structure and low molecular weight (334 Da) has been successful in crossing BBB and Aβ detection [110]. Despite the promising results of this method, further studies are still needed for making this technique noninvasive; looks for this purpose and increasing the redshift and quantum yield can be effective.

Quantum Dots

Quantum dots are an emerging group of nanomaterials with fluorescent properties [111, 112]. These semiconductor nanoparticles due to their unique properties, including minimal photobleaching, optimal stability, high signal-to-noise ratio, and broad absorption spectrum with very narrow but size-dependent tunable emission spectrum, have received a lot of attention. Long-term tracking and simultaneous visualization of multiple physiological and pathological molecular events are possible by using nano-scale crystals [113, 114]. Due to the involvement of various biomarkers in AD pathology, simultaneous multiple labeling subject is important for AD diagnosis. In the case of AD disease In addition to detecting biomarkers, determining their ratio can also be very useful to rule out other differential diagnoses [115].

FUTURE PERSPECTIVES

The multifactorial nature of AD disease has led to the failure of conventional treatment methods for this disease. Unfortunately, mismatches between the research and clinical phase of studies have been very disappointing, but the use of nanotechnology-based methods in recent years has opened the windows of hope for patients and researchers. Although the pathophysiology of AD is not known and there are many unanswered questions but it seems that the use of nanotechnology methods for disease management is inevitable. There is a pressing need to investigate nanotechnology-based methods for early diagnosis and treatment of AD because this is the best way to delay the disease. The main challenge in using nanotechnology-based therapies is the issue of toxicity, which must be seriously and thoroughly examined. Nowadays, many efforts to synthesize biocompatible nanoparticles could be promising to solve the toxicity problem of nanomedicine. Despite many studies, the clinical use of this technology in AD therapy still has a long way to go. Nanotechnology methods for AD treatment, aspects of pharmacokinetic and toxicological should be considered. Magnetic nanoparticles have a lot of potentials to work in this field. Dendrimers and nanoemulsions have also received a great deal of attention as drug delivery systems for this disease, because these systems, due to their special nature and properties, can cross the BBB.

CONCLUSION

Although scientific communities around the world in recent years have taken valuable and significant steps to eradicate many chronic diseases, they have not been very successful in the case of Alzheimer's. It seems the multifactorial nature of AD is the most important cause. Today use of nanomedicine, because it provides the targeted delivery of drugs to brain lesions, has been proposed as a promising option for the treatment of AD. Although patients with mild-to - moderate AD treatment by intranasal nanoparticles of APH-1105 enter phase 2 clinical trials but Use of Nano- based methods for AD treatment needs much more researches. More potent and non-toxic nanomedicine formulations are required for successful drug delivery to treat patients with CNS disorders like AD.

CONSENT FOR PUBLICATION

Not applicable.

CONFLICT OF INTEREST

The authors declare no conflict of interest, financial or otherwise.

ACKNOWLEDGEMENTS

Declared none.

REFERENCES

[1]   Fong TG, Tulebaev SR, Inouye SK. Delirium in elderly adults: diagnosis, prevention and treatment. Nat Rev Neurol 2009; 5(4): 210-20.
        [http://dx.doi.org/10.1038/nrneurol.2009.24] [PMID: 19347026]

[2]   Guo T, Zhang D, Zeng Y, Huang TY, Xu H, Zhao Y. Molecular and cellular mechanisms underlying the pathogenesis of Alzheimer's disease. Mol Neurodegener 2020; 15(1): 40.
        [http://dx.doi.org/10.1186/s13024-020-00391-7] [PMID: 32677986]

[3]   Jack CR Jr, Holtzman DM. Biomarker modeling of Alzheimer's disease. Neuron 2013; 80(6): 1347-58.
        [http://dx.doi.org/10.1016/j.neuron.2013.12.003] [PMID: 24360540]

[4]   Swerdlow RH. Pathogenesis of Alzheimer's disease. Clin Interv Aging 2007; 2(3): 347-59.
        [PMID: 18044185]

[5]   Harilal S, Jose J, Parambi DGT, *et al.* Advancements in nanotherapeutics for Alzheimer's disease: current perspectives. J Pharm Pharmacol 2019; 71(9): 1370-83.
        [http://dx.doi.org/10.1111/jphp.13132] [PMID: 31304982]

[6]   Gupta J, Fatima MT, Islam Z, Khan RH, Uversky VN, Salahuddin P. Nanoparticle formulations in the diagnosis and therapy of Alzheimer's disease. Int J Biol Macromol 2019; 130: 515-26.
        [http://dx.doi.org/10.1016/j.ijbiomac.2019.02.156] [PMID: 30826404]

[7]   Babazadeh A, Mohammadi Vahed F, Jafari SM. Nanocarrier-mediated brain delivery of bioactives for treatment/prevention of neurodegenerative diseases. J Control Release 2020; 321: 211-21.
        [http://dx.doi.org/10.1016/j.jconrel.2020.02.015] [PMID: 32035189]

[8]   Alyautdin R, Khalin I, Nafeeza MI, Haron MH, Kuznetsov D. Nanoscale drug delivery systems and the blood-brain barrier. Int J Nanomedicine 2014; 9: 795-811.
        [PMID: 24550672]

[9]   Wortmann M. Importance of national plans for Alzheimer's disease and dementia. Alzheimers Res Ther 2013; 5(5): 40.

[http://dx.doi.org/10.1186/alzrt205] [PMID: 24007939]

[10]   Gerenu G, Liu K, Chojnacki JE, *et al.* Curcumin/melatonin hybrid 5-(4-hydroxy-phenyl)--oxo-pentanoic acid [2-(5-methoxy-1H-indol-3-yl)-ethyl]-amide ameliorates AD-like pathology in the APP/PS1 mouse model. ACS Chem Neurosci 2015; 6(8): 1393-9.
       [http://dx.doi.org/10.1021/acschemneuro.5b00082] [PMID: 25893520]

[11]   Revett TJ, Baker GB, Jhamandas J, Kar S. Glutamate system, *et al.* amyloid β peptides and tau protein: functional interrelationships and relevance to Alzheimer disease pathology. J Psychiatry Neurosci 2013; 38(1): 6.
       [http://dx.doi.org/10.1503/jpn.110190] [PMID: 22894822]

[12]   Farlow MR, Miller ML, Pejovic V. Treatment options in Alzheimer's disease: maximizing benefit, managing expectations. Dement Geriatr Cogn Disord 2008; 25(5): 408-22.
       [http://dx.doi.org/10.1159/000122962] [PMID: 18391487]

[13]   Birks JS. Cholinesterase inhibitors for Alzheimer's disease. Cochrane Database Syst Rev 2006; (1): CD005593.
       [http://dx.doi.org/10.1002/14651858.CD005593]

[14]   Hansen RA, Gartlehner G, Webb AP, Morgan LC, Moore CG, Jonas DE. Efficacy and safety of donepezil, galantamine, and rivastigmine for the treatment of Alzheimer's disease: a systematic review and meta-analysis. Clin Interv Aging 2008; 3(2): 211-25.
       [PMID: 18686744]

[15]   WAGSTAFF A and, MCTAVISH D. A review of its pharmacodynamic and pharmacokinetic properties, and therapeutic efficacy in alzheimers-disease Drugs Aging 1994; 5(2): 95-5.

[16]   Mullard A. BACE inhibitor bust in Alzheimer trial. Nat Rev Drug Discov 2017; 16: p. (3)155.
       [http://dx.doi.org/10.1038/nrd.2017.43]

[17]   Becker RE, Greig NH, Giacobini E. Why do so many drugs for Alzheimer's disease fail in development? Time for new methods and new practices? J Alzheimers Dis 2008; 15(2): 303-25.
       [http://dx.doi.org/10.3233/JAD-2008-15213] [PMID: 18953116]

[18]   Sharma S, Singh A. Nanotechnology based targeted drug delivery: current status and future prospects for drug development. Drug Discovery and Development–Present and Future, published by infotech open science open minds 2011; 427-63.
       [http://dx.doi.org/10.5772/28902]

[19]   Parveen S, Sahoo SK. Nanomedicine: clinical applications of polyethylene glycol conjugated proteins and drugs. Clin Pharmacokinet 2006; 45(10): 965-88.
       [http://dx.doi.org/10.2165/00003088-200645100-00002] [PMID: 16984211]

[20]   Brambilla D, Verpillot R, Taverna M, *et al.* New method based on capillary electrophoresis with laser-induced fluorescence detection (CE-LIF) to monitor interaction between nanoparticles and the amyloid-β peptide. Anal Chem 2010; 82(24): 10083-9.
       [http://dx.doi.org/10.1021/ac102045x] [PMID: 21086977]

[21]   Bianco A, Kostarelos K, Partidos CD, Prato M. Biomedical applications of functionalised carbon nanotubes. Chem Commun (Camb) 2005; (5): 571-7.
       [http://dx.doi.org/10.1039/b410943k] [PMID: 15672140]

[22]   Kumari A, Yadav SK, Yadav SC. Biodegradable polymeric nanoparticles based drug delivery systems. Colloids Surf B Biointerfaces 2010; 75(1): 1-18.
       [http://dx.doi.org/10.1016/j.colsurfb.2009.09.001] [PMID: 19782542]

[23]   Karthivashan G, Ganesan P, Park SY, Kim JS, Choi DK. Therapeutic strategies and nano-drug delivery applications in management of ageing Alzheimer's disease. Drug Deliv 2018; 25(1): 307-20.
       [http://dx.doi.org/10.1080/10717544.2018.1428243] [PMID: 29350055]

[24]   Zhang F, Lin YA, Kannan S, Kannan RM. Targeting specific cells in the brain with nanomedicines for CNS therapies. J Control Release 2016; 240: 212-26.

[http://dx.doi.org/10.1016/j.jconrel.2015.12.013] [PMID: 26686078]

[25]   Garcia-Chica J, D Paraiso WK, Tanabe S, *et al.* An overview of nanomedicines for neuron targeting. Nanomedicine (Lond) 2020; 15(16): 1617-36.
[http://dx.doi.org/10.2217/nnm-2020-0088] [PMID: 32618490]

[26]   Su S, Kang PM. Systemic review of biodegradable nanomaterials in nanomedicine. Nanomaterials (Basel) 2020; 10(4): 656.
[http://dx.doi.org/10.3390/nano10040656] [PMID: 32244653]

[27]   Sato Y, Sakurai Y, Kajimoto K, *et al.* Innovative technologies in nanomedicines: from passive targeting to active targeting/from controlled pharmacokinetics to controlled intracellular pharmacokinetics. Macromol Biosci 2017; 17(1): 1600179.
[http://dx.doi.org/10.1002/mabi.201600179] [PMID: 27797146]

[28]   Fisher A. Cholinergic modulation of amyloid precursor protein processing with emphasis on M1 muscarinic receptor: perspectives and challenges in treatment of Alzheimer's disease. J Neurochem 2012; 120 (Suppl. 1): 22-33.
[http://dx.doi.org/10.1111/j.1471-4159.2011.07507.x] [PMID: 22122190]

[29]   Jones AT. Gateways and tools for drug delivery: endocytic pathways and the cellular dynamics of cell penetrating peptides. Int J Pharm 2008; 354(1-2): 34-8.
[http://dx.doi.org/10.1016/j.ijpharm.2007.10.046] [PMID: 18068916]

[30]   Zanganeh S, Spitler R, Erfanzadeh M, Alkilany AM, Mahmoudi M. Protein corona: Opportunities and challenges. Int J Biochem Cell Biol 2016; 75: 143-7.
[http://dx.doi.org/10.1016/j.biocel.2016.01.005] [PMID: 26783938]

[31]   Derakhshankhah H, Sajadimajd S, Jafari S, *et al.* Novel therapeutic strategies for Alzheimer's disease: Implications from cell-based therapy and nanotherapy. Nanomedicine 2020; 24: 102149.
[http://dx.doi.org/10.1016/j.nano.2020.102149] [PMID: 31927133]

[32]   Stegemann S, Leveiller F, Franchi D, de Jong H, Lindén H. When poor solubility becomes an issue: from early stage to proof of concept. Eur J Pharm Sci 2007; 31(5): 249-61.
[http://dx.doi.org/10.1016/j.ejps.2007.05.110] [PMID: 17616376]

[33]   Sainsbury F, Zeng B, Middelberg AP. Towards designer nanoemulsions for precision delivery of therapeutics. Curr Opin Chem Eng 2014; 4: 11-7.
[http://dx.doi.org/10.1016/j.coche.2013.12.007]

[34]   Aprahamian M, Michel C, Humbert W, Devissaguet JP, Damge C. Transmucosal passage of polyalkylcyanoacrylate nanocapsules as a new drug carrier in the small intestine. Biol Cell 1987; 61(1-2): 69-76.
[http://dx.doi.org/10.1111/j.1768-322X.1987.tb00571.x] [PMID: 2833966]

[35]   Mathison S, Nagilla R, Kompella UB. Nasal route for direct delivery of solutes to the central nervous system: fact or fiction? J Drug Target 1998; 5(6): 415-41.
[http://dx.doi.org/10.3109/10611869808997870] [PMID: 9783675]

[36]   De Rosa G, Salzano G, Caraglia M, Abbruzzese A. Nanotechnologies: a strategy to overcome blood-brain barrier. Curr Drug Metab 2012; 13(1): 61-9.
[http://dx.doi.org/10.2174/138920012798356943] [PMID: 22292810]

[37]   Fonseca-Santos B, Gremião MPD, Chorilli M. Nanotechnology-based drug delivery systems for the treatment of Alzheimer's disease. Int J Nanomedicine 2015; 10: 4981-5003.
[http://dx.doi.org/10.2147/IJN.S87148] [PMID: 26345528]

[38]   Loureiro JA, Gomes B, Coelho MA, do Carmo Pereira M, Rocha S. Targeting nanoparticles across the blood-brain barrier with monoclonal antibodies. Nanomedicine (Lond) 2014; 9(5): 709-22.
[http://dx.doi.org/10.2217/nnm.14.27] [PMID: 24827845]

[39]   Markoutsa E, Papadia K, Giannou AD, *et al.* Mono and dually decorated nanoliposomes for brain targeting, *in vitro* and in vivo studies. Pharm Res 2014; 31(5): 1275-89.

[http://dx.doi.org/10.1007/s11095-013-1249-3] [PMID: 24338512]

[40]   Mufamadi MS, Choonara YE, Kumar P, *et al.* Ligand-functionalized nanoliposomes for targeted delivery of galantamine. Int J Pharm 2013; 448(1): 267-81.
[http://dx.doi.org/10.1016/j.ijpharm.2013.03.037] [PMID: 23535346]

[41]   Rocha S. Targeted drug delivery across the blood brain barrier in Alzheimer's disease. Curr Pharm Des 2013; 19(37): 6635-46.
[http://dx.doi.org/10.2174/13816128113199990613] [PMID: 23621533]

[42]   Gregori M, Salvati E. Functionalization with TAT-peptide enhances blood-brain barrier crossing *in vitro* of nanoliposomes carrying a curcumin-derivative to bind amyloid-β peptide. J Nanomed Nanotechnol 2013; 4:3.
[http://dx.doi.org/10.4172/2157-7439.1000171]

[43]   Härtig W, Paulke BR, Varga C, Seeger J, Harkany T, Kacza J. Electron microscopic analysis of nanoparticles delivering thioflavin-T after intrahippocampal injection in mouse: implications for targeting β-amyloid in Alzheimer's disease. Neurosci Lett 2003; 338(2): 174-6.
[http://dx.doi.org/10.1016/S0304-3940(02)01399-X] [PMID: 12566180]

[44]   Sun W, Xie C, Wang H, Hu Y. Specific role of polysorbate 80 coating on the targeting of nanoparticles to the brain. Biomaterials 2004; 25(15): 3065-71.
[http://dx.doi.org/10.1016/j.biomaterials.2003.09.087] [PMID: 14967540]

[45]   Zhang C, Wan X, Zheng X, *et al.* Dual-functional nanoparticles targeting amyloid plaques in the brains of Alzheimer's disease mice. Biomaterials 2014; 35(1): 456-65.
[http://dx.doi.org/10.1016/j.biomaterials.2013.09.063] [PMID: 24099709]

[46]   Zhang C, Chen J, Feng C, *et al.* Intranasal nanoparticles of basic fibroblast growth factor for brain delivery to treat Alzheimer's disease. Int J Pharm 2014; 461(1-2): 192-202.
[http://dx.doi.org/10.1016/j.ijpharm.2013.11.049] [PMID: 24300213]

[47]   Yang Z-Z, Zhang YQ, Wang ZZ, Wu K, Lou JN, Qi XR. Enhanced brain distribution and pharmacodynamics of rivastigmine by liposomes following intranasal administration. Int J Pharm 2013; 452(1-2): 344-54.
[http://dx.doi.org/10.1016/j.ijpharm.2013.05.009] [PMID: 23680731]

[48]   Wilson B, Samanta MK, Santhi K, Kumar KP, Ramasamy M, Suresh B. Chitosan nanoparticles as a new delivery system for the anti-Alzheimer drug tacrine. Nanomedicine 2010; 6(1): 144-52.
[http://dx.doi.org/10.1016/j.nano.2009.04.001] [PMID: 19446656]

[49]   Swerdlow RH, Burns JM, Khan SMJBeBA-MBoD. The Alzheimer's disease mitochondrial cascade hypothesis: progress and perspectives. 2014; 1842(8): 1219-31.
[http://dx.doi.org/10.1016/j.bbadis.2013.09.010]

[50]   Dey A, Bhattacharya R, Mukherjee A, Pandey DK, *et al.* Natural products against Alzheimer's disease: Pharmaco-therapeutics and biotechnological interventions. Biotechnol Adv 2017; 35(2): 178-216.

[51]   Daulatzai MAJAjond. Fundamental role of pan-inflammation and oxidative-nitrosative pathways in neuropathogenesis of Alzheimer's disease. 2016; 5(1): 1.

[52]   Heller A, Brockhoff G, Goepferich A, *et al.* Targeting drugs to mitochondria. Eur J Pharm Biopharm 2012; 82(1): 1-18.
[http://dx.doi.org/10.1016/j.ejpb.2012.05.014]

[53]   Md S, Bhattmisra Sk, zeeshan F, *et al.* Nano-carrier enabled drug delivery systems for nose to brain targeting for the treatment of neurodegenerative disorders. J Drug Deliv Sci Technol 2018; 43: 295-310.
[http://dx.doi.org/10.1016/j.jddst.2017.09.022]

[54]   Contestabile AJBbr. The history of the cholinergic hypothesis. Behav Brain Res 2011 Aug; 221(2): 334-40.
[http://dx.doi.org/10.1016/j.bbr.2009.12.044]

[55] Wen MM, El-Salamouni NS, El-Refaie WM, *et al.* Nanotechnology-based drug delivery systems for Alzheimer's disease management: Technical, industrial, and clinical challenges. J Control Release 2017; 245: 95-107.

[56] Sanabria-Castro A, Alvarado-Echeverría I, Monge-Bonilla CJAon. Monge-Bonilla, Molecular pathogenesis of Alzheimer's disease: an update. Ann Neurosci 2017 May; 24(1): 46-54.
[http://dx.doi.org/10.1159/000464422]

[57] Van Giau V, An SSA, Hulme JP. Mitochondrial therapeutic interventions in Alzheimer's disease. J Neurol Sci 2018; 395: 62-70.
[http://dx.doi.org/10.1016/j.jns.2018.09.033] [PMID: 30292965]

[58] Kulkarni PV, Roney CA, Antich PP, Bonte FJ, Raghu AV, Aminabhavi TM. Quinoline--butylcyanoacrylate-based nanoparticles for brain targeting for the diagnosis of Alzheimer's disease. Wiley Interdiscip Rev Nanomed Nanobiotechnol 2010; 2(1): 35-47.
[http://dx.doi.org/10.1002/wnan.59] [PMID: 20049829]

[59] Wilson B, Samanta MK, Muthu MS, Vinothapooshan G. Design and evaluation of chitosan nanoparticles as novel drug carrier for the delivery of rivastigmine to treat Alzheimer's disease. Ther Deliv 2011; 2(5): 599-609.
[http://dx.doi.org/10.4155/tde.11.21] [PMID: 22833977]

[60] Jaruszewski KM, Ramakrishnan S, Poduslo JF, Kandimalla KK. Chitosan enhances the stability and targeting of immuno-nanovehicles to cerebro-vascular deposits of Alzheimer's disease amyloid protein. Nanomedicine 2012; 8(2): 250-60.
[http://dx.doi.org/10.1016/j.nano.2011.06.008] [PMID: 21704598]

[61] Ahmad J, Akhter S, Rizwanullah M, *et al.* Nanotechnology based Theranostic approaches in Alzheimer's disease management: current status and future perspective. Curr Alzheimer Res 2017; 14(11): 1164-81.
[http://dx.doi.org/10.2174/1567205014666170508121031] [PMID: 28482786]

[62] Brambilla D, Verpillot R, Le Droumaguet B, *et al.* PEGylated nanoparticles bind to and alter amyloid-beta peptide conformation: toward engineering of functional nanomedicines for Alzheimer's disease. ACS Nano 2012; 6(7): 5897-908.
[http://dx.doi.org/10.1021/nn300489k] [PMID: 22686577]

[63] Mathew A, Fukuda T, Nagaoka Y, *et al.* Curcumin loaded-PLGA nanoparticles conjugated with Tet-1 peptide for potential use in Alzheimer's disease. PLoS One 2012; 7(3): e32616.
[http://dx.doi.org/10.1371/journal.pone.0032616] [PMID: 22403681]

[64] Reddy PH, Manczak M, Yin X, *et al.* Protective effects of Indian spice curcumin against amyloid-β in Alzheimer's disease. J Alzheimers Dis 2018; 61(3): 843-66.
[http://dx.doi.org/10.3233/JAD-170512] [PMID: 29332042]

[65] Lim GP, Chu T, Yang F, Beech W, Frautschy SA, Cole GM. The curry spice curcumin reduces oxidative damage and amyloid pathology in an Alzheimer transgenic mouse. J Neurosci 2001; 21(21): 8370-7.
[http://dx.doi.org/10.1523/JNEUROSCI.21-21-08370.2001] [PMID: 11606625]

[66] Patil R, Gangalum PR, Wagner S, *et al.* Curcumin targeted, polymalic acid-based MRI contrast agent for the detection of Aβ plaques in Alzheimer's disease. Macromol Biosci 2015; 15(9): 1212-7.
[http://dx.doi.org/10.1002/mabi.201500062] [PMID: 26036700]

[67] Elnaggar YS, Etman SM, Abdelmonsif DA, Abdallah OY, *et al.* Intranasal piperine-loaded chitosan nanoparticles as brain-targeted therapy in Alzheimer's disease: optimization, biological efficacy, and potential toxicity. J Pharm Sci 2015; 104(10): 3544-56.
[http://dx.doi.org/10.1002/jps.24557]

[68] Jose J, Charyulu RN. Prolonged drug delivery system of an antifungal drug by association with polyamidoamine dendrimers. Int J Pharm Investig 2016; 6(2): 123-7.

[http://dx.doi.org/10.4103/2230-973X.177833] [PMID: 27051632]

[69]   Patel DA, Henry JE, Good TA. Attenuation of β-amyloid-induced toxicity by sialic-acid-conjugated dendrimers: role of sialic acid attachment. Brain Res 2007; 1161: 95-105.
[http://dx.doi.org/10.1016/j.brainres.2007.05.055] [PMID: 17604005]

[70]   Igartúa DE, Martinez CS, Temprana CF, Alonso SDV, Prieto MJ. PAMAM dendrimers as a carbamazepine delivery system for neurodegenerative diseases: A biophysical and nanotoxicological characterization. Int J Pharm 2018; 544(1): 191-202.
[http://dx.doi.org/10.1016/j.ijpharm.2018.04.032] [PMID: 29678547]

[71]   Aso E, Martinsson I, Appelhans D, *et al.* Poly(propylene imine) dendrimers with histidine-maltose shell as novel type of nanoparticles for synapse and memory protection. Nanomedicine 2019; 17: 198-209.
[http://dx.doi.org/10.1016/j.nano.2019.01.010] [PMID: 30708052]

[72]   Bozzuto G, Molinari A. Liposomes as nanomedical devices. Int J Nanomedicine 2015; 10: 975-99.
[http://dx.doi.org/10.2147/IJN.S68861] [PMID: 25678787]

[73]   Naseri N, Valizadeh H, Zakeri-Milani P. Solid lipid nanoparticles and nanostructured lipid carriers: structure, preparation and application. Adv Pharm Bull 2015; 5(3): 305-13.
[http://dx.doi.org/10.15171/apb.2015.043] [PMID: 26504751]

[74]   Mourtas S, Lazar AN, Markoutsa E, Duyckaerts C, Antimisiaris SG. Multifunctional nanoliposomes with curcumin-lipid derivative and brain targeting functionality with potential applications for Alzheimer disease. Eur J Med Chem 2014; 80: 175-83.
[http://dx.doi.org/10.1016/j.ejmech.2014.04.050] [PMID: 24780594]

[75]   Bernardi A, Frozza RL, Meneghetti A, *et al.* Indomethacin-loaded lipid-core nanocapsules reduce the damage triggered by Aβ1-42 in Alzheimer's disease models. Int J Nanomedicine 2012; 7: 4927-42.
[http://dx.doi.org/10.2147/IJN.S35333] [PMID: 23028221]

[76]   Song Q, Huang M, Yao L, *et al.* Lipoprotein-based nanoparticles rescue the memory loss of mice with Alzheimer's disease by accelerating the clearance of amyloid-beta. ACS Nano 2014; 8(3): 2345-59.
[http://dx.doi.org/10.1021/nn4058215] [PMID: 24527692]

[77]   Muntimadugu E, Dhommati R, Jain A, Challa VG, Shaheen M, Khan W. Intranasal delivery of nanoparticle encapsulated tarenflurbil: A potential brain targeting strategy for Alzheimer's disease. Eur J Pharm Sci 2016; 92: 224-34.
[http://dx.doi.org/10.1016/j.ejps.2016.05.012] [PMID: 27185298]

[78]   Eriksen JL, Sagi SA, Smith TE, *et al.* NSAIDs and enantiomers of flurbiprofen target γ-secretase and lower Abeta 42 in vivo. J Clin Invest 2003; 112(3): 440-9.
[http://dx.doi.org/10.1172/JCI18162] [PMID: 12897211]

[79]   Vellas B. Tarenflurbil for Alzheimer's disease: a "shot on goal" that missed. Lancet Neurol 2010; 9(3): 235-7.
[http://dx.doi.org/10.1016/S1474-4422(10)70030-2] [PMID: 20170836]

[80]   Loureiro JA, Andrade S, Duarte A, *et al.* Resveratrol and grape extract-loaded solid lipid nanoparticles for the treatment of Alzheimer's disease. Molecules 2017; 22(2): 277.
[http://dx.doi.org/10.3390/molecules22020277] [PMID: 28208831]

[81]   Karami Z, Saghatchi Zanjani MR, Hamidi M. Nanoemulsions in CNS drug delivery: recent developments, impacts and challenges. Drug Discov Today 2019; 24(5): 1104-15.
[http://dx.doi.org/10.1016/j.drudis.2019.03.021] [PMID: 30914298]

[82]   Sood S, Jain K, Gowthamarajan K. Intranasal delivery of curcumin–/INS; donepezil nanoemulsion for brain targeting in Alzheimer's disease. J Neurol Sci 2013; 333: e316-7.
[http://dx.doi.org/10.1016/j.jns.2013.07.1182]

[83]   Ferreira LM, Cervi VF, Gehrcke M, *et al.* Ketoprofen-loaded pomegranate seed oil nanoemulsion stabilized by pullulan: Selective antiglioma formulation for intravenous administration. Colloids Surf

B Biointerfaces 2015; 130: 272-7.
[http://dx.doi.org/10.1016/j.colsurfb.2015.04.023] [PMID: 25935266]

[84]   Hussein J, El-Bana M, Refaat E, El-Naggar ME, *et al.* Synthesis of carvacrol-based nanoemulsion for treating neurodegenerative disorders in experimental diabetes. J Funct Foods 2017; 37: 441-8.
[http://dx.doi.org/10.1016/j.jff.2017.08.011]

[85]   Md S, Gan SY, Haw YH, Ho CL, Wong S, Choudhury H. *In vitro* neuroprotective effects of naringenin nanoemulsion against β-amyloid toxicity through the regulation of amyloidogenesis and tau phosphorylation. Int J Biol Macromol 2018; 118(Pt A): 1211-9.
[http://dx.doi.org/10.1016/j.ijbiomac.2018.06.190] [PMID: 30001606]

[86]   Haimov E, Harel Y, Polani S, *et al.* Metal-based nanoparticles as carriers of mTHPC drug for effective photodynamic therapy. Nanoscale imaging, sensing, and actuation for biomedical applications XVI. International Society for Optics and Photonics 2019.
[http://dx.doi.org/10.1117/12.2508456]

[87]   Das S, Dowding JM, Klump KE, McGinnis JF, Self W, Seal S. Cerium oxide nanoparticles: applications and prospects in nanomedicine. Nanomedicine (Lond) 2013; 8(9): 1483-508.
[http://dx.doi.org/10.2217/nnm.13.133] [PMID: 23987111]

[88]   Zhang C, Zheng X, Wan X, *et al.* The potential use of H102 peptide-loaded dual-functional nanoparticles in the treatment of Alzheimer's disease. J Control Release 2014; 192: 317-24.
[http://dx.doi.org/10.1016/j.jconrel.2014.07.050] [PMID: 25102404]

[89]   Karami Z, Hamidi M. Cubosomes: remarkable drug delivery potential. Drug Discov Today 2016; 21(5): 789-801.
[http://dx.doi.org/10.1016/j.drudis.2016.01.004] [PMID: 26780385]

[90]   Elnaggar YS, Etman SM, Abdelmonsif DA, Abdallah OY. Novel piperine-loaded Tween-integrated monoolein cubosomes as brain-targeted oral nanomedicine in Alzheimer's disease: pharmaceutical, biological, and toxicological studies. Int J Nanomedicine 2015; 10: 5459-73.
[http://dx.doi.org/10.2147/IJN.S87336] [PMID: 26346130]

[91]   Do TD, Ul Amin F, Noh Y, Kim MO, Yoon J. Guidance of magnetic nanocontainers for treating Alzheimer's disease using an electromagnetic, targeted drug-delivery actuator. J Biomed Nanotechnol 2016; 12(3): 569-74.
[http://dx.doi.org/10.1166/jbn.2016.2193] [PMID: 27280254]

[92]   Sivasankarapillai VS, Jose J, Shanavas MS, Marathakam A, Uddin MS, Mathew B. Silicon Quantum dots: promising theranostic probes for the future. Curr Drug Targets 2019; 20(12): 1255-63.
[http://dx.doi.org/10.2174/1389450120666190405152315] [PMID: 30961492]

[93]   Arruebo M, Valladares M, González-Fernández Á. Antibody-conjugated nanoparticles for biomedical applications. J Nanomaterials 2009; (1687-4110):
[http://dx.doi.org/10.1155/2009/439389]

[94]   Carradori D, Balducci C, Re F, *et al.* Antibody-functionalized polymer nanoparticle leading to memory recovery in Alzheimer's disease-like transgenic mouse model. Nanomedicine 2018; 14(2): 609-18.
[http://dx.doi.org/10.1016/j.nano.2017.12.006] [PMID: 29248676]

[95]   Tamba B, *et al.* Tailored surface silica nanoparticles for blood-brain barrier penetration: Preparation and in vivo investigation. Arab J Chem 2018; 11(6): 981-90.
[http://dx.doi.org/10.1016/j.arabjc.2018.03.019]

[96]   Kamigaito O. What can be improved by nanometer composites? Journal of the Japan Society of Powder and Powder Metallurgy 1991; 38(3): 315-21.
[http://dx.doi.org/10.2497/jjspm.38.315]

[97]   Chen Q, Du Y, Zhang K, *et al.* Tau-targeted multifunctional nanocomposite for combinational therapy of alzheimer's disease. ACS Nano 2018; 12(2): 1321-38.

[http://dx.doi.org/10.1021/acsnano.7b07625] [PMID: 29364648]

[98]   Humpel , C.J.T.i.b . Identifying and validating biomarkers for Alzheimer's disease. Trends Biotechnol 2011; 29(1): 26-32.
[http://dx.doi.org/10.1016/j.tibtech.2010.09.007]

[99]   Blennow K, *et al.* Cerebrospinal fluid and plasma biomarkers in Alzheimer disease. 2010; 6(3): 131-44.
[http://dx.doi.org/10.1038/nrneurol.2010.4]

[100] Visser PJ, *et al.* Prevalence and prognostic value of CSF markers of Alzheimer's disease pathology in patients with subjective cognitive impairment or mild cognitive impairment in the DESCRIPA study: a prospective cohort study. 2009; 8(7): 619-27.
[http://dx.doi.org/10.1016/S1474-4422(09)70139-5]

[101] Georganopoulou DG, Chang L, Nam JM, *et al.* Nanoparticle-based detection in cerebral spinal fluid of a soluble pathogenic biomarker for Alzheimer's disease. Proc Natl Acad Sci USA 2005; 102(7): 2273-6.
[http://dx.doi.org/10.1073/pnas.0409336102] [PMID: 15695586]

[102] Nazem A, Mansoori GA. Nanotechnology for Alzheimer's disease detection and treatment. Insciences J 2011; 1(4): 169-93.
[http://dx.doi.org/10.5640/insc.0104169]

[103] Nazem A, Mansoori GAJIJ. Nanotechnology for Alzheimer's disease detection and treatment. 2011; 1(4): 169-93.
[http://dx.doi.org/10.5640/insc.0104169]

[104] Kruman II, Kumaravel TS, Lohani A, *et al.* Folic acid deficiency and homocysteine impair DNA repair in hippocampal neurons and sensitize them to amyloid toxicity in experimental models of Alzheimer's disease. J Neurosci 2002; 22(5): 1752-62.
[http://dx.doi.org/10.1523/JNEUROSCI.22-05-01752.2002] [PMID: 11880504]

[105] Haes AJ, Chang L, Klein WL, Van Duyne RP. Detection of a biomarker for Alzheimer's disease from synthetic and clinical samples using a nanoscale optical biosensor. J Am Chem Soc 2005; 127(7): 2264-71.
[http://dx.doi.org/10.1021/ja044087q] [PMID: 15713105]

[106] Skaat H, Margel S. Synthesis of fluorescent-maghemite nanoparticles as multimodal imaging agents for amyloid-β fibrils detection and removal by a magnetic field. Biochem Biophys Res Commun 2009; 386(4): 645-9.
[http://dx.doi.org/10.1016/j.bbrc.2009.06.110] [PMID: 19559008]

[107] Hofmann-Amtenbrink M, Hofmann H, Montet X. Superparamagnetic nanoparticles - a tool for early diagnostics. Swiss Med Wkly 2010; 140: w13081. [ARTICLE].
[http://dx.doi.org/10.4414/smw.2010.13081] [PMID: 20853192]

[108] Yang J, Wadghiri YZ, Hoang DM, *et al.* Detection of amyloid plaques targeted by USPIO-Aβ1-42 in Alzheimer's disease transgenic mice using magnetic resonance microimaging. Neuroimage 2011; 55(4): 1600-9.
[http://dx.doi.org/10.1016/j.neuroimage.2011.01.023] [PMID: 21255656]

[109] Nesterov EE, Skoch J, Hyman BT, Klunk WE, Bacskai BJ, Swager TM. In vivo optical imaging of amyloid aggregates in brain: design of fluorescent markers. Angew Chem Int Ed 2005; 44(34): 5452-6.
[http://dx.doi.org/10.1002/anie.200500845] [PMID: 16059955]

[110] Willets KA, Ostroverkhova O, He M, Twieg RJ, Moerner WE. Novel fluorophores for single-molecule imaging. J Am Chem Soc 2003; 125(5): 1174-5.
[http://dx.doi.org/10.1021/ja029100q] [PMID: 12553812]

[111] Dubertret B, Skourides P, Norris DJ, Noireaux V, Brivanlou AH, Libchaber A. In vivo imaging of

quantum dots encapsulated in phospholipid micelles. Science 2002; 298(5599): 1759-62.
[http://dx.doi.org/10.1126/science.1077194] [PMID: 12459582]

[112] Zhang T, Stilwell JL, Gerion D, *et al.* Cellular effect of high doses of silica-coated quantum dot profiled with high throughput gene expression analysis and high content cellomics measurements. Nano Lett 2006; 6(4): 800-8.
[http://dx.doi.org/10.1021/nl0603350] [PMID: 16608287]

[113] Jain KK. Nanotechnology in clinical laboratory diagnostics. Clin Chim Acta 2005; 358(1-2): 37-54.
[http://dx.doi.org/10.1016/j.cccn.2005.03.014] [PMID: 15890325]

[114] Jaiswal JK, Mattoussi H, Mauro JM, Simon SM. Long-term multiple color imaging of live cells using quantum dot bioconjugates. Nat Biotechnol 2003; 21(1): 47-51.
[http://dx.doi.org/10.1038/nbt767] [PMID: 12459736]

[115] Nazem A, Mansoori GA. Nanotechnology solutions for Alzheimer's disease: advances in research tools, diagnostic methods and therapeutic agents. J Alzheimers Dis 2008; 13(2): 199-223.
[http://dx.doi.org/10.3233/JAD-2008-13210] [PMID: 18376062]

CHAPTER 5

Polyphenol Compounds as Potential Therapeutic Agents in Alzheimer's Disease

Érika Paiva de Moura[1], Alex France Messias Monteiro[1], Natan Dias Fernandes[1], Herbert Igor Rodrigues de Medeiros[1], Igor José dos Santos Nascimentos[2], Marcus Tullius Scotti[1], Francisco Jaime Bezerra Mendonça Júnior[3], Edeildo Ferreira da Silva-Júnior[*, 2] and Luciana Scotti[1,4]

[1] *Postgraduate Program in Natural and Bioactive Synthetic Products, Health Sciences Center, Federal University of Paraíba, Castelo Branco Street, João Pessoa, PB , Brazil*

[2] *Chemistry and Biotechnology Institute, Federal University of Alagoas, 57072-970, Maceió, AL , Brazil*

[3] *Laboratory of Synthesis and Drug Delivery, State University of Paraíba, Horácio Trajano de Oliveira Street, João Pessoa, PB , Brazil*

[4] *Teaching and Research Management - University Hospital of the Federal University of Paraíba, João Pessoa, PB , Brazil*

Abstract: According to the World Health Organization (WHO), dementia is a syndrome that affects thoughts, memory, the ability to perform day-to-day activities, and behavior. Alzheimer's disease (AD) corresponds to almost 70% of dementia cases, affecting mainly the elderly over 60 years old, causing physical, psychological, social, and economic impacts. Whether of natural or synthetic origin, the polyphenols and their derivatives have great versatility in terms of biological activity, as can be seen in the literature, exhibiting different properties, such as anti-inflammatory, anti-tumor, anti-viral, anti-microbial, and others. Among therapeutic alternatives are polyphenols and their derivatives as a molecular class broadly studied against neurodegenerative diseases, including AD. This chapter consists of a literary review of some polyphenols and derivatives with proven activity against AD, thus showing their importance among the other molecular classes, when it comes to proposing new bioactive agents against AD. Many targets are studied for this disease, since the pathogenesis of AD requires clarification and approved drugs only delay the evolution of the disease, such as donepezil hydrochloride, galantamine, among others. In addition to encouraging new studies by relating polyphenols and derivatives targeting AD, this work can assist research groups by providing some recent studies that have proven this relationship. At the end of this research, it is possible to realize the importance and applicability of these compounds in AD.

* **Corresponding author Edeildo Ferreira da Silva-Júnior:** Laboratory of Medicinal Chemistry - LMC, Chemistry and Biotechnology Institute, Federal University of Alagoas, Brazil; Tel: (+55)-87-99610-8311; Email: edeildo.junior@esenfar.ufal.br

Dr. José Juan Antonio Ibarra Arias (Eds.)

Keywords: Alzheimer's Disease, Amyloid-Beta, Aβ Amyloid Fibrils, BACE, Polyphenols .

INTRODUCTION

Alzheimer's disease (AD) is a neurodegenerative disorder that is becoming more common with the aging of the world's population, where it is estimated that one in every 85 people will have AD, in 2050 [1]. The disease is characterized by the progressive deterioration of cognitive functions due to brain atrophy, resulting from the death of neurons and degeneration of synapses in the hippocampus [2]. Thus, individuals with AD may experience memory loss, difficulty completing daily tasks, confusion with time or place, and learning problems [3].

Apart from the reduction in the brain volume, AD has other hallmarks, such as extracellular plaques containing amyloid-beta (Aβ) and intracellular neurofibrillary tangles containing hyperphosphorylated Tau protein [4]. Aβ is formed through the cleavage of amyloid precursor protein (APP) by three endoproteases, being α-secretase, β-secretase (BACE), and γ-secretase. These are responsible for cutting APP in different positions and thus producing Aβ peptides, ranging in length (23, 40, 42, or 56 amino acids) [5, 6]. As α-secretase cleavage does not produce complete Aβ, this protease does not participate in the development of AD, whereas BACE and γ-secretase are associated with the disease by producing various Aβ isoforms which can aggregate and form senile plaques [7]. In normal individuals, Aβ peptides removed from APP by actions of BACE and γ-secretase are released into the extracellular medium and rapidly degraded, while in elderly individuals the metabolic ability to degrade Aβ is decreased and Aβ peptides may be accumulated, inducing the formation of Aβ amyloid fibrils (fAβ) that can originate senile plaques and, consequently, cause neurotoxicity and induction of the Tau pathology [8].

Phosphorylation associated with Tau protein pathology has been the focus of recent researches, in which the most studied Tau kinases are proline-directed kinases, such as glycogen synthase kinase-3β (GSK-3β), mitogen-activated protein kinase (MAPK), among others [9]. However, the main Tau kinase is the active GSK-3β which, when increased in AD patients' brains, triggers irregular patterns of the Tau phosphorylation, which contributes to the disease progression [10]. One of the biological functions of Tau consists of its connection to the microtubule and thus regulating axonal transports. However, Tau phosphorylation can lead to its aggregation and as a consequence, it decreases the levels of soluble functional Tau and hinders the axonal transport, promoting neurodegeneration [11].

Recently, extensive studies suggest that soluble forms of A β produced during fibrillization, known as A β oligomers (A β O), are toxic species in AD capable of triggering a harmful cascade, damaging neurons and synapses [12]. A β O can cause toxicity by directly interacting with membranes and receptors and can be organized in different structures, such as dimers, trimers, tetramers, pentamers, decamers, dodecamers, among other forms [13]. Although pathological hallmarks of AD are considered to be the increase in the amount of A β monomers and the development of amyloid plaques, more research reports which A β O is a type of A β most strongly associated the severity of dementia in humans [14].

AD is also related to decreased brain levels of the neurotransmitter acetylcholine (ACh), where there is a failure in cholinergic neurotransmission that affects the brain's cognitive processes [15]. Thus, the recovery of ACh levels may be determinant for the treatment of AD, in which the regulation of these levels occurs by two different enzymes, being acetylcholinesterase (AChE) and butyrylcholinesterase (BuChE) [16]. AChE is found at high concentrations in the brain, while BuChE is distributed throughout the body, acting as cholinesterases (ChEs) extremely efficient since they cleave more than 10,000 ACh molecules *per* second [17]. In the brain of a healthy individual, AChE predominates over BuChE, but during the development of AD, AChE activity declines while BuChE activity increases [18].

The literature reports that several therapeutic targets have been investigated for the treatment of AD, such as proteins (Aβ and Tau), enzymes (BACE, γ-secretase, ChEs, and GSK-3β), receptors (cholinergic, dopaminergic, glutamatergic, among others), and processes/pathway (oxidative stress, neurogenesis, excitotoxicity, among others) involved in the pathogenesis of AD [19, 20]. The complicated pathogenesis of AD leads to the failure of several promising drug candidates in clinical trials, making it difficult to develop new drugs [21]. Currently, approved anti-AD drugs are related to memory recovery, where the AChE inhibition improves the cholinergic defect [22]. Thus, the current drug therapy for AD only attenuates the symptoms and does not interfere in the mechanisms related to disease progression. In this context, the search for efficient therapeutic approaches is an unmet need [23, 24].

Polyphenols are a class of natural, synthetic, or semi-synthetic organic compounds that are characterized by the presence of one or more aromatic rings with one or more hydroxyl groups [25]. According to their chemical structures, they can be classified into flavonoids, such as isoflavones, flavonols, flavones, neoflavonoids, chalcones, and others; also, nonflavonoids, such as stilbenoids, phenolic acids, and phenolic amides are polyphenols [26]. Polyphenols can be found in vegetables, fruits, roots, bark, leaves from different plants, coffee, red wine, tea, among other sources [27]. Animal and human studies have shown that the anti-inflammatory and antioxidant properties of various polyphenols may potentially prevent or serve as a treatment against cardiovascular disease, obesity, cancer, and neurodegenerative disorders [28]. Then, this chapter aims to seek research that evaluated the effects of polyphenols and derivatives on targets related to the development of AD.

POLYPHENOLS WITH ANTI-ALZHEIMER ACTIVITIES

Anti-Amyloid Aggregation Agents

A study published by Inbar *et al.* (2008) describes a simple and parallel method to quickly trace molecules concerning their ability to associate with Aβ fibrils related to AD [29]. The biological assay was performed by depositing these fibrils in a 96-well plate where different concentrations of the molecules were added into each well. The authors removed excess solution from the compounds, blocked the surface of the wells with bovine serum albumin (BSA), and incubated the wells with a solution of the anti-Aβ IgG. Thus, they tested the association of 30 commercial substances in which eight compounds showed excellent activity against the peptide with concentrations below 20 μM, where the results can be seen in Table **1**.

Table 1. **Results for eight molecules screened using a *β*-amyloid fibril inhibition assay.**

Molecule name	IC$_{50}$ [μM]
Acridine orange	1.2
Diamino acridine	4.6
Crystal violet	4.0
Thioflavin T	5.0
Dopamine	5.8
(*S*)-Naproxen	1.4
Tetracycline	8.2
Tannic acid	0.39

The results obtained by Imbar *et al.* (2008) allowed them to observe that two polyphenols bound with good efficiency to the fibrillar surface and are known to have low toxicity, being tannic acid and dopamine, with IC$_{50}$ values of 0.39 and 5.8 μM, respectively. Besides, tetracycline, a compound with multiple hydroxyl groups, was also able to efficiently bind to A β fibrils (8.2 μM). According to the authors, further studies toward elucidating the structure-activity relationship (SAR) of various similar compounds are currently in progress [29].

Churches *et al.* (2014) investigated the biological activity of polyphenolic compounds as Aβ aggregation inhibitors by employing different tests, such as thioflavin T (ThT) assay, transmission electron microscopy (TEM), dynamic light scattering, and size exclusion chromatography. From the 24 compounds tested in the ThT assay, maritimetin, luteolin, and transilitin exhibited inhibitory activity comparable to the epigallocatechin gallate (EGCG, used as control), a potent inhibitor of Aβ aggregation. TEM analysis was used to verify all inhibitory results from the ThT assay, where the results indicated that both maritimetin and luteolin allowed a reduction in the formation of fibrils, while transilitin caused a significant reduction in the frequency and size of Aβ aggregates. therefore, according to the data from ThT and TEM, transilitin was chosen to study its effects on the Aβ aggregation pathway. Dynamic light scattering and size exclusion chromatography results proved that transilitin stabilized an early oligomeric species and inhibited the formation of fibrils. In order to understand how phenolic compounds, such as transilitin, modulate the Aβ aggregation, molecular docking studies were performed in Surflex-Dock v. 2.6 within Sybyl-X v. 2.0 using a crystalline structure of oligomeric Aβ (PDB ID: 3MOQ), where it was observed that transilitin interacts with residues Ile^{32} and Leu^{34} *via* H-bonds, in addition to favorable hydrophobic interactions with the oligomeric interface. The authors concluded that transilitin inhibits Aβ fibrillation by binding and stabilizing oligomers [30].

Tomaselli *et al.* (2019) performed biophysical and *in vivo* studies, allowing to identify a new polyphenol PP04 (Fig. **1**), neutralizing Aβ oligomerization in memory impairment mediated by oligomers *In vitro* and neuroinflammation in an acute mouse model of AD [31]. In this study, six molecules were synthesized and the authors analyzed the ability of the molecules to inhibit Aβ through the acute mouse model induced by AβO. Additionally, the authors carried out molecular docking studies involving the Aβ$_{42}$ monomer (PDB ID: 1IYT) and amyloid fibril of the complete Aβ$_{1-42}$ peptide (PDB ID: 5OQV).

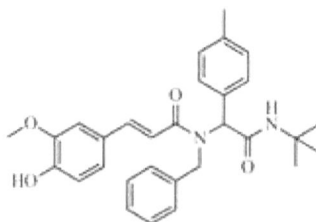

Fig. (1). Chemical structure of PP04.

As the conclusion of the study, the authors believe that the polyphenolic compound PP04 efficiently interferes in all stages of $A\beta_{42}$ *in vitro* aggregation kinetics, targets $A\beta Os$, and neutralizes *in vivo* memory impairment mediated by $A\beta O$ and the hippocampus microgliosis in naive C57BL/6 mice. This aspect is particularly interesting since $A\beta Os$ are considered the most potent toxic species and neuroinflammation has reappeared as a vital therapeutic target.

Orteca *et al.* (2018) studied curcumin derivatives against $A\beta$-fibrillar aggregates to combat neurodegenerative diseases, for this purpose *in silico* studies involving molecular docking were performed using AutoDock and molecular dynamics simulations by using GROMOS software using the PDB ID: 2LMN, which corresponds to the $A\beta$ fibrils formed by the peptide of 40 residues. *In vitro* tests showed that these derivatives are capable of interfering with $A\beta$ fibrils, in which the compound K2F21 stands out as the best candidate for diagnostic/therapeutic purposes, due to its high stability under physiological conditions. This compound quickly binds to fibrillar aggregates, causing a strong depolymerizing activity, high cytoprotection effects against oxidative stress, and low cytotoxicity [32].

Chaudhury *et al.* (2018) investigated the effects of EGCG against the aggregation of human γB-crystallin, believed to be one of the key reasons for age-onset cataract and in the formation of amyloid-like fibrils. Thus, the turbidity assay was performed and as a result, it was observed that there was an increase in the absorbance of γB-crystallin in the absence of EGCG and a decrease in this absorbance in the presence of EGCG, indicating that EGCG prevented the aggregation process. With the ThT assay, it was found that EGCG inhibited the formation of γB-crystallin fibrils, as there was less intensity of ThT fluorescence. To verify the effects of EGCG in the tertiary structure of γB-crystallin, ANS fluorescence spectroscopy was used. The authors noticed that there was a reduction in ANS fluorescence when γB-crystallin was incubated with EGCG and the extent of the blue shift is lower when compared to the protein fibrils alone. This means that the surface hydrophobicity of the protein has decreased and therefore, EGCG prevented the exposure of hydrophobic sites from the protein [33].

Regarding the results of the circular dichroism spectroscopy by Chaudhury *et al.* (2018), the native human γB-crystallin sample showed a predominance of β-sheet secondary structure in conditions of elevated temperature and low pH, where the increase in the content of this secondary structure is related to fibrillation protein. When the protein was incubated in the presence of EGCG, a much lower extent of fibril formation was observed, meaning that EGCG acted as an efficient inhibitor. Theoretical studies including molecular dynamics have demonstrated aggregation-prone regions and a flexible region near to that may instigate the aggregation process. However, as EGCG can bind to the flexible region, incubation with polyphenols is likely to prevent protein aggregation by preventing the formation of extended β-sheet. Finally, it was possible to verify the inhibitory potency of EGCG against the aggregation of γB-crystallin and provide additional information for the identification of potent compounds against aggregation/fibrillation [33].

Eggers *et al.* (2019) studied the neuroprotective and anti-aggregative properties of cannflavin A (Fig. **2**), a geranylated flavonoid from the *Cannabis sativa* L., against Aβ and compared with the effects of the flavonoids, mimulone and diplacone. In the thiazolyl blue tetrazolium bromide (MTT) assay used to verify neuronal viability, it was possible to observe that cannflavin A increased viability by 40%, with concentrations ranging from 1 to 10 µM. However, in higher concentrations of cannflavin A, neurotoxicity is observed, while mimulone and diplacone showed neurotoxicity in a dose-dependent manner. To determine the effect of the selected flavonoids on the formation of Aβ fibrils, the ThT test was conducted and it was found that cannflavin A and diplacone inhibited ThT fluorescence, which means that they have an anti-fibrillar effect. The TEM test revealed that each of the flavonoids allowed the reduction of Aβ aggregates, where diplacone had the greatest inhibitory effect among the flavonoids. The results of the molecular docking performed in CLC Drug Discovery Workbench 2.0 demonstrated that cannflavin A had a higher binding affinity towards the oligomer Aβ (PDB ID: 2BEG). Therefore, the authors concluded that cannflavin A has, in low concentrations, neuroprotective and anti-aggregative effects upon the neurotoxicity caused by Aβ, but in high concentrations, it could generate toxicity [34].

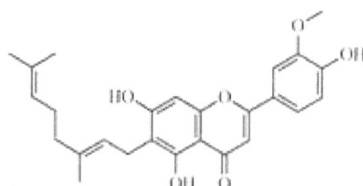

Fig. (2). Chemical structure of cannflavin A.

Huang *et al.* (2019) proposed a binary modulator (Aβ-segment KLVFF and (-) - epigallocatechin-3-gallate, KLVFF/EGCG) to inhibit Aβ aggregation. The ThT assay demonstrated that KLVFF and EGCG individually inhibited Aβ42 aggregation. However, the association of these two compounds decreased the fluorescence of ThT to a low level, indicating that there was a synergistic inhibitory effect. In the analysis of circular dichroism spectroscopy, it was possible to verify that the formation of stable aggregates occurs due to the co-

contribution of KLVFF and EGCG. Using AFM, the authors observed that the $A\beta_{42}$ morphology, in the presence of the KLVFF/EGCG complex, was amorphous and did not have nanofibrils, demonstrating that the complex had an excellent inhibitory effect. The authors proposed the molecular interactions between $A\beta_{42}$ and the complex through liquid-state nuclear magnetic resonance (NMR) spectroscopy and molecular dynamics simulations, where KLVFF could selectively bind to the key segment of the $A\beta_{42}$ chains while EGCG performs H-bonds with the main chain and side groups from the $A\beta_{42}$ due to their hydroxyl groups. Besides, MTT assay and DCFH-DA (2′,7′-dichlorodihydrofluorescein diacetate) tests were performed and the results indicated that the inhibitors reduced cytotoxicity when interacting with $A\beta_{42}$ and decreased the production of reactive oxygen species. It is concluded that the strategy of developing peptide breaker/polyphenol binary inhibitors can be an effective approach for the prevention and treatment of AD [35].

Kurnik *et al.* (2018) performed a time-resolved fluorescence resonance energy transfer (FRET) -based high-throughput screen (HTS) involving 746,000 compounds to identify α-synuclein (αSN) aggregation inhibitors, a protein widely expressed in the brain that spontaneously forms amyloid fibrils after a lag time ranging from hours to days. Hits compounds were defined as compounds reducing the FRET signal by > 48%. Then, the authors performed an orthogonal assay with ThT that reduced the number of hits compounds to 136. These compounds were further triaged by a tubulin polymerization assay resulting in 58 hits molecules. To characterize the action mechanisms of the hits, they were evaluated in three medium-throughput plate reader assays, being inhibition of anionic surfactant sodium dodecyl sulfate (SDS)-induced fibrillation (assay 1), inhibition of shaking induced fibrillation in the absence of SDS (assay 2), and inhibition of vesicle permeabilization (assay 3). All 58 hits exhibited significant inhibitory effects in both trials 1 and 2, and some compounds were comparable to or superior to the EGCG control in inhibiting vesicle permeabilization in trial 3 [36].

As the three trials performed by Kunik *et al.* (2018) did not correlate closely, the compounds were evaluated by CNS multiparameter optimization (MPO) function, which produces a desirable score for essential drug-related attributes. Thus, nine compounds were selected using a simple linear weighting of three secondary assays and the CNS-MPO scale. Among pyrazolone and (4-hydroxynaphthalen-1-yl) sulfonamide analogs, six sulfonamides were found to be the most promising since they exhibited the highest CNS-MPO scores, and all showed the most positive results in the screenings and orthogonal assays, except the compound C3. In conclusion, the authors conducted an HTS that resulted in the identification of structurally diverse compounds that have the potential to elucidate the mechanisms related to the pathophysiology of αSN, contributing with important information for future therapies [36].

Oliveri (2019) summarized the available data concerning inhibitors of α-synuclein aggregation, providing insights into their mechanism of action and pharmacology. Among these, benzoic acid derivatives such as protocatechuic acid (PCA) and gallic acid (GA) (Fig. 3) have been found as promising candidates. The literature reports that PCA inhibits aggregation and destabilizes αSyn fibrils at a ratio of 1:1, as confirmed by Western blot analysis, fluorescence assay, transmission electron microscopy (TEM), and thioflavin-T (ThT). Some authors suggest that PCA binds to oligomers or preformed fibrils rather than monomers. Concerning GA, several biochemical and biophysical techniques have indicated that this molecule inhibits αSyn and A53T αSyn fibrillization and toxicity. The presence of three vicinal hydroxyl groups in GA increases its ability to inhibit αSyn fibrillization compared to dihydroxybenzoic or monohydroxybenzoic acids [37].

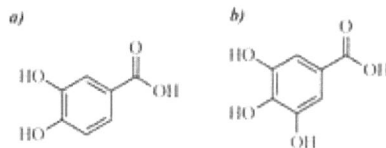

Fig. (3). Molecular structures of protocatechuic acid (*a*) and gallic acid (*b*).

Hirohata *et al.* (2007) showed that flavonoids, especially myricetin, exhibited an anti-amyloidogenic effect *in vitro* by preferential and reversible binding to the fAβ structure. In this study, the incubation of Aβ with fresh or oxidized flavonoids significantly reduced the fluorescence of ThT. Aβ and fAβ incubated with fresh or oxidized flavonoids also resulted in reduced ThT fluorescence. The authors observed that incubation of fAβ (1-40) with fresh or oxidized myricetin and quercetin decreased significantly ThT fluorescence compared to controls. Lastly, among the flavonoids tested, myricetin exhibited the most potent destabilizing activity.

In the affinity analysis of flavonoids for Aβ/fAβ by surface plasmon resonance (SPR), all compounds examined (morine, quercetin, myricetin, catechin or epi-catechin, 1-25 μM) reacted very weakly with the immobilized Aβ monomer. There was also no increase in response after adding flavan-3-ol compounds (catechin and epicatechin, 1-50 μM) to immobilized fAβ. However, after the addition of flavonol compounds in lower concentrations (myricetin 0.1-6 μM and quercetin 0.1-8 mM) to fAβ immobilized, different dissociation and association reactions were observed. Myricetin was more strongly bound to fAβ, while very weak interaction occurs with Aβ monomer. A significant decrease in the fAβ extension rate was observed when myricetin at a concentration > 0.5 μM was injected. Maximum inhibition (approximately 30%) was achieved by injection of myricetin at a concentration > 5 μM. These data showed that the binding of myricetin to fibrils competed with the binding of the Aβ monomer to the ends of the fibril, inhibiting the subsequent polymerization of Aβ. Thus, the authors demonstrated that flavonoids, especially myricetin, exhibited anti-amyloidogenic effect *in vitro* by preferential and reversible binding to fAβ, rather than Aβ monomers [38].

Siposova *et al.* (2019) reported the promising anti-amyloid activity of rottlerin (Fig. **4**), a natural polyphenol, focusing on its effect on the fibrillation of three different amyloidogenic proteins, namely insulin, lysozyme, and $A\beta_{1-40}$ peptide. The study demonstrated that the formation of amyloid fibril from any of the three proteins was inhibited by low micromolar concentrations of rottlerin, due to the interaction of this polyphenol with the amyloidogenic regions of the proteins. Rottlerin was also effective in disassembling preformed fibrils, suggesting that the interactions of rottlerin with fibrils were able to disrupt the fibril stabilizing bonds with β-sheets. The IC_{50} and DC_{50} values were determined ranging from 1.3 to 36.4 μM and from 15.6 to 25.8 μM, respectively [39].

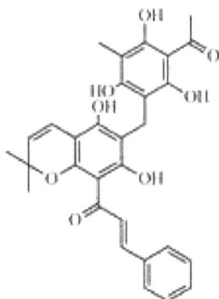

Fig. (4). Chemical structure of rottlerin.

The strongest inhibitory effect of rottlerin was observed on the $A\beta_{1-40}$ peptide. Additionally, the cytotoxicity assay performed on *Neuro 2a* cells, indicated that over time cell morphology changed; however, rottlerin affected cell viability only at concentrations above 50 μM. These results were justified because the conformation of rottlerin can easily adjust and, therefore, fit well in the binding sites of amyloidogenic proteins. Thus, the results of the study suggest that chemical modifications in rottlerin may be tested in the future, as a promising strategy for modulating the aggregation of amyloidogenic proteins [39].

In the work described by Patel *et al.* (2018) the therapeutic potential of ascorbic acid against amyloid disorders has been reported by using microscopic, spectroscopic, and computational techniques. The ability of ascorbic acid to interact with lysozyme and thus prevent its conversion to amyloid fibrils has been described, indicating a therapeutic potential for this vitamin. The authors verified that ascorbic acid inhibited hen egg-white lysozyme (HEWL) fibrillation probably by binding to the protein *via* H-bonds. Besides, ascorbic acid interacted with the partially unfolded form of the enzyme and inhibited its fibrillation at the beginning of the fibrillation pathway and significantly reduced cytotoxicity. In summary, ascorbic acid has the potential to strongly inhibit HEWL fibrillation with a 10-fold molar excess of lysozyme concentration, considering that ascorbic acid forms a stable complex with HEWL. Specifically, ascorbic acid binds to the beta-domain prone to the aggregation of natively folded monomeric lysozyme and/or partially unfolded monomers and stabilizes them, preventing them from undergoing conformational changes that lead to the formation of the amyloid fibril [40].

Chong *et al.* (2017) reported the anti-amyloid effect of flavonoid morine, on the fibrillation of HEWL, through computational and spectrofluorometric studies. Surprisingly, morine inhibited the formation of HEWL amyloid fibrils, binding to the region of the aggregation slit of the HEWL domain, stabilizing the molecule in its native state. After binding, excess morine adhered to and coated the surface of the HEWL protein, minimizing the interaction between the protein's surface and water molecules. The presence of morine prolonged the delayed fibrillation phase and showed a significant decrease in the fluorescence of ThT over the period of formation of the amyloid fibril [41]. The IC_{50} value found for morine was equal to 13.1 µM, a value that confirms its significant inhibitory activity, in comparison with other molecules, such as bis-(indolyl)arylmethanes, erythrosine B, nordic hydroguaiaretic acid, and myricetin, which have IC_{50} values of 12.3, 13.3, 26.3, and 28.2 µM respectively. In summary, the work demonstrated that after the addition of morine (in low concentration) with HEWL, only a very small portion of morine molecules was available to interact directly with HEWL in a site-specific manner, resulting in greater exposure of the partially unfolded protein surface, which in turn would lead to self-association and fibrillation. On the other hand, when morine was present in excess (at higher concentrations), it interacted freely with the protein surface, displacing the water molecules, thus reducing the interaction between the water molecules [41].

Computational studies also play an important role in determining interactions with amyloidogenic proteins. Wang *et al.* (2017) applied molecular dynamics simulations upon four phenolic compounds (EGCG, resveratrol, curcumin, and vanillin (Fig. **5**) to investigate the effect on $A\beta_{17-36}$ aggregation. The results showed that the inhibition of the formation of $A\beta_{17-36}$ fibrils is more effective by using EGCG, followed by resveratrol, curcumin, and vanillin [42].

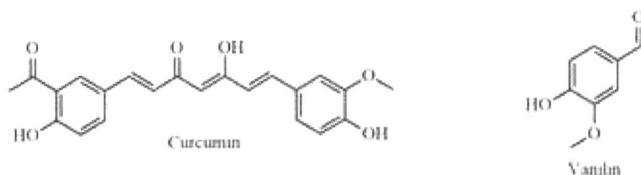

Fig. (5). Chemical structures of curcumin and vanillin.

In the work of Huy *et al.* (2013), the anti-amyloidogenic activity of 15 synthetic vitamin K3 analogs (VK3) is described. As a result, many VK3 analogs, such as VK3–9, VK3–10, and VK3–6, inhibited $A\beta_{1-40}$ aggregation. However, only VK3-9 was able to protect cells against $A\beta_{1-40}$-induced toxicity. The effective dose of VK3–9 was approximately 0.1 μM, which is as effective as amyloidogenic compounds, such as curcumin. Also, VK3-2, VK3-6, VK3-8, VK3-9, VK3-10, VK3-199, VK3-221, and VK3-224 inhibited $A\beta_{1-40}$ aggregation. Specifically, among these, VK3, VK3-6, VK3-9, and VK3-10 were the most effective analogs for inhibiting $A\beta_{1-40}$ aggregation. On the other hand, some analogs, such as VK3-1, VK3-4, VK3-5, VK3-233-2d, and especially VK3-4 and VK3-5, improved the aggregation of $A\beta_{1-40}$. Thus, although most analogs did not have protective effects for cells against $A\beta$, the VK3-9 analog can effectively inhibit $A\beta$ aggregation, reduce free radicals produced by $A\beta$ and also protect cells against toxicity induced by $A\beta$ [43].

Ulicna *et al.* (2018) reported the inhibitory effects of three different groups of heterodimers (tacrine, acridone, and coumarin) on the fibrillation of HEWL *in vitro*. Interestingly, the ability of heterodimers to interfere with the aggregation of amyloid lysozyme was performed using the ThT fluorescence assay, atomic force microscopy, and the coupling method. The chemical composition and structural configuration of the tacrine/acridone-coumarin heterodimers were divided into three groups, such as *I* tacrine-coumarin derivatives with the halogenylenediamine ligand which consisted of 6 to 9 alkyl groups; *II* tacrine-coumarin derivatives with alkylenepolamine linkers, consisting of 4, 6, 8 alkyl groups; *III* acridone-coumarin derivatives with alkylenediamine linkers, consisting of 6-9 alkyl groups. As a result, acridone had a very low inhibitory effect, however, tacrine and coumarin along with the alkylenediamine ligand were the most effective inhibitors of lysozyme fibrillation when compared to the alkylenepolamine ligand. The IC_{50} values for tacrine and coumarin were 396.8 and 322.2 μM, respectively; and the heterodimers studied had no cytotoxic effects on human neuroblastoma cells [44].

Simões *et al.* (2016) described the synthesis of a new series of molecules that inhibit amyloid formation by protein transthyretin (TTR) based on a virtual screening protocol. Thus, virtual screening was carried out on the ZINC database, containing approximately 2,259,573 compounds. In total, 12 compounds were selected, where 11 of them were posteriorly eliminated considering the authors' criteria. As a consequence, these criteria resulted in one promising compound, ZINC00041425. The authors observed that the best compound in this screening had a furan nucleus. Then, it was decided to optimize the ZINC00041425, generating the AT09-B00 and AT09-B06 analogs that demonstrated similar interactions. Finally, the optimized compounds showed IC_{50} values against TTR amyloid fibril formation lower than the *hit* screening compounds (7.10 μM for ZINC00041425; 2.02 μM for AT09-B00, and 2.49 μM for AT09-B06), being more active than the tafamidis, employed as a standard compound ($IC_{50} = 3.10 \pm 0.15$ μM) [45].

Wang *et al.* (2012) investigated the effects of α-mangostin, a polyphenolic xanthone derivative from mangosteen, on Aβ-aggregation, and the cytotoxicity induced by Aβ oligomers [46]. The α-mangostin was commercially obtained and Aβ_{1-40} and Aβ_{1-42} oligomers were prepared according to the procedures described by Kayed 2003, with mild modifications [47]. Also, the cell viability of primary culture cerebral cortical neurons treated with Aβ oligomers and α-mangostin was measured by CCK-8 kit. In order to understand how the α-mangostin interacts with the structure of Aβ, the software Molecular Operating Environment (MOE) was employed to perform the automatic docking, using a nuclear magnetic resonance (NMR) structure of Aβ in the solution obtained from protein data bank (PDB ID: 1BA4). Moreover, molecular dynamics simulations were performed in Amber 11 software. The authors verified that the α-mangostin had significant protective effects against neurotoxicity induced by Aβ_{1-40} or Aβ_{1-42} oligomers with EC_{50} values of 3.89 and 4.14 nM, respectively. Also, the mechanism of action is probably associated with its ability to redirect Aβ aggregation cascades, confirmed by dissociating toxic β-sheet-rich aggregated Aβ oligomers and later-stage fibrils and inhibition of fibril formation process. Additionally, docking molecular simulations revealed that α-mangostin interacted with Lys[16] and Asp[23] amino acids *via* H-bond interactions; Phe[19] and Glu[22] residues *via* π-stacking and van der Waals interactions, respectively. Finally, the authors suggest that α-mangostin represents a potential candidate for treating AD [46].

Yuan *et al.* (2016) studied the pomegranate extract (PE) to identify its compounds, to verify if these compounds can cross the blood-brain barrier (BBB), and if they have effects on Aβ_{1-42} fibrillation. Besides, *in vivo* assays were performed to evaluate the ability of PE and some compounds to abrogate Aβ_{1-42} induced neurotoxicity and paralysis in *Caenorhabditis elegans*. In total, 21 compounds were identified and isolated, including four derivatives of microbial metabolites, urolithin analogs (Fig. **6**). Then, the authors performed an *in silico* study showing that none of them presents the criteria to cross BBB, except for urolithine derivatives [48]. Also, the authors showed that PE reduced $A\beta$ fibrillation (35.9 and 76.4% at 10 and 100 μg/mL concentrations, respectively) and the isolated compounds, as well as the urolithins, reduced $A\beta$ fibrillation ranging from 6.5 to 65.4% at 10 μM concentration, and from 20.2 to 76.3% at 100 μM concentration. However, in the tests for protection of neurotoxicity and paralysis induced by $A\beta_{1-42}$ in *C. elegans*, the compounds were not effective, except for the urolithin analog (mUB). Finally, the authors concluded that further studies are needed to prove the neuroprotective effects of urolithins using animal models [48].

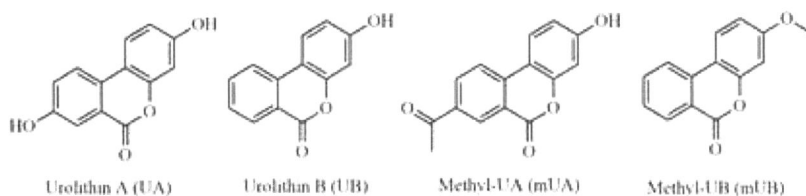

Urolithin A (UA)   Urolithin B (UB)   Methyl-UA (mUA)   Methyl-UB (mUB)

Fig. (6). Derivatives of urolithins.

Dong *et al.* (2016) applied molecular dynamics simulations upon three benzothiazole-derived CQ_i compounds (Fig. **7**) and Cu^{2+}-$A\beta_{40}$ monomers to evaluate the effects of inhibition and related disaggregation mechanisms, such as changes in the $A\beta_{40}$ peptide configuration at different physiological conditions. The results showed that the compounds present binding energy of 100 KJ/mol and they are preferentially placed into a hydrophobic cavity. However, in the model with lower binding energy, the molecules are only partially placed into the $A\beta_{40}$ peptide structure. These results are important because the binding energies are entirely related to the effectiveness of inhibiting modes and $A\beta_{40}$ aggregates formation. These results may be useful in the discovery of new anti-AD compounds. Concerning binding energies, the authors showed that the compound CQ_3 was the most effective, being proposed that CQ_3 has the greatest disaggregation capacity of the $A\beta_{40}$ peptide mainly due to the formation of a hydrogen bond with the Asp^{23} residue. In sense, this important parameter should be taken into account in the design of new drugs. It was also observed that the increase in the polarity of the compounds results in increased interactions with the *N*-terminal portion from the $A\beta_{40}$ structure, inhibiting the aggregation induced by Cu^{2+} ions. Also, benzothiazole in combination with the phenolic ring overcomes the activation barrier, assuming a more flexible character, which allows the bond, obtaining maximum interactions with the residues of this region at the $A\beta_{40}$ peptide structure. In conclusion, these structural characteristics of CQ_i compounds provide new courses in the design of drugs anti-AD [49].

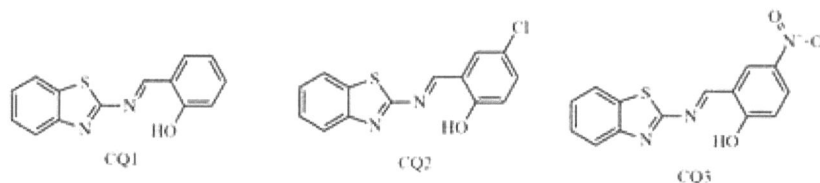

Fig. (7). Benzothiazoles studied by Dong *et al..*

Simoni *et al.* (2016) synthesized a series of five piperidin-4-one derivatives aimed to investigate anti-aggregating activity at the mitochondrial level. The results showed that compound A was the most effective in preventing the self-assembly process from Aβ42 (53% inhibition). This fact is mainly related to the hydroxyl substituents at the aromatic ring, a catechol, considering that any modification in the catechol group led to the loss of anti-aggregating efficacy. In addition, this compound showed no toxicity. The study also revealed through molecular dynamics simulations that polyamine is fundamental for the process of molecular recognition in the anti-aggregating and neuroprotective effects. Thus, compound A can be a lead compound addressed to neuroprotective effects [50].

Cholinesterase inhibitors

Neagu *et al.* (2015) performed the inhibitory activity against AChE and tyrosinase from the extracts of *Alchemilla vulgaris* and *Filipendula ulmaria*, where the study determined the total amount of polyphenols (88 - 112.3 µg/mL), flavones (360 - 862 µg/mL), and proanthocyanidins (77.6 - 130 µg/mL) in the extract of these plants [51]. According to the study published by the authors, one of the hypotheses that try to explain the pathogenesis of AD is based on AChE inhibition. To determine polyphenols content, the authors used the Folin-Ciocalteau method [52], where this amount was expressed in microgram of gallic acid equivalents (GAE) *per* milliliter of extract and the activity against AChE was measured using methods described by Ingkaninan *et al.* (2003) [53]. Polyphenols were isolated by HPLC, according to the methods described by Cristea *et al.* (2009) [54]. Finally, these compounds are shown in Table **2**.

Table 2. Compounds isolated from the extracts of *Alchemilla vulgaris* and *Filipendula ulmaria*.

Entry	Name	Entry	Name
01	Epicatechin	10	Ferulic acid
02	Quercetin	11	Rosmarinic acid
03	Rutin	12	Chlorogenic acid
04	Myricetin	13	Quercetin 3-β-D-glucoside

(Table 2) cont.....

Entry	Name	Entry	Name
05	Kaempferol	14	Luteolin
06	Ellagic acid	15	Genistein
07	Gallic acid	16	Daidzein
08	*P*-karmic acid	17	Caffeic acid
09	Synapic acid	-	-

Continuing the study of Neagu *et al.* (2015), after the determination of the polyphenolic structures, the biological activity of aqueous and ethanol extracts containing the substances described in Table **2** was carried out, thus it was possible to evaluate the activity upon AChE [51], as shown in Table **3**.

Table 3. AChE inhibition (%).

Species	-	AChE inhibition (%) 3 mg/mL
Alchemilla vulgaris	Aqueous extract	84.56 ± 5.64*
-	70% Ethanolic extract	96.50 ± 4.93*
Filipendula ulmaria	Aqueous extract	77.03 ± 3.82*
-	70% Ethanolic extract	98.30 ± 3.91*

The data represent the means ± SD of triplicate samples of three independent experiments. $p < 0.05$, compared the activity of the ethanolic extracts with that of the aqueous extracts. $p < 0.01$, compared the activity of the ethanolic extracts with that of the aqueous extracts.

According to the data presented, the authors concluded that these two lesser-known medicinal plants in Romania, *A. vulgaris* and *F. ulmaria*, presented cholinergic effects, considering that their extracts had high AChE inhibitory effects, exhibiting 98.3% ± 3.91 for *F. ulmaria*, and 96.5% ± 2.93 for *A. vulgaris*, in addition to considerable amounts of flavones (*F. ulmaria*), polyphenols, and proanthocyanidins (*A. vulgaris*) that make them extremely useful in the treatment of degenerative diseases, especially where the cholinergic hypothesis is involved.

Zengin *et al.* (2017) conducted a study involving extracts of *Achillea phrygia* and *Bupleurum croceum* plants where phenolic components were separated by high-performance liquid chromatography (HPLC) [55], whose phenolic composition was determined through a small variation of the method Folin-Ciocalteau. The components found in the extracts are shown in Table **4**.

Table 4. Molecules described in Zengin *et al*. study.

Entry	Name	Entry	Name
01	Gallic acid	13	Benzoic acid
02	Protocatechuic acid	14	*o*-Coumaric acid
03	(+)-Catechin	15	Rutin
04	*p*-Hydroxybenzoic acid	16	Hesperidin
05	Chlorogenic acid	17	Rosmarinic acid
06	Caffeic acid	18	Eriodictyol
07	(-)-Epicatechin	19	*trans*-Cinnamic acid
08	Syringic acid	20	Quercetin
09	Vanillin	21	Luteolin
10	*p*-Coumaric acid	22	Kaempferol
11	Ferulic acid	23	Apigenin
12	Synapinic acid	-	-

The authors report that anti-cholinesterase activity found from *A. phrygia* and *B. croceum* herbs is moderate and their activities could be attributed to the presence of quercetin and apigenin, polyphenols previously reported as modest inhibitors of cholinesterases. Molecular docking studies were also performed with these molecules using the proteins, such as AChE (PDB ID: 4X3C) and BuChE (PDB ID: 4BDS), in which energies values were calculated by using Glide software. The authors observed that quercetin interacted efficaciously to the enzymatic cavity of AChE by forming two hydrogen bonds with Tyr [70] and His [440], and also interacted with the cavity of BuChE by forming three H-bonds with Glu [197], Leu [286], and Ser [287] residues. Regarding apigenin, two H-bonds are performed with residues present into the enzymatic cavity of AChE and BuChE [55]. These compounds showed an affinity toward studied proteins, thus contributing to the proposition of new bioactives and inspiring new studies involving these plants against AD.

Kuppusamy *et al.* (2017) combined computational approaches with *in vitro* assays of some polyphenols and derivatives against AChE. Molecular docking studies to perform analyzes of conformational sites and docking parameters, such as binding energy, inhibition constant, and intermolecular energy, were determined by using AutoDock v. 4.2. In sense, these molecular docking studies were carried out with diosmin, silibinin, scopoletin, taxifolin, and tricetin, where this last compound exhibited higher binding energy upon AChE than donepezil. Finally, *In vitro* tests with scopoletin showed the best result, with an IC_{50} value of 10.18 ± 0.68 μM [56].

Orhan *et al.* (2018) analyzed the inhibitory potential of xanthohumol, naringenin, and acyl phloroglucinol derivatives obtained from *Humulus lupulus* L. against AChE and BuChE. For this, they tested fourteen compounds in an ELISA microplate assay and performed the TLC-DPPH test to verify the ability to eliminate free radicals. Besides, molecular modeling was performed, which included absorption, distribution, metabolism, and excretion of compounds and molecular docking studies with the selected compounds, and finally *in vitro* tests. The results of the analysis of cholinesterase inhibitory activity demonstrated that xanthohumol and 3-hydroxyxanthohumol showed moderate inhibitory activity against AChE compared to the drug galantamine and were also active against BuChE. In the antioxidant activity test, eight compounds exhibited high antioxidant properties than rutin [57]. The pharmacokinetic study conducted by Orhan *et al.* (2018) indicated that all compounds had acceptable drug-like properties. Finally, the docking data obtained in Schrödinger's Induced Fit (IFD) protocol revealed that xanthohumol and 3-hydroxyxanthohumol are energetically favorable against AChE (PDB ID: 4M0E0) and BuChE (PDB ID: 4TPK). Regarding the types of interactions performed by these compounds, the authors found that both performed H-bonds and π-π stacking interactions with the AChE amino acids Phe[297] and Trp[286]. Furthermore, aromatic xanthohumol rings performed strong π-π stacking interactions with BuChE amino acids Trp[82], Trp[231], and Ala[328]. Thus, the authors concluded that xanthohumol is a promising compound that may be important in the discovery of new cholinesterase inhibitors [57].

In a study performed by Zengin *et al.* (2018) conducted their research using the total phenolic components from the extract of *Silene salsuginea* where these components were separated by HPLC [58]. In sense, it was possible to isolate 23 compounds, which were subjected to biological tests and *in silico* studies with the proteins AChE (PDB ID: 4EY6), BuChE (PDB ID: 1P0P) and others, whose affinity energy values are presented in Table **5**.

Table 5. Affinity energy of phenolic components of extracts from *Silene salsuginea* toward AChE and BuChE.

Name	AChE [kcal/mol]	BuChE [kcal/mol]
Gallic acid	−4.89	−4.66
Protocatechuic acid	−5.32	−5.13
(+)-Catechin	−8.86	−8.56
p- Hydroxybenzoic acid	−5.18	−5.27

(Table 5) cont.....

Name	AChE [kcal/mol]	BuChE [kcal/mol]
Chlorogenic acid	−7.37	−6.16
Caffeic acid	−6.38	−6.29
Epicatechin	−8.88	−9.35
Syringic acid	−5.76	−5.27
Vanillin	−5.50	-5.30
p-Coumaric acid	−6.35	−6.09
Ferulic acid	−6.45	−6.19
Synapic acid	−6.36	−6.28
Benzoic acid	−5.30	−4.76
o-Coumaric acid	−6.96	−6.10
Rutin	−4.25	−3.58
Hesperidin	−3.18	−4.67
Rosmarinic acid	−6.95	−5.92
Eriodictyol	−8.63	−8.50
Cinnamic acid	−6.19	−5.63
Quercetin	−8.64	−8.53
Luteolin	−8.53	−8.32
Kaempferol	−8.68	−8.27
Apigenin	−8.31	−8.46
Control **	−9.27	−5.90

Continuing the study by Zengin *et al.* (2018), the biological test corroborated the computational data previously obtained, where the extracts showed activity against the studied proteins [58], as shown in Table **6**.

Table 6. Enzymatic inhibitory effects of *Silene salsuginea* extracts.

Extracts	AChE	BuChE
	GALAE/g extract	
Ethyl acetate	3.09 ± 0.08	4.83 ± 0.04
Methanol	3.03 ± 0.08	4.53 ± 0.32
Water	n.a.	0.34 ± 0.03

± SD (n = 3). n.a.: not active; GALAE: galantamine equivalent. Different superscript letters in the same columns indicate a significant difference between the extracts (p < 0.05).

It was concluded that ethyl acetate and methanol extracts were almost equally effective in inhibiting AChE (3.09 and 3.03 mg of GALAE/g, respectively), while ethyl acetate extract was the most effective inhibitor of BuChE (4.83 mg of GALAE/g), followed by the methanol extract (4.53 mg of GALAE/g of extract). On the other hand, the aqueous extract was the least active cholinesterase inhibitor, showing no inhibition against the AChE enzyme.

Ferlemi *et al.* (2015) investigated the protective effects of consuming the infusion of *Rosmarinus officinalis* L. leaves in rats, observing changes in their behavior and learning, as well as its activity against cholinesterase isoforms (ChEs) [59]. Thus, they sought to determine the content of metabolites in the infusion by using Total Phenolics Content (TPC) through Folin-Ciocalteau reagent method [60], Total Flavonoid Content (TFC) according to aluminum chloride colorimetric method [61], and Total Monomeric Anthocyanin Content (TMAC) by the pH-differential method of reference [62]. The observed results were 12.7 ± 0.008 mg GAE/100 mL of TPC in the *R. officinalis* tea, 1.44 ± 0.069 mg QE/100 mL of TFC, and 0.61 ± 0.038 mg 3-cyanidin glucoside equivalents/100 mL of TMAC. The authors were able to identify 16 compounds by LC/DAD/ESI-MS analysis, demonstrated in Table 7.

Table 7. Compounds identified in the extract of *Rosmarinus officinalis*.

Entry	Name	Entry	Name
01	Caffeic acid-*O*-hexoside derivative	09	Cirsimaritin
02	Quinic acid	10	Rosmanol isomer
03	Caffeic acid derivative	11	Rosmanol isomer
04	Isorhamnetin-3-*O*-hexoside	12	Rosmanol isomer
05	Homaplantaginin	13	Rosmadial
06	Luteolin-7-*O*-glucuronide	14	Notohamosin R
07	Rosmarinic acid	15	Carnosol
08	Luteolin-acetyl-glucuronide	16	Carnosic acid

Ferlemi *et al.* (2015) divided the rats into two groups (one group treated with *R. officinalis* tea and another group that received water) and verified behavioral changes through passive avoidance, elevated plus maze, and forced swimming tests. The results indicated that the tea had anxiolytic and anti-depressant effects in the rats; however, it did not affect the learning ability [59]. Ellman's colorimetric method [63] was used to determine the effect of *R. officinalis* infusion on ChE activity, being observed that ChE isoforms activity was significantly decreased in the brain and liver of treated mice. The authors also performed the molecular docking of the four main compounds identified (rosmarinic acid, luteolin-7-*O*-glucuronide, caffeic acid, and carnosic acid) with recombinant human acetylcholinesterase (rhAChE) (PDB ID: 4EY7) using Schrödinger's suite 2012. 2 and GlideXP. Docking data allowed to verify that all four compounds showed favorable binding scores similar to approved drugs' scores. In addition, the four compounds interacted with the same residues that galantamine and donepezil. Thus, it was possible to notice in the study that *R. officinalis* contains several polyphenols and has benefits due to antidepressant/anxiolytic-like and anticholinesterase effects observed in the results of *in vivo* and *in vitro* tests and *in silico* prediction.

Ślusarczyk *et al.* (2019) analyzed the inhibitory activity of polyphenols (isoflavones and stilbenes) from *Belamcandae chinensis rhizoma* against ChE isoforms. Using HPLC and NMR spectroscopy, they isolated and identified 11 metabolites from the rhizome [64], which can be visualized in Table **8**.

Table **8. Compounds isolated from *Belamcandae chinensis rhizome*.**

Isoflavonoids	Stilbenes	Xanthone glucoside
Irisflorentin	Piceatannol	Mangiferin
Tectorigenin	Resveratrol	-
Iristectorigenin B	-	-
Irigenin	-	-
Irilin D	-	-
Iridin	-	-
Tectoridin	-	-
Iristectorin B	-	-

The identified compounds were tested against AChE and BuChE in ELISA microtiter assay, where piceatannol inhibited both cholinesterases (AChE: 67% inhibition at 100 µg/mL, IC_{50} = 53.42 µg/mL, and BuChE: 91% inhibition at 100 µg/mL, IC_{50} = 18.2 µg/mL), while irilin D and resveratrol were active only against BuChE (IC_{50} of irilin D = 109.53 µg/mL and IC_{50} of resveratrol = 78.07 µg/mL). Irisflorentin was a weak inhibitor against AChE (36% inhibition at 100 µg/mL). The other compounds showed inhibition below 40%. Thus, the authors selected the compounds considered active (irilin B, piceatannol, and resveratrol) to perform molecular docking in the AutoDock Vina software, using the enzymes AChE (PDB ID: 1EVE) and BuChE (PDB ID: 1P0I). The results indicated that H-bond interactions were performed with residues from the catalytic triad and peripheral anionic site, thus compounds blocked the access to key residues involved in the normal functioning of the enzyme, which explains the inhibitory activity of the compounds. Thus, irilin B, piceatannol, and resveratrol showed anticholinesterase activity *In vitro* and *in silico*, being promising for future research involving the planning of substances against cognitive impairment [64].

Nicolaou *et al.* (2010) performed a study involving the synthesis of resveratrol-derived polyphenol compounds hopeanol and hopeahainol A (Fig. **8**) that are natural products isolated from *Hopea hainanensis*. Additionally, the biological evaluation of their activities upon AChE was performed. In sense, inhibition assays were performed at times of 10, 20, 30, 40, 50, and 60 minutes and confirmed the acetylcholinesterase inhibitory activity of the synthetic hopeahainol (+)-3 (Fig. **8**) with IC_{50} values of 4.92 ± 0.09; 4.78 ± 1.07; 3.67 ± 0.39; 3.63 ± 0.03; 3.82 ± 0.09; 3.96 ± 0.25 µM, respectively. The authors concluded that the hopeahainol A is a potent AChE inhibitor and that the results of their study are expected to facilitate synthesis and biological evaluation of related compounds as potential drugs for the treatment of diseases such as AD [65].

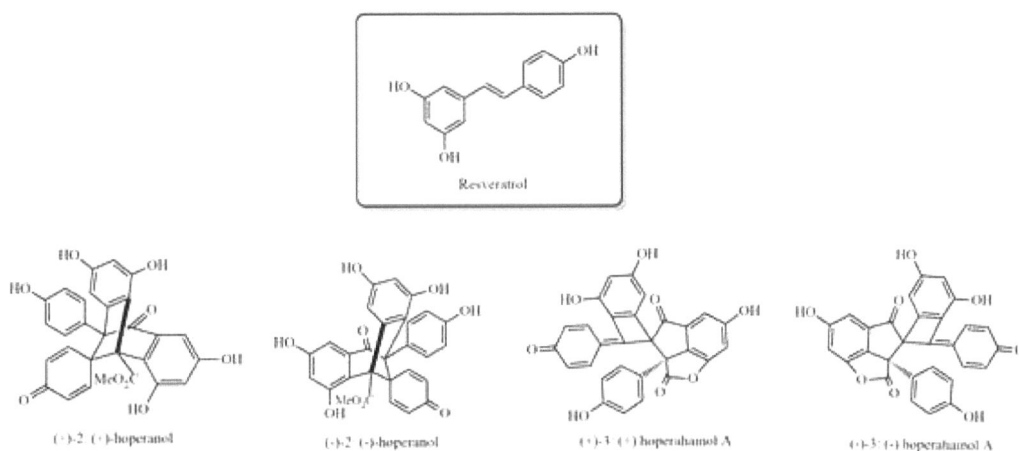

Fig. (8). Compounds synthesized by Nicolaou *et al.*.

Orostachys japonicus A. Berger (*O. japonicus*) is an herb from the *Crassulaceae* and is broadly found in China, Japan, and Korea. It is also used to treat cancer, fever, and gingivitis in folk medicine [66 - 68]. Additionally, some studies have reported other activities, such as protection of neuronal cells from apoptosis, calpain inhibition, anti-oxidant, and antitumoral effects against human breast adenocarcinoma (MCF-7), human leukemia (HP-60), and human lung adenocarcinoma (A549) cancer cells [67, 69, 70]. Catechins are flavanols that can be found in many plant species, including *O. japonicus*, existing as several different isomers [71 - 73]. In sense, Kim and collaborators (2016) evaluated the activity of an in-house library containing *O. japonicus* compounds, such as (-) - epicatechin, (+)-catechin, (−)-epicatechin-3-*O*-gallate, (+)-catechin-3-*O*-gallate, gallic acid, (-)-epicatechin-3,5-O-digallate, and methyl gallate against AChE and BuChE enzymes. Additionally, AutoDock v. 4.2 software was used to provide insight into the interactions between ligand and target (BuChE, PDB ID: 1P0I). Also, molecular dynamics simulations were performed by using Gromacs v. 4.6.5. From the enzymatic assays, it was observed that all compounds had inhibitory activity ranging from 31.4 ± 3.7 to 145.0 ± 2.5 µg/mL and from 17.8 ± 3.8 to 182.2 ± 1.9 µg/mL on AChE and BuChE, respectively. Moreover, it was verified that the (-)-epicatechin-3,5-O-digallate has the most promising activity on both AChE and BuChE enzymes, with a K_i value of 15.4 ± 2.5 µg/mL [74].

The (-)-epicatechin-3,5-O-digallate was selective towards BuChE than AChE, being 3-fold more active. Besides, enzymatic kinetic studies performed by the generation of Lineweaver-Burk and Dixon plots demonstrated that gallic acid is a competitive inhibitor for BuChE. Moreover, it suppresses the BuChE activity and prevents the neurons from reactive oxygen species (ROS) damage. The molecular docking simulation revealed that (-)-epicatechin-3,5-O-digallate interacts with 21 amino acid residues, including H-bond interactions with Gln[67], Thr[120], Glu[197], Pro[285], and Tyr[440]. Also, it was observed that it displays an excellent affinity towards BuChE, with an energy value of -12.21 Kcal/mol. Finally, molecular dynamics simulations demonstrated that (-)-epicatechin-3,5-O-digallate remains in a distance that can be hydrolyzed by the action of His[484] at the active site. Furthermore, it was obtained that it has a potential energy value of -1.18×10^6 KJ/mol over the trajectory period of the simulation. In summary, the authors concluded that the compound (-)-epicatechin 3,5-O-digallate is a promising inhibitor of the human BuChE [74].

BACE-1 Inhibitors

Omar *et al.* (2018) conducted a study involving the chemical components of olives (*Olea europaea* L.) against the enzymes BACE-1, cholinesterases, histone deacetylase, and tyrosinase for the proposal of new bioactive agents to combat AD, whose compounds used were acetyl thicololine iodide, *S*-butyrylthiocholine chloride, caffeic acid, dimethyl sulfoxide (DMSO), *L*-3,4-dihydroxyphenylalanine (L-DOPA), 2-nitrobenzoic acid, epigallocatechin gallate (EGCG), ethylenediamine tetraacetic acid (EDTA), galantamine hydrobromide, 1-glutamine, hydroxytyrosol, kojic acid, luteolin, oleuropein, penicillin-streptomycin, quercetin, rutin, tris-HCl buffer, and trypan blue. As a conclusion, the authors published that tested compounds showed inhibitory action against the rate-limiting enzyme BACE-1, consequently, they can reduce the production of Aβ and the pathogenesis of AD [75].

Cox *et al.* (2015) studied the biological activity of some polyphenolic substances

(Table **9**) against β-mediated amyloid precursor protein processing. This study identified a single flavanol, (-)-epicatechin, to be effective in reducing the production and pathology of Aβ in wild type neurons and a transgenic model of AD and that this is most likely by modulating BACE-1 activity. Deeming that Aβ toxicity is almost certainly initiated in pre-symptomatic stages of AD, any potential benefit of an intervention with (-)-epicatechin would likely be achieved through a risk reduction strategy, rather than as a treatment [76].

Table 9. Molecules studied by Cox *et al.*

Flavonoid group	Compound name
Flavonol	Fisetin A
	Kaempferol
	Kaempferol
Flavone	Quercetin
	Apigenin
	Apigenin glycoside
	Coumarin
	Diosmetin
	Hyperoside
Flavanone	Sinensetin B
	Hesperetin
Anthocyanin	Narirutin
	Cyanidin chloride
	Delphinidin chloride
Flavanol	Pelargonidin chloride A
	(+)-Catechin
	(-)-Epicatechin
	Epicatechin gallate
	Epigallocatechin B

Faghih *et al.* (2015) studied the P7C3 molecule (Fig. **9**) [77], a derivative of dibromocarbazole discovered by MacMillan *et al.* (2011) and that can protect newborn neurons from apoptotic cell death [78]. It is metabolically stable, orally bioavailable, nontoxic, and capable of crossing the blood-brain barrier (BBB). Faghih *et al.* (2015) performed molecular dynamics simulations using the GROMACS v. 4.5.5 package to verify the ability of P7C3 to prevent the onset or to slow the progression of AD [77]. The authors observed that the penetration of

P7C3 in the I-1 simulation system consisting of Aβ protofibril structure (PDB ID: 2BEG) in an aqueous solution resulted in the destabilization of Asp[23]–Lys[28] salt bridge, due to the formation of H-bond interactions between P7C3 and Asp[23]/Lys[28]. The results of interactions between BACE (PDB ID: 1W51) and P7C3 indicated that P7C3 adopted a favorable binding mode in complexes with flap-closed BACE. Besides, the P7C3 hydroxyl performs an important H-bond interaction with the Asp[228] residue from the BACE enzyme. Finally, P7C3 bound more strongly to BACE compared to Aβ, suggesting that P7C3 could treat AD mainly by BACE inhibition [77].

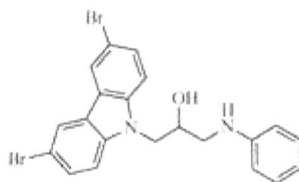

Fig. (9). Chemical structure of P7C3.

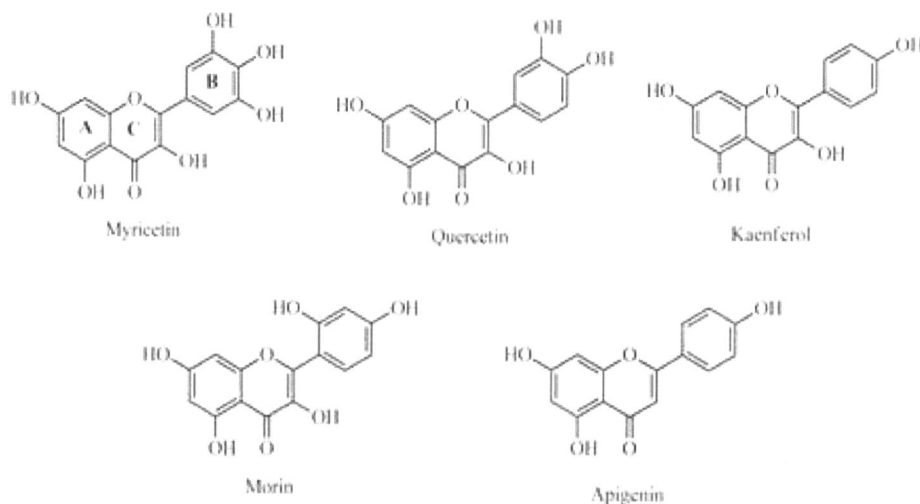

Fig. (10). Molecular structures of the five flavonoids.

In the study performed by Shimmyo *et al.* (2008), the promising activity of five natural flavonoids for the treatment of AD is reported. For this, molecular docking was performed in the Discovery Studio platform using five flavonoids of interest, being myricetin, quercetin, kaempferol, morine, and apigenin (Fig. **10**); toward BACE-1 (PDB ID: 2B8L). In addition, *In vitro* tests were carried out against neuronal cells of rats and the ELISA method was used. Flavonoids acted as BACE-1 inhibitors, reducing Aβ levels. The IC$_{50}$ values upon BACE-1 were found to be 2.8, 5.4, 14.7, 21.7, and 38.5 µM for myricetin, quercetin, kaempferol, morine, and apigenin, respectively. Finally, each flavonoid inhibited BACE-1, significantly reducing the levels of $A\beta_{1-40}$ and $A\beta_{1-42}$ [79].

Neuroprotectors

Singh *et al.* (2008) compiled the activities of several polyphenolic compounds present in foods (fruits, vegetables, and beverages) associated with neuroprotection, more precisely against AD, highlighting the main challenges in this area. Thus, it was shown that among the main derivatives, flavonoids, phenolic acids, and stilbenes are related to the neuroprotective activity. Among these classes of molecules, the EGCG is the main constituent of green tea and has demonstrated neuroprotective activity in several studies. Although it is promising, some challenges should be overcome, such as studies that determine their absorption, bioavailability, and ability to cross the BBB, as well as tests on humans who suffer irreversible and intense neuronal loss [80].

Tandon *et al.* (2020) evaluated the therapeutic potential of curcumin (see Figure 4, C21), a polyphenol component from saffron (*Curcuma longa*) in inhibition of bisphenol-A (BPA) induced neurotoxicity. The oligosphere growth kinetics test was performed to verify the effects of BPA and curcumin on the size and number of oligospheres. The authors noted that there was a decrease in the quantity and size of oligospheres with BPA treatment, while an increase was seen in cultures treated with BPA and curcumin, compared to the group exposed to BPA only. The effects of BPA and curcumin on the proliferation and differentiation of oligodendrocyte progenitor cells (OPC) were verified through immunocytochemistry, where curcumin significantly improved the growth and differentiation of OPC. The authors also performed a neuron-oligodendrocyte co-culture and found that curcumin increased the potential for myelination of oligodendrocytes. The results of fluoromyelin staining indicated that curcumin attenuated BPA-induced neutrality in the myelination of the hippocampus of the rat brain [81]. Also, immunohistochemical analysis was performed by Tandon *et al.* (2020) using the rat hippocampus, and it was found that curcumin causes neuroprotection by allowing the proliferation of OPCs. Besides, immunohistochemical analysis has shown that curcumin may be involved in preventing apoptosis caused by BPA. Real-time PCR (Rt-PCR) and western blot analysis suggested that curcumin increased levels of myelination markers in the hippocampus of BPA-treated mice. To study the influence of curcumin and BPA on learning and memory, two-way conditioned avoidance test and passive avoidance test were performed on rats. The results showed that BPA decreased learning and memory capacity. However, the group treated with curcumin and BPA showed a significant increase in these cognitive functions, where curcumin had a neuroprotective function by mitigating the harmful effects of BPA. The authors also performed an *in silico* analysis to verify the interaction of BPA and curcumin with Notch pathway molecules and found that curcumin showed a higher binding affinity with the proteins Notch1, Hes1, and Mib1 when compared to BPA. In conclusion, curcumin protected groups exposed to BPA *in vitro* and *in vivo* analyzes and that curcumin neuroprotection involves the Notch pathway [81].

Hirata *et al.* (2020) developed a new hybrid curcumin-oxindol compound, GIF-2165X-G1, and evaluated its effect on the oxidative stress on *HT22* cells compared to curcumin and the oxindol derivative GIF-0726-r. The results of the glutamate-induced cell death and viability tests demonstrated that GIF-2165X-G1 is more protective against oxidative stress than curcumin and GIF-0726-r, in addition to being less toxic than curcumin. Therefore, GIF-2165X-G1, GIF-072--r, and curcumin (Fig. **11**) were also tested for activation of antioxidant response element (ARE) and expression of Heme Oxygenase 1 (HO-1). The first increased the ARE activity in a dose-dependent manner, the second slightly induced the activity and the third demonstrated a strong activation of ARE at the concentration of 10 µM and then there was a decrease in this activity by 25 µM. In the test that evaluated the effect of the compounds on the content of intracellular Fe^{2+} ions, it was observed that the three had similar Fe^{2+} ions chelating activity. Using a 6-hydroxydopamine (6-OHDA) mouse model of Parkinson's disease (PD) it was possible to verify that GIF-2165X-G1 prevented the loss of dopaminergic neurons. Thus, it was found that the compound GIF-2165X-G1 has antioxidant activity and exhibit neuroprotective activity through reactive oxygen species scavenging *via* multifunctional mechanisms [82].

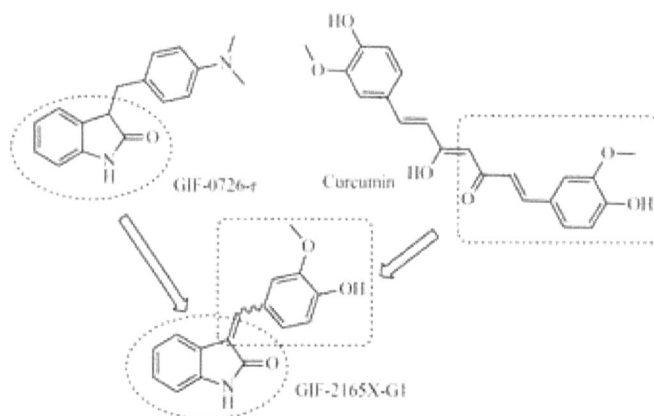

Fig. (11). Chemical structures of GIF-0726-r, GIF-2165X-G1, and curcumin.

Catalogna *et al.* (2019) reviewed the polypharmacological properties of resveratrol (*trans*-3,4',5-trihydroxystilbene) (RSV, Fig. (**12**)), a polyphenolic natural product found in more than 70 species of plants, one of which is the species *Polygonum cuspidatum*. The authors observed that RSV can have neuroprotective actions since the literature reports that it activates sirtuins, particularly SIRT1, an enzyme responsible for the deacetylation of proteins involved in cell regulation. Thus, it can protect neurons against apoptosis, oxidative stress, and inflammation. It has also been reported that Aβ-induced neuron death has been reduced by RSV in a dose-dependent manner from 15 to 40 µM concentrations. This can happen because of RSV's ability to bind to Aβ_{1-42}, interfering with its aggregation, changing the oligomeric conformation of Aβ_{1-42}, and attenuating its cytotoxicity [83].

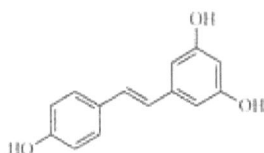

Fig. (12). Chemical structure of resveratrol.

Bahri *et al.* (2016) reviewed the literature for *in vivo* and *in vitro* data on the effectiveness of carnosic acid (CA, Fig. (**13**)), a phenolic diterpene, facing several diseases including AD. One of the studies referenced by the authors demonstrated that CA at a dose of 3 mg/kg in an experimental rat model for AD protected neurons in the hippocampus against cell death. Another study found that CA at the same dose could ameliorate the spatial and learning memory deficits. Data from other research showed that 30 µM of CA in SH-SY5Y human neuroblastoma cells reduced Aβ secretion *via* tumor necrosis factor-α-converting enzyme activation (α-secretase TACE) without effect on BACE. Still, the same work found that treatment with Aβ caused apoptosis in SH-SY5Y cells through the activation of caspases 3-8 and 9, in addition to the cleavage of Poly (ADP-ribose) polymerase (PARP). However, pretreatment with 10 µM CA caused partial inhibition of apoptosis and reduced Aβ oligomers. Thus, it can be seen that there is evidence in the literature demonstrating that CA is a natural product that can be useful in the search for compounds against AD [84].

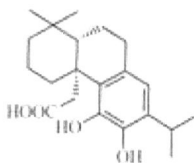

Fig. (13). Chemical structure of carnosic acid.

Due to the important role of flavonoids, extracts containing this class of molecules can be used in the treatment of neurodegenerative diseases, such as AD, where these could exert neuroprotective effects related to the reduction of pro-inflammatory cytokines. Candiracci *et al.* (2012) studied the ability of an extract of raw multifloral honey (HFE) rich in flavonoids (luteolin, quercetin, apigenin, kaempferol, isorhamnetin, acacetin, tamarixetin, chrysin, and galangin, (Fig. **14**) to reduce induced pro-inflammatory mediators in the microglia activated for lipopolysaccharide (LPS). As result, it was shown that HFE decreases the production of iNOS mRNA and protein in N13 microglia after LPS stimulation; decreased TNF-α and IL-1β dose-dependent; and inhibits the production of ROS (reactive oxygen intermediates) after LPS induction. The authors suggest that the antioxidant and anti-inflammatory effects of HFE could be used to prevent neurodegenerative diseases [85].

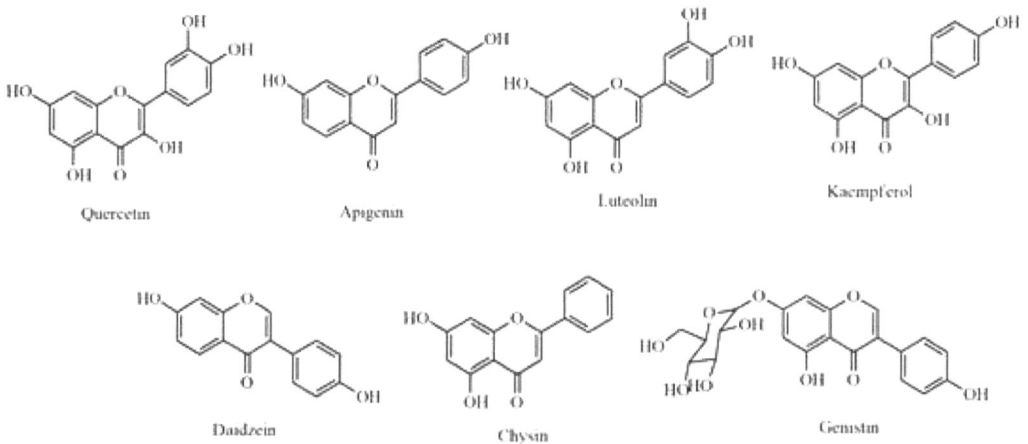

Fig. (14). Flavonoids found in the raw multifloral honey extract.

Based on a previous study that showed neuroprotective activity of chloroform extracts from *Laurus nobilis* leaves in three cell lines, Pacifico *et al.* (2014) investigated the alcoholic extract obtained from the powdered leaves pre-extracted with chloroform. In total, 13 compounds (Fig. **15**) were identified from the LnM-2c fraction, which showed the most promising results, with anti-radical effects in the 2,2'-diphenyl-1-picrylhydrazyl (DPPH) and 2,2'-azino-bis-3-ethylbenzothiazoline)-6-sulfonic acid (ABTS$^{·+}$) assays, with values of 11.8 and 0.96 µg/mL, respectively. Additionally, it was verified the absence of cytotoxicity in the MTT assay; a decrease in the oxidant activity of the amyloidogenic fragment; and inhibition of ROS production by 56%, after one-hour exposure. Finally, the authors highlighted the action of flavonoids in the neuroprotection effect and which they could be promising in both treatments and prevention of AD [86].

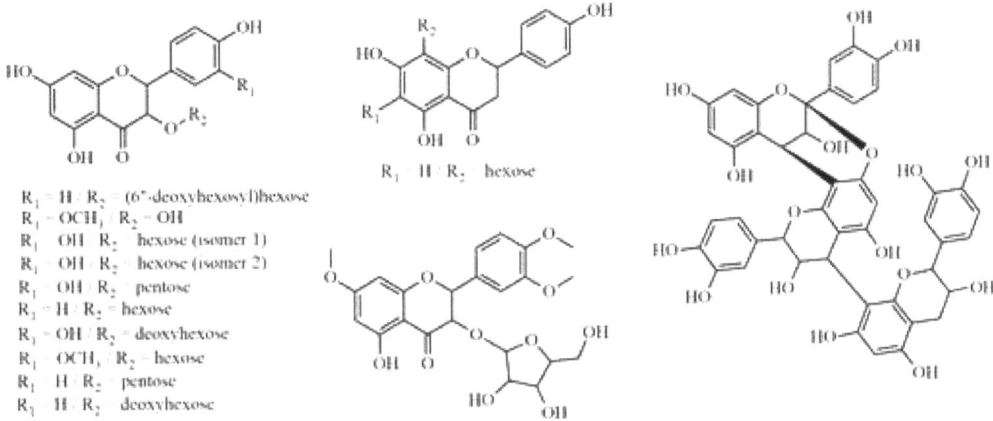

Fig. (15). Compounds identified from the LnM-2c fraction.

Pérez *et al.* (2019) synthesized four derivatives of 5-aminosalicylic acid and evaluated them for anti-inflammatory, antioxidant activity and as inhibitors of myeloperoxidase (MPO), an enzyme that plays a crucial role in inflammation and oxidative stress. The results of the topic TPA-induced mouse ear edema method indicated that the compound C1 had better anti-inflammatory activity than compounds C2, C3, and C4. Also, the *O*-dianisidine method was used to verify the inhibitory activity of the compounds against MPO, where it was observed that compounds C2, C3, and C4 demonstrated better inhibitory effects than the compound C1. Moreover, DPPH and ABTS$^{·+}$ methods revealed that compounds C3 and C4 have good antioxidant properties. Finally, the molecular docking performed in AutoDock v. 4.2.0 software demonstrated that all four compounds tested interact with amino acid residues reported in the main active MPO site (PDB ID: 1DNU). However, compounds C3 and C4 fit into the MPO site with better values of free energy than the other compounds. Thus, the authors concluded that these compounds C1-C4 may be useful in the treatment of diseases with inflammatory processes, such as AD [87].

In the study carried out by Lecanu *et al.* (2005), the synthesis and neuroprotective activity of (4-ethyl-piperaz-1-yl)-phenylmethanone derivatives were reported, because this molecule exhibits strong neuroprotective properties against Aβ_{1-42}. For this, natural compounds were researched in a database, using procaine as a substructure, in order to select a common chemical nucleus carrying the activity and, from there, to develop and synthesize a series of analogs [88]. Notwithstanding this information, three active compounds that had the substructure of 4-ethyl-1-benzoyl-piperazine were chosen, named SP015, SP016, and SP017 which can be extracted from plants belonging the *Asteraceae* family, *Inula britanica*, and *Artemisia glabella*. Among the selected natural compounds,

SP017 showed a greater protective effect on mitochondrial tissue, SP015 exhibited a neuroprotective effect at 1 and 10 μM concentrations and SP016 only showed a significant effect only at low concentrations of $A\beta_{1-42}$ (0.1 μM). However, SP015 and SP017 unexpectedly caused a free radical-generating effect. From these results, a molecule containing 4-ethyl-1-benzoyl-piperazine, called SP008 was then synthesized. SP008 demonstrated significant neuroprotective properties against $A\beta_{1-42}$, exhibiting strong anti-amyloid activity, and showing no toxicity at higher concentrations, compared to SP017 [88].

Another class of polyphenols that may be promising against AD are derived from chalcones, as shown in the study that investigated the neuroprotective activity of Xanthohumol (2′,4′,4-trihydroxy-6′-methoxy-3′-prenylchalcone, Xn, (Fig. **16**) performed by Yao *et al.* (2015) [89]. The authors synthesized the compound Xn, as well as a derivative with the absence of α,β-unsaturated ketone (TH-Xn, Figure 16), and evaluated its cytoprotective effect, as well as its effects underlying mechanism in a cell line neuron-like rat pheochromocytoma (PC12). Then, it was shown that the compound Xn can prevent neuronal damage induced by oxidative stress, inhibiting ROS *In vitro*. The results also show that the cytoprotective of TH-Xn is due to the induction of endogenous molecules, this because it presents an α,β-unsaturated ketone in its structure so that no induction of Nrf2 occurs on PC12 cells in the absence of this chemical group (TH-Xn). Moreover, the authors concluded that Xn is an activator of Keakp1-Nrf2, providing important information in the development of new compounds with neuroprotective effects [89].

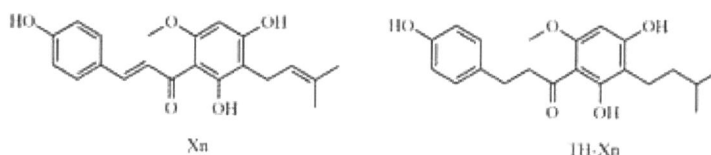

Fig. (16). Molecular structures of compounds Xn and TH-Xn.

Kinases Inhibitors and Anti-Tau Pathology Agents

Carvalho *et al.* (2017) performed the screening of quercetin protein targets based on the sequential arrangement of ligand similarity search, binding site comparison, and inverse docking. SHAFTS (SHApe-Feature Similarity) algorithm was used to identify quercetin-similar ligands associated with protein complexes in the PDB. Thus, 34 molecules sharing the typical flavonoid scaffold were obtained, except for six molecules. Also, the authors noted that the corresponding co-crystallized protein set comprised of 25 mammalian proteins, where proteins from plants and other organisms were disregarded for the subsequent analysis. The protein candidates selected by SHAFTS were subjected to binding site comparison against a representative set of proteins contained in the PDB, using the LIBRA software. The results indicated that three main classes of proteins, along with the previous set, could be distinguished, such as sulfotransferases, tankyrases-poly (ADP-ribosyl) transferases, and kinases [90].

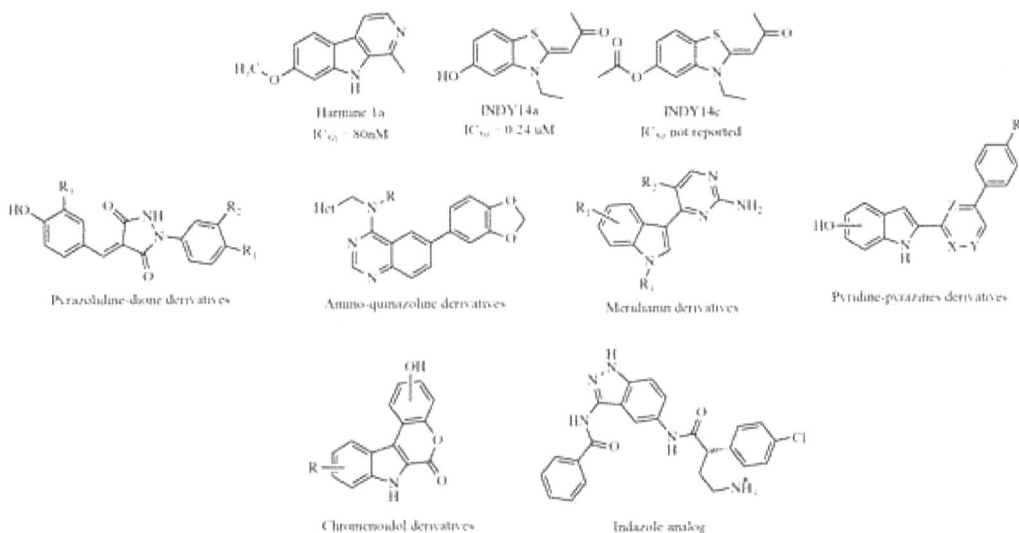

Fig. (17). Promising compounds against DYRK1A. *R, R1, R2, and R3 are non-hydrogen atom.*

From the previous screening stages performed by Carvalho *et al.* (2017), a list of 74 protein target candidates was identified, where these were subsequently subjected to structure-based screening against quercetin using idTarget. The authors observed, through the docking poses obtained, that quercetin reached a known binding site of every protein, which was generally a catalytic or an allosteric site and that the predicted affinity ranged from 15.1 to 494.1 nM. Molecular dynamics simulations were performed in Molecular Operating Environment (MOE) software and the following quercetin-protein complexes were analyzed: glycogen synthase kinase-3 beta (GSK-3b) (PDB ID: 1J1B), mitogen-activated protein kinase 14 (MAPK14) (PDB ID: 3S3I), phosphatidylinositol 3-kinase gamma (PI3K) (PDB ID: 1E8W), and poly [ADPribose] polymerase 1 (PARP) (PDB ID: 3GJW). In this analysis, it was observed that the hydroxyl groups of quercetin participated in interactions with proteins, in which groups at positions 3', 4', and 7 were generally in direct contact with the targets. Thus, these results indicate that quercetin may be a multi-kinase and multi-PARP inhibitor [90].

Several studies have suggested that a kinase present in eukaryotes known as Dual Specificity Tyrosine Phosphorylation Regulated Kinase 1A (DYRK1A) may be an exploitable target in the search for new agents for the treatment of neurodegenerative diseases, such as AD. Smith *et al.* (2012), through a literature review, highlighted the main advances related to the development of new DYRK1A antagonists, showing that harmine (Fig. **17**) and polyphenol EGCG are promising natural derivatives. Besides this, the phenolic compound INDY (Fig. **17**) proved to be one of the best inhibitors; however, it has limited application because it does not present selectivity for other kinases. Another promising compound, acetylated-INDY derivative (Fig. **17**), showed a potential therapeutic application against neurodegenerative diseases. Besides, other scaffolds have been identified as promising, such as pyrazolidine-dione, amino-quinazoline, meridianin, pyridine-pyrazines, and chromenoidol derivatives (Fig. **17**), as well as an indazole analog (Fig. **17**), which, through its crystalline structure, showed the good interactions with the active site from DYRK1A, providing new horizons in the discovery of new compounds against AD [91].

Chua *et al.* (2017) evaluated the effects of altenusin (Fig. **18**), a polyphenolic compound isolated from Penicillium sp., on heparin-induced Tau fibrillization and phosphorylation *in vitro* and *in vivo*. *In vitro* results indicated that altenusin inhibited aggregation of Tau protein into paired helical filaments, in which it is associated with the stabilization of Tau dimers and other oligomers. Also, it was verified the reduction of the capacity for Tau phosphorylation in affected cells, resulting in a decrease in the pathology progress. However, treatment of Tau transgenic mice did not improve neuropathology and functional deficits. Therefore, the authors concluded that altenusin inhibits *in vitro* fibrillization of tau and reduces induced Tau pathology in neurons [92].

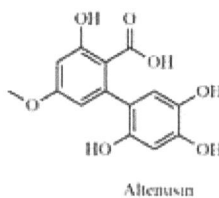

Fig. (18). Chemical structure of Altenusin.

Yao *et al* . (2013) performed multiple experiments to elucidate the binding properties and aggregation inhibitory activity of tannic acid (Fig. **19**) , a plant polyphenol, on Tau peptide R3, in comparison with its moiety, gallic acid [93]. The results showed that TA has an affinity for the R3 peptide from Tau. In contrast, gallic acid does not demonstrate to have a good affinity for Tau. It was shown that tannic acid showed the ability to inhibit the aggregation of Tau R3 peptide *In vitro* with an IC $_{50}$ value of 3.5 μM, while gallic acid presented 92 μM. The authors suggest, through molecular docking studies, that the high affinity of tannic acid may be due to structural differences concerning gallic acid, such as hydrophobic interactions promoted by the central glucose ring, aromatic rings, as well as hydrophobic center and size of phenolic groups. The main factor related to the recognition of tannic acid by R3 Tau peptide is the formation of a hairpin structure, in which it has fundamental importance for inhibiting polymerization of Tau. The authors concluded that this study offers a new strategy for designing new Tau aggregation inhibitors useful for the treatment of AD [93].

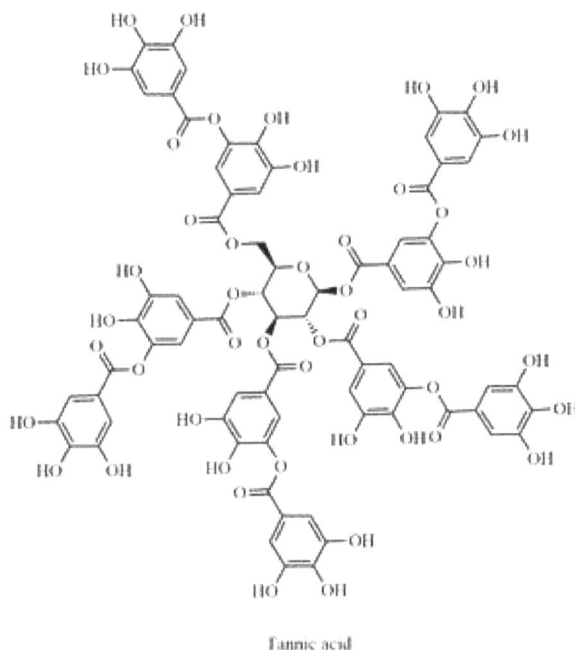

Fig. (19). Chemical structure of tannic acid.

Guéroux *et al.* (2015) report the promising activity of 10 polyphenols (procyanidins) present in wine, in the regression of AD. Specifically, the work sought to understand the mode of action of polyphenols on the Tau protein, at a molecular level that could explain its benefits on the progression of the disease. The authors used NMR and molecular modeling techniques to study the fixation of the 10 flavan-3-ols (procyanidins) (Fig. **20**) in the proline-rich domains from the Tau because these domains are considered potent targets capable of fixing a variety of compounds, especially polyphenols. The authors concluded that the polyphenols could slow the progression of AD by fixing the proline-rich region of the Tau protein, protecting it against uncontrolled attacks of kinases, and avoiding the disorganization of microtubules and the subsequent cell death [94].

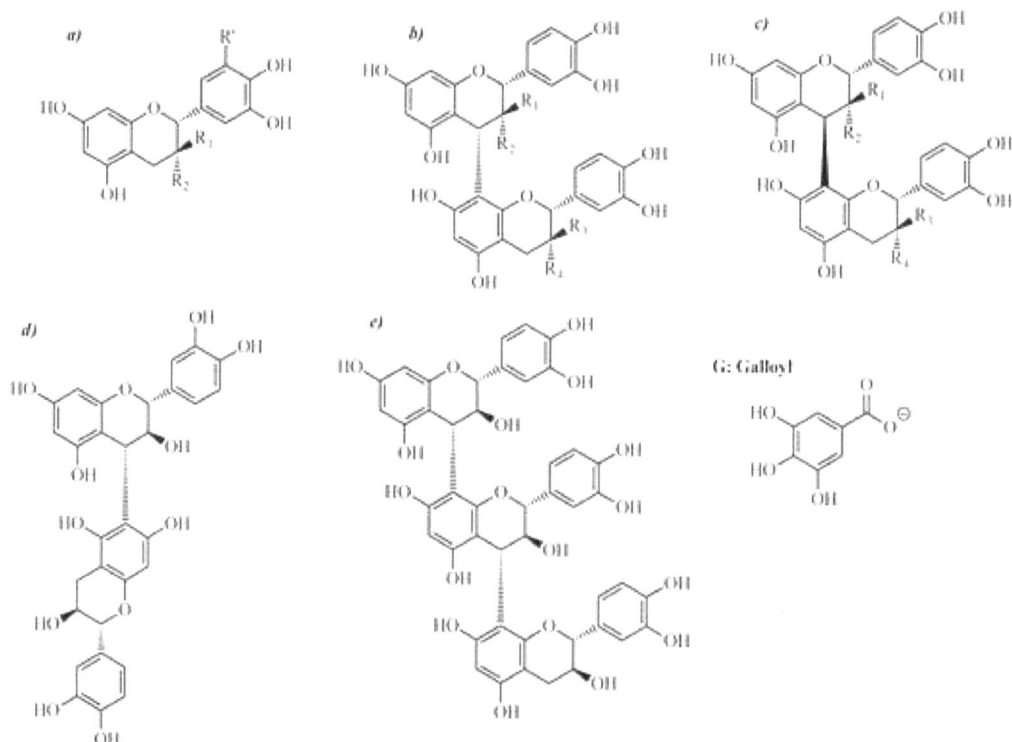

Fig. (20). *Chemical structures of procyanidins studied by Guéroux* et al . a) EC, Epicatechin (R_1=R'=H, R_2=OH); ECG, Epicatechin gallate (R_1=H, R'=OH, R_2=G); EGCG, Epigallocatechin gallate (R_1=H, R'=H, R_2=G). b) B3, Catechin 4α-8 Catechin (R_1/R_3=OH, R_2/R_4=H); B4, Catechin 4α-8 Epicatechin (R_1/R_4=OH, R_2/R_3=H); B3G, Catechin 4α-8 Catechin gallate (R_1=OH, R_3=G, R_2/R_4=H). c) B2, Epicatechin 4β-8 Epicatechin (R_1/R_3=H, R_2/R_4=OH); B1, Epicatechin 4β-8 Catechin (R_2/R_3=OH, R_1/R_4=H). d) B6, Catechin 4α-6 Catechin.e) C2, Catechin 4α-8 Catechin 4α-8 Catechin.

Multi-Target Agents Against Alzheimer's Disease

Chakraborty and Basu (2017) developed a multi-target virtual screening protocol to locate potent phytochemicals against AD that could inhibit BACE-1 and demonstrate antioxidant and anti-amyloidogenic activity. They found narirutin (Fig. **21**), a flavanone glycoside present in orange juice. Both the antioxidant activity and the ability to inhibit BACE-1 and aggregation of Aβ from this compound were analyzed through *In vitro* assays. The result of steady-state and time-resolved fluorescence indicated that the gradual addition of narirutin causes a continuous increase in BACE-1 fluorescence, because of the conformational transition from BACE-1 from the open to the closed form, induced by narirutin. The authors also performed an assay using specific dyes for amyloid, ANS, and ThT, to observe the formation of Aβ and its inhibition by narirutin. ThT presents an increased intrinsic fluorescence when it binds to the nucleus of fibril Aβ, so when the authors verified the decrease in the intensity of this fluorescence in the co-incubation of Aβ with 10 and 20 μm of narirutin, they realized that there was a reduction in the degree of amyloid formation. The atomic force microscope (AFM) study also confirmed the potential that narirutin has to inhibit Aβ aggregation. ABTS^{+} scavenging assay revealed that narirutin has moderate concentration-dependent antioxidant activity. Therefore, it was concluded that narirutin has multi-potent activity, as it is an inhibitor of BACE-1 and Aβ aggregation, in addition to having antioxidant activity, which can be a therapeutic option for AD [95].

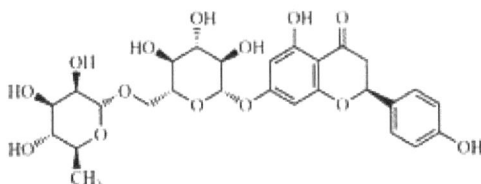

Fig. (21). Chemical structure of narirutin.

Pugazhendhi *et al.* (2018) evaluated the anticholinesterase, anti-amyloidogenic, and antioxidant activities of the fruit extract of *Terminalia chebula* (TCF) and bark extract of *Terminalia arjuna* (TAB) [96]. The results of the activities of the extracts against BuChE can be found in Table **10**.

Table 10. Butyrylcholinesterase inhibitory activities of different solvent extracts of *T. chebula* and *T. arjuna*.

Solvent extract	IC$_{50}$ (µg/mL)
Terminalia chebula	-
Donepezil	6.78 ± 1.17
Petroleum ether	n.d.
Chloroform	n.d.
Ethyl acetate	238.16 ± 3.94
Methanol	81.44 ± 1.23
Water	115.58 ± 10.67
Terminalia arjuna	-
Donepezil	6.78 ± 1.17
Petroleum ether	n.d.
Chloroform	n.d.
Ethyl acetate	170.17 ± 50
Methanol	52.16 ± 3.94

Solvent extract	IC_{50} (µg/mL)
Water	90.23 ± 3.94

*The results were expressed as Mean ± SD (n = 3); n.d. = Not determined; * P < 0.05.*

The authors observed that of different solvent extracts, the methanolic extract of TFC and TAB exhibited significant BuChE inhibitory activity with an IC_{50} value of 81.44 ± 1.23 and 52.16 ± 3.94 µg/mL, respectively. Therefore, they used chromatography to fractionate the compounds present in the methanolic extracts, and thus 10 fractions were obtained. Table **11** presents the compounds of the fractions F7 of TFC and F5 of TAB.

Table 11. LC-MS analysis of the fraction of column F7 of TFC and F5 of TAB.

Terminalia arjuna		*Terminalia chebula*	
Entry	**Compound**	**Entry**	**Compound**
01	Arjunic acid	17	Pyrogallol
02	Arjunetin	18	Gallic acid
03	Morintanic acid	19	Chebulagic Acid
04	7-Methyl gallate	20	Chebulinic Acid
05	Gallic acid	21	Punic acid
06	Methoxycoumarin	22	Punicalagin
07	Trimethoxycoumarin	23	Corilagin
08	Coumestrol	24	Benzoic acid
09	Coumaroylquinic acid	25	Dicoumarol
10	Coumaric acid	26	Coumaroylquinic Acid
11	Coumarin	27	Coumarin
12	Ellagic acid	28	7-Methyl gallate
13	Quercetin	29	Luteolin
14	Quercetin methyl ether	30	Hydroxyluteoline
15	3-Methyl ellagic acid	31	Luteolin Gluteoside
16	Naringenin	32	Luteoliflavan glucoside

Finishing the study by Pugazhendhi *et al.* (2018), the authors concluded that the methanolic extract of TCF and TAB, with a substantial quantity of polyphenols, exhibited potent activity against BuChE and AChE enzymes. In their work, it was possible to verify that the compound 7-methyl-gallic acid acts as a neuroprotective constituent of extracts of TFC and TAB. *In vitro* analysis demonstrated that 7-methyl-gallic acid exhibited a double cholinergic effect and also had potent antioxidant activity and anti-amyloidogenic effect by effectively inhibiting Aβ fibrillogenesis [96].

Chakraborty *et al.* (2016) performed a multi-target screening as a therapeutic strategy for AD and obtained hesperidin (Fig. **22**), a flavanone glycoside. To verify the interaction of hesperidin with BACE-1, studies were carried out involving steady-state and time-resolved fluorescence, determination of binding affinity and inhibitory activity, and molecular modeling. Fluorescence spectroscopy revealed that hesperidin causes a conformational change in BACE-1 by binding to the active site of this protein, which changes from open to closed conformation. The Benesie-Hildebrand equation was used to determine the equilibrium constant (K) and, thus, verify the hesperidin affinity for BACE-1, where the constant resulted in a value of 3.02 x 10^5, indicating a strong link between hesperidin and the active protein site. The inhibitory activity was confirmed with the BACE-1 assay kit, as it demonstrated that only 500 nM of hesperidin causes 100% inhibition of BACE-1 activity [97]. The results of molecular docking performed using the FlexX software by Chakraborty *et al.* (2016) showed that hesperidin performed 15 H-bond interactions with various residues from the BACE-1 active site (PDB ID: 3TPP), which explains its high binding affinity. Besides, the authors also carried out tests to verify the ability of hesperidin to inhibit Aβ aggregation, being ANS, and ThT tests. The results indicated complete inhibition of aggregation of Aβ fibrils. To check the antioxidant activity of hesperidin, ABTS$^{\cdot+}$ scavenging assay was performed and it was observed that hesperidin showed a moderate capacity to eliminate free radicals. Thus, the multi-target screening, performed by the authors, selected hesperidin as an excellent compound against AD, as it exhibits strong binding affinity with BACE-1, demonstrates high inhibition of Aβ aggregation, and antioxidant activity [97].

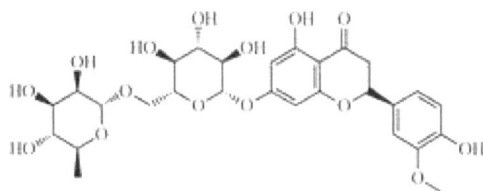

Fig. (22). Chemical structure of hesperidin.

Khan *et al.* (2019) verified the chemical and biological effects of different extracts (methanol, ethyl acetate, *n*-hexane, aqueous) from *Caragana ambigua*. Thus, they determined total phenolic content (TPC) using the Folin-Ciocalteau method and total flavonoid content (TFC) through aluminum chloride colorimetric assay, while the profile of secondary metabolites was obtained by UHPLC/MS. The authors observed that ethyl acetate extract had a higher content of phenolic compounds and flavonoids compared to the other three extracts. To check the antioxidant potential of the extracts, different tests were performed (DPPH, ABTS$^{·+}$, FRAP, CUPRAC, phosphomolybdenum, and metal chelating assays), where the ethyl acetate extract was the one that demonstrated the greatest ability to eliminate free radicals. This extract also showed better results in the FRAP and CUPRAC tests, being the most active. Besides, through the phosphomolybdenum assay, the ethyl acetate extract showed the highest total antioxidant capacity [98]. The extracts of *C. ambigua* in the study by Khan *et al.* (2019) were also tested *in vitro* against cholinesterases (AChE and BuChE) and other enzymes. The results of these tests indicated that all the extracts of *C. ambigua* (except aqueous extract) had the potential for inhibition against the two cholinesterase enzymes. Thus, the authors performed an *in silico* analysis with three of the dominant compounds of the extracts (isobergaptene, jujubasaponin IV, and phellodensin D), where the compounds were subjected to molecular docking in AutoDock v. 4 software using the enzymes AChE (PDB ID: 4EY6) and BuChE (PDB ID: 1P0P). Among the three compounds, phellodensin D (Fig. **23**) showed a higher binding affinity with cholinesterases. Therefore, with all these results, the authors found that the ethyl acetate extract exhibited the highest content of bioactive compounds and the greatest antioxidant potential (except metal chelation) and that through this study it was possible to verify the pharmacological potential of *C. ambigua*, which presented molecules that can be useful in industrial applications [98].

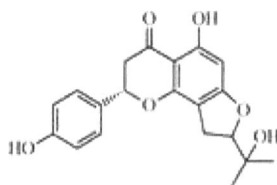

Fig. (23). Chemical structure of phellodensin D.

Chang *et al.* (2019) researched the effect of caffeic acid (see Fig. **5**) in the development of AD and related mechanisms in high-fat (HF) diet-induced hyperinsulinemic rats. For this, the Morris water maze test was performed in hyperinsulinemic rats to verify the memory decline, resulting in escape latency time for rats in the group treated with caffeic acid significantly less than that from the group not treated with caffeic acid. Moreover, the authors studied the effect of caffeic acid on antioxidant substances such as superoxide dismutase (SOD), catalase, and glutathione of the hippocampus and the cortex in hyperinsulinemic rats. They found that the HF group showed a decrease in SOD and catalase activities compared to the normal group, while the caffeic acid-treated group exhibited an increase in SOD activity by 36.6% in the hippocampus and 22.8% in the cortex and a decrease in the catalase activity of 13.3% in the hippocampus, and 12.8% in the cortex [99]. Continuing the results of the aforementioned research, the western blot method confirmed that hyperinsulinemic rats administered with caffeic acid showed decreased phosphorylated-Tau protein expression compared to the HF group, and the level of inactive GSK3β expression in the hippocampus was regulated positively, 38.3% when compared to that of the HF group. Regarding the effect of caffeic acid on proteins related to Aβ degradation, the authors observed that the expression levels of the BACE enzyme were decreased in the hippocampus and cortex of the group of HF rats treated with caffeic acid, in 26 and 31.6%, respectively, compared to HF group rats. Therefore, caffeic acid also promoted a decrease in APP in the hippocampus and cortex in HF diet rats. Therefore, it was concluded that caffeic acid can protect neurons against oxidative stress by improving the activity of antioxidant substances, it can decrease the neurotoxicity induced by the accumulation of Aβ by attenuating the expression of BACE, among other benefits, being a compound that has potential in preventing progression of AD [99].

Mazumder and Choudhury (2019) tested *in silico* the polyphenols (+)-catechin, (-)-epicatechin, (-)-epigallocatechin, and (-)-epigallocatechin gallate against the main targets of AD. For this, they performed the molecular docking of these compounds in the Molegro Virtual Docker (MVD) software using the proteins: BACE (PDB ID: 1FKN), γ-secretase (PDB ID: 5FN2), GSK-3β (PDB ID: 1Q5K), AChE (PDB ID: 4EY6), and BuChE (PDB ID: 4BDS) [100]. The results of the docking of polyphenols are shown in Table **12** in comparison with those of the protein inhibitors described, being elenbecestat, semagacestat, AR-A014418, galantamine, and tacrine.

Table 12. MolDock score of different ligands with the enzymes.

Enzyme	Ligand	MolDock score
BACE	*Elenbecestat*	-146.205
-	-(-)-Epigallocatechin	-91.72
-	(-)-Epicatechin	-99.986
-	(-)-Epigallocatechin gallate	-141.686
-	(+)-Catechin	-108.556
γ-Secretase	*Segamacestat*	-79.525
-	(-)-Epigallocatechin	-83.671
-	(-)-Epicatechin	-75.59
-	(-)-Epigallocatechin gallate	-128.822

Enzyme	Ligand	MolDock score
-	(+)-Catechin	-104.63
GSK-3β	AR-A014418	-103.617
-	(-)-Epigallocatechin	-79.114
-	(-)-Epicatechin	-87.606
-	(-)-Epigallocatechin gallate	-114.384
-	(+)-Catechin	-90.040
AChE	Galantamine	-125.778
-	(-)-Epigallocatechin	-128.042
-	(-)-Epicatechin	-123.617
-	(-)-Epigallocatechin gallate	-176.664
-	(+)-Catechin	-132.426
BuChE	Tacrine	-82.874
-	(-)-Epigallocatechin	-121.804
-	(-)-Epicatechin	-104.442
-	(-)-Epigallocatechin gallate	-150.744
-	(+)-Catechin	-124.505

Molecular docking showed that, against BACE, the known inhibitor elenbecestat had the highest MolDock score (-146.205), while among polyphenols, (-) - epigallocatechin gallate showed the maximum score (-141.686), similar to elenbecestat. The H-bond scores for all ligands were comparable, showing that receptor inhibition is largely caused by hydrogen bonding interactions. Against γ-secretase, (-)-epigallocatechin gallate had the highest MolDock score (-128.822) while (-)-epicatechin showed the highest H-bond score. Although the MolDock score of segamacestat (-79.525) was comparable to (-)-epigallocatechin (-83.671) and (-)-epicatechin (-75.590), the inhibitor had a much lower H-bond score. H-bond scores for (-)-epicatechin and (-)-epigallocatechin gallate were therefore 3.49 times and 2.83 times higher, respectively, compared to the known inhibitor. The lower H-bond scores and the MolDoc scores of the inhibitor may be due to a smaller number of hydrogen bond donor and hydrogen bond acceptor groups present in their structure, compared to polyphenols. Thus, it is shown that the inhibition of the receptor by polyphenols is mostly due to more interactions in hydrogen bonds, concerning the known inhibitor Segamacestat [100].

Regarding GSK-3β, the authors noted that (-) -epigallocatechin gallate had the highest MolDoc score (-114.384) and H-bond score. The higher scores for (---epigallocatechin gallate can be attributed to the greater number of groups accepting hydrogen bonds and donors of hydrogen bonds in polyphenol. The results of the ligands against AChE showed that although the (-)-epigallocatechin gallate had the highest MolDoc score (-176.664), the H-bond score was much lower, suggesting that the MolDock score is mostly due to weaker interactions. However, the (-)-epigallocatechin showed the highest H-bond score, which was 2.61 times higher than the known inhibitor galantamine. The other polyphenols (-)-epigallocatechin and (+)-catechin) also showed good H-bond scores. Considering the H-bond, the authors considered (+)-catechin, (-)-epicatechin, and (-)-epigallocatechin as the most promising inhibitors of the enzyme, compared to (-)-epigallocatechin gallate and the known inhibitor. At the BuChE active site, the (-)-epigallocatechin gallate had the highest MolDock score (-150.744), however, (+)-catechin showed the highest H-bond score of all ligands. The H-bond scores of (-)-epigallocatechin gallate and (+)-catechin were 9.04 times and 9.91 times higher than the known inhibitor tacrine. Thus, it was found that the inhibition of BuChE by polyphenols is greatly affected by the interactions of hydrogen bonds, and can be attributed to the fact that Tacrine has only three groups that can form these bonds [100].

According to the data presented, the authors concluded that the polyphenols: (+)-catechin, (-)-epicatechin, (-)-epigallocatechin, and (-)-epigallocatechin gallate potentially inhibit the enzymes AChE, BuChE, BACE, γ-secretase, and GSK-3β. Thus, it is expected that this inhibition will increase the level of ACh in the brain and thus improve the progression of AD. As a result, polyphenols can become alternative candidates to the drugs currently used in the treatment of AD, and further *in vivo* and *in vitro* studies are necessary to establish their effectiveness [100].

Rodrigues *et al.* (2018) studied the neuroprotective *in vitro* effects of extracts from the leaves and roots of *Polygonum maritimum*. For this, the authors made extracts using methanol and dichloromethane from leaves and roots of this species, which were characterized by UHPLC/MS-MS. The extracts were subsequently evaluated regarding the inhibition of AChE and BuChE, the ability to attenuate the hydrogen peroxide-induced cytotoxicity on neuroblastoma cells and lipopolysaccharide (LPS)-induced neuroinflammation on microglia cells. UHPLC analysis allowed the identification of 15 bioactive polyphenols, including flavonol glycosides, flavan-3-ols, and proanthocyanidins [101]. Concerning *in vitro* tests, methanol extracts showed greater activity in AChE (leaves: $IC_{50} = 0.27$ mg/mL; roots: $IC_{50} = 0.17$ mg/mL), and BuChE (leaves: $IC_{50} = 0.62$ mg/mL; roots: $IC_{50} = 0.61$ mg/mL). Besides, methanol extracts resulted in the highest increase in SH-SY5Y cellular viability after co-application with H_2O_2 and the highest decrease of NO production on LPS-stimulated microglia. The authors performed molecular docking simulations using Induced Fit Docking (IFD) protocol and they observed that all the compounds were capable of binding to AChE and

BuChE proteins, where the procyanidin was observed as the highest potent compound with average IFD scores of −9.54 and −10.38 Kcal/mol in the catalytic domains of AChE and BuChE, respectively. Thus, the study data indicate that the methanol extracts of P. maritimum may be a promising source of bioactive polyphenols with cognitive enhancement and anti-aging properties [101].

Mahomoodally *et al.* (2019) investigated the enzyme inhibitory potential, antioxidant activity, and phenolic content of different extracts of *Hypericum lanuginosum*. For this, they performed liquid chromatography/high-resolution mass spectrometry (LC-HRMS) analysis that identified 21 phenolic compounds, including flavonoids, bioflavonoids, phenolic acids, and acylquinic acids. They found that quinic acid, 3-caffeoylquinic acid (Fig. **24**), and other molecules are major compounds in the extracts. The enzyme inhibitory activity results indicated that the ethyl acetate extract was the most effective inhibitor of AChE (5.03 mg galantamine equivalent (GALAE)/g extract) and BuChE (6.02 mg GALAE). Regarding the antioxidant activity, the aqueous extract showed a better antioxidant effect in five out of the six assays conducted and the authors believe this is because the aqueous extract has a greater amount of phenolic and flavonoid compounds than other extracts [102]. In summary, *H. lanuginosum* could be considered as a source of phenolic compounds with antioxidant properties and cholinesterases inhibitory activities, where further *in vivo* investigation is needed.

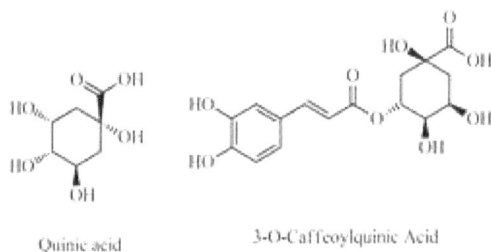

Fig. (24). Molecular structures of two compounds found in the extracts.

CONCLUSION

Based on the results of several studies in this chapter, we conclude that polyphenols are promising molecules in the treatment of Alzheimer's disease (AD) due to their ability to interact with the different targets of the disease. It is noted that AD has complex pathogenesis and, consequently, it is essential to search for compounds capable of acting in various mechanisms related to the disease. Thus, it was possible to identify, in this work, some polyphenols that stood out for their action on the various targets and processes involved in the development of AD. As an example, the phenolic compound known as 7-methy-

-gallic acid inhibited the enzymes, AChE and BuChE, *in vitro*, and therefore, improved the cholinergic deficit, blocked the aggregation of A β and disaggregated preformed fibrils, in addition to having antioxidant activity. Another compound that demonstrated good results *in vitro* was the flavonoid narirutin presenting strong A β aggregation inhibitory potential, moderate antioxidant activity, and considered a high-affinity BACE-1 inhibitor. Still, several studies have reported that epigallocatechin gallate inhibits A β aggregation and the BACE-1 enzyme *In vitro* and *in silico* , thus, it can reduce the production of A β peptides and amyloid plaques. Therefore, it is believed that multipotent polyphenols can become new alternatives in drug development, given that there is still no cure for AD, so their use shows a viable path for therapy.

CONSENT FOR PUBLICATION

Not applicable.

CONFLICT OF INTEREST

The authors declare no conflict of interest, financial or otherwise.

ACKNOWLEDGEMENTS

The authors thank Coordenação de Aperfeiçoamento Pessoal de Nível Superior (CAPES), National Council for Scientific and Technological Development (CNPq), Fundação de Amparo à Pesquisa do Estado de Alagoas (FAPEAL), and Financier of Studies and Projects (FINEP) for their financial support to the Brazilian Post-Graduate Programs.

REFERENCES

[1]   Cummings J, Lee G, Ritter A, Zhong K. Alzheimer's disease drug development pipeline: 2018. Alzheimers Dement (N Y) 2018; 4: 195-214.
[http://dx.doi.org/10.1016/j.trci.2018.03.009] [PMID: 29955663]

[2]   Cheignon C, Tomas M, Bonnefont-Rousselot D, Faller P, Hureau C, Collin F. Oxidative stress and the amyloid beta peptide in Alzheimer's disease. Redox Biol 2018; 14: 450-64.
[http://dx.doi.org/10.1016/j.redox.2017.10.014] [PMID: 29080524]

[3]   Association A. 2019 Alzheimer's disease facts and figures. Alzheimers Dement 2019; 15: 321-87.
[http://dx.doi.org/10.1016/j.jalz.2019.01.010]

[4]   Honig LS, Vellas B, Woodward M, *et al.* Trial of solanezumab for mild dementia due to Alzheimer's disease. N Engl J Med 2018; 378(4): 321-30.
[http://dx.doi.org/10.1056/NEJMoa1705971] [PMID: 29365294]

[5]   Vyas S, Kothari SL, Kachhwaha S. Nootropic medicinal plants: Therapeutic alternatives for Alzheimer's disease. J Herb Med 2019; 17–18100291.
[http://dx.doi.org/10.1016/j.hermed.2019.100291]

[6]   Kumar D, Ganeshpurkar A, Kumar D, Modi G, Gupta SK, Singh SK. Secretase inhibitors for the treatment of Alzheimer's disease: Long road ahead. Eur J Med Chem 2018; 148: 436-52.
        [http://dx.doi.org/10.1016/j.ejmech.2018.02.035] [PMID: 29477076]

[7]   MacLeod R, Hillert EK, Cameron RT, Baillie GS. The role and therapeutic targeting of α-, β- and γ-secretase in Alzheimer's disease. Future Sci OA 2015; 1(3): FSO11.
        [http://dx.doi.org/10.4155/fso.15.9] [PMID: 28031886]

[8]   Kametani F, Hasegawa M. Reconsideration of amyloid hypothesis and tau hypothesis in Alzheimer's disease. Front Neurosci 2018; 12: 25.
        [http://dx.doi.org/10.3389/fnins.2018.00025] [PMID: 29440986]

[9]   Cuadrado A, Kügler S, Lastres-Becker I. Pharmacological targeting of GSK-3 and NRF2 provides neuroprotection in a preclinical model of tauopathy. Redox Biol 2018; 14: 522-34.
        [http://dx.doi.org/10.1016/j.redox.2017.10.010] [PMID: 29121589]

[10]   Ivashko-Pachima Y, Gozes I. NAP protects against tau hyperphosphorylation through GSK3. Curr Pharm Des 2018; 24(33): 3868-77.
        [http://dx.doi.org/10.2174/1381612824666181112105954] [PMID: 30417779]

[11]   Gao Y-L, Wang N, Sun F-R, Cao X-P, Zhang W, Yu J-T. Tau in neurodegenerative disease. Ann Transl Med 2018; 6(10): 175-5.
        [http://dx.doi.org/10.21037/atm.2018.04.23] [PMID: 29951497]

[12]   Mroczko B, Groblewska M, Litman-Zawadzka A, Kornhuber J, Lewczuk P. Amyloid β oligomers (AβOs) in Alzheimer's disease. J Neural Transm (Vienna) 2018; 125(2): 177-91.
        [http://dx.doi.org/10.1007/s00702-017-1820-x] [PMID: 29196815]

[13]   Sengupta U, Nilson AN, Kayed R. The role of amyloid-β oligomers in toxicity, propagation, and immunotherapy. EBioMedicine 2016; 6: 42-9.
        [http://dx.doi.org/10.1016/j.ebiom.2016.03.035] [PMID: 27211547]

[14]   Brody AH, Strittmatter SM. Synaptotoxic signaling by amyloid Beta oligomers in alzheimer's disease through prion protein and mGluR5. 1st ed. Elsevier Inc. 2018; Vol. 82.

[15]   Knez D, Coquelle N, Pišlar A, *et al.* Multi-target-directed ligands for treating Alzheimer's disease: Butyrylcholinesterase inhibitors displaying antioxidant and neuroprotective activities. Eur J Med Chem 2018; 156: 598-617.
        [http://dx.doi.org/10.1016/j.ejmech.2018.07.033] [PMID: 30031971]

[16]   Jiang Y, Gao H. Pharmacophore-based drug design for the identification of novel butyrylcholinesterase inhibitors against Alzheimer's disease. Phytomedicine 2019; 54: 278-90.
        [http://dx.doi.org/10.1016/j.phymed.2018.09.199] [PMID: 30668379]

[17]   Kumar A, Pintus F, Di Petrillo A, *et al.* Novel 2-pheynlbenzofuran derivatives as selective butyrylcholinesterase inhibitors for Alzheimer's disease. Sci Rep 2018; 8(1): 4424.
        [http://dx.doi.org/10.1038/s41598-018-22747-2] [PMID: 29535344]

[18]   Jalili-Baleh L, Babaei E, Abdpour S, *et al.* A review on flavonoid-based scaffolds as multi-targe--directed ligands (MTDLs) for Alzheimer's disease. Eur J Med Chem 2018; 152: 570-89.
        [http://dx.doi.org/10.1016/j.ejmech.2018.05.004] [PMID: 29763806]

[19]   Gong CX, Liu F, Iqbal K. Multifactorial hypothesis and multi-targets for alzheimer's disease. J Alzheimers Dis 2018; 64(s1): S107-17.
        [http://dx.doi.org/10.3233/JAD-179921] [PMID: 29562523]

[20]   Lauretti E, Dincer O, Praticò D. Glycogen synthase kinase-3 signaling in Alzheimer's disease. Biochim Biophys Acta Mol Cell Res 2020; 1867(5)118664.
        [http://dx.doi.org/10.1016/j.bbamcr.2020.118664] [PMID: 32006534]

[21]   Zhang P, Xu S, Zhu Z, Xu J. Multi-target design strategies for the improved treatment of Alzheimer's disease. Eur J Med Chem 2019; 176: 228-47.

[http://dx.doi.org/10.1016/j.ejmech.2019.05.020] [PMID: 31103902]

[22] Hiremathad A, Keri RS, Esteves AR, Cardoso SM, Chaves S, Santos MA. Novel Tacrine-Hydroxyphenylbenzimidazole hybrids as potential multitarget drug candidates for Alzheimer's disease. Eur J Med Chem 2018; 148: 255-67.
[http://dx.doi.org/10.1016/j.ejmech.2018.02.023] [PMID: 29466775]

[23] Coimbra JRM, Marques DFF, Baptista SJ, *et al.* Highlights in BACE1 inhibitors for Alzheimer's disease treatment. Front Chem 2018; 6: 178.
[http://dx.doi.org/10.3389/fchem.2018.00178] [PMID: 29881722]

[24] Piton M, Hirtz C, Desmetz C, *et al.* Alzheimer's disease: Advances in drug development. J Alzheimers Dis 2018; 65(1): 3-13.
[http://dx.doi.org/10.3233/JAD-180145] [PMID: 30040716]

[25] Tresserra-Rimbau A, Lamuela-Raventos RM, Moreno JJ. Polyphenols, food and pharma. Current knowledge and directions for future research. Biochem Pharmacol 2018; 156: 186-95.
[http://dx.doi.org/10.1016/j.bcp.2018.07.050] [PMID: 30086286]

[26] Yahfoufi N, Alsadi N, Jambi M, Matar C. The immunomodulatory and anti-inflammatory role of polyphenols. Nutrients 2018; 10(11): 1-23.
[http://dx.doi.org/10.3390/nu10111618] [PMID: 30400131]

[27] Gorzynik-Debicka M, Przychodzen P, Cappello F, *et al.* Potential health benefits of olive oil and plant polyphenols. Int J Mol Sci 2018; 19(3)E686.
[http://dx.doi.org/10.3390/ijms19030686] [PMID: 29495598]

[28] Cory H, Passarelli S, Szeto J, Tamez M, Mattei J. The role of polyphenols in human health and food systems: A mini-review. Front Nutr 2018; 5: 87.
[http://dx.doi.org/10.3389/fnut.2018.00087] [PMID: 30298133]

[29] Inbar P, Bautista MR, Takayama SA, Yang J. Assay to screen for molecules that associate with Alzheimer's related β-amyloid fibrils. Anal Chem 2008; 80(9): 3502-6.
[http://dx.doi.org/10.1021/ac702592f] [PMID: 18380488]

[30] Churches QI, Caine J, Cavanagh K, *et al.* Naturally occurring polyphenolic inhibitors of amyloid beta aggregation. Bioorg Med Chem Lett 2014; 24(14): 3108-12.
[http://dx.doi.org/10.1016/j.bmcl.2014.05.008] [PMID: 24878198]

[31] Tomaselli S, La Vitola P, Pagano K, *et al.* Biophysical and *in Vivo* studies identify a new natural-based polyphenol, counteracting Aβ oligomerization *in vitro* and Aβ oligomer-mediated memory impairment and neuroinflammation in an acute mouse model of alzheimer's disease. ACS Chem Neurosci 2019; 10(11): 4462-75.
[http://dx.doi.org/10.1021/acschemneuro.9b00241] [PMID: 31603646]

[32] Orteca G, Tavanti F, Bednarikova Z, *et al.* Curcumin derivatives and Aβ-fibrillar aggregates: An interactions' study for diagnostic/therapeutic purposes in neurodegenerative diseases. Bioorg Med Chem 2018; 26(14): 4288-300.
[http://dx.doi.org/10.1016/j.bmc.2018.07.027] [PMID: 30031653]

[33] Chaudhury S, Dutta A, Bag S, Biswas P, Das AK, Dasgupta S. Probing the inhibitory potency of epigallocatechin gallate against human γB-crystallin aggregation: Spectroscopic, microscopic and simulation studies. Spectrochim Acta A Mol Biomol Spectrosc 2018; 192: 318-27.
[http://dx.doi.org/10.1016/j.saa.2017.11.036] [PMID: 29172128]

[34] Eggers C, Fujitani M, Kato R, Smid S. Novel cannabis flavonoid, cannflavin A displays both a hormetic and neuroprotective profile against amyloid β-mediated neurotoxicity in PC12 cells: Comparison with geranylated flavonoids, mimulone and diplacone. Biochem Pharmacol 2019; 169113609
[http://dx.doi.org/10.1016/j.bcp.2019.08.011] [PMID: 31437460]

[35] Huang Q, Zhao Q, Peng J, *et al.* Peptide-Polyphenol (KLVFF/EGCG) binary modulators for inhibiting

aggregation and neurotoxicity of amyloid-β peptide. ACS Omega 2019; 4: 4233-42.
[http://dx.doi.org/10.1021/acsomega.8b02797]

[36]   Kurnik M, Sahin C, Andersen CB, *et al.* Potent α-synuclein aggregation inhibitors, identified by high-throughput screening, mainly target the monomeric state. Cell Chem Biol 2018; 25(11): 1389-1402.e9.
[http://dx.doi.org/10.1016/j.chembiol.2018.08.005] [PMID: 30197194]

[37]   Oliveri V. Toward the discovery and development of effective modulators of α-synuclein amyloid aggregation. Eur J Med Chem 2019; 167: 10-36.
[http://dx.doi.org/10.1016/j.ejmech.2019.01.045] [PMID: 30743095]

[38]   Hirohata M, Hasegawa K, Tsutsumi-Yasuhara S, *et al.* The anti-amyloidogenic effect is exerted against Alzheimer's β-amyloid fibrils *in vitro* by preferential and reversible binding of flavonoids to the amyloid fibril structure. Biochemistry 2007; 46(7): 1888-99.
[http://dx.doi.org/10.1021/bi061540x] [PMID: 17253770]

[39]   Siposova K, Kozar T, Huntosova V, Tomkova S, Musatov A. Inhibition of amyloid fibril formation and disassembly of pre-formed fibrils by natural polyphenol rottlerin. Biochim Biophys Acta Proteins Proteomics 2019; 1867(3): 259-74.
[http://dx.doi.org/10.1016/j.bbapap.2018.10.002] [PMID: 30316862]

[40]   Patel P, Parmar K, Patel D, Kumar S, Trivedi M, Das M. Inhibition of amyloid fibril formation of lysozyme by ascorbic acid and a probable mechanism of action. Int J Biol Macromol 2018; 114: 666-78.
[http://dx.doi.org/10.1016/j.ijbiomac.2018.03.152] [PMID: 29596935]

[41]   Chong X, Sun L, Sun Y, *et al.* Insights into the mechanism of how Morin suppresses amyloid fibrillation of hen egg white lysozyme. Int J Biol Macromol 2017; 101: 321-5.
[http://dx.doi.org/10.1016/j.ijbiomac.2017.03.107] [PMID: 28341174]

[42]   Wang Y, Latshaw DC, Hall CK. Aggregation of Aβ(17-36) in the presence of naturally occurring phenolic inhibitors using coarse-grained simulations. J Mol Biol 2017; 429(24): 3893-908.
[http://dx.doi.org/10.1016/j.jmb.2017.10.006] [PMID: 29031698]

[43]   Huy PDQ, Yu YC, Ngo ST, *et al. In silico* and *in vitro* characterization of anti-amyloidogenic activity of vitamin K3 analogues for Alzheimer's disease. Biochim Biophys Acta 2013; 1830(4): 2960-9.
[http://dx.doi.org/10.1016/j.bbagen.2012.12.026] [PMID: 23295971]

[44]   Ulicna K, Bednarikova Z, Hsu WT, *et al.* Lysozyme amyloid fibrillization in presence of tacrine/acridone-coumarin heterodimers. Colloids Surf B Biointerfaces 2018; 166: 108-18.
[http://dx.doi.org/10.1016/j.colsurfb.2018.03.010] [PMID: 29550545]

[45]   Simões CJV, Almeida ZL, Costa D, *et al.* A novel bis-furan scaffold for transthyretin stabilization and amyloid inhibition. Eur J Med Chem 2016; 121: 823-40.
[http://dx.doi.org/10.1016/j.ejmech.2016.02.074] [PMID: 27020050]

[46]   Wang Y, Xia Z, Xu JR, *et al.* A-mangostin, a polyphenolic xanthone derivative from mangosteen, attenuates β-amyloid oligomers-induced neurotoxicity by inhibiting amyloid aggregation. Neuropharmacology 2012; 62(2): 871-81.
[http://dx.doi.org/10.1016/j.neuropharm.2011.09.016] [PMID: 21958557]

[47]   Kayed R. Common structure of soluble amyloid oligomers implies common mechanism of pathogenesis. Science 2003; 300(80): 9-486.
[http://dx.doi.org/10.1126/science.1079469]

[48]   Yuan T, Ma H, Liu W, *et al.* Pomegranate's neuroprotective effects against alzheimer's disease are mediated by urolithins, its ellagitannin-gut microbial derived metabolites. ACS Chem Neurosci 2016; 7(1): 26-33.
[http://dx.doi.org/10.1021/acschemneuro.5b00260] [PMID: 26559394]

[49]   Dong M, Li H, Hu D, Zhao W, Zhu X, Ai H. Molecular dynamics study on the inhibition mechanisms of drugs CQ1-3 for alzheimer amyloid-β40 aggregation induced by Cu_{2+}. ACS Chem Neurosci 2016;

7(5): 599-614.
[http://dx.doi.org/10.1021/acschemneuro.5b00343] [PMID: 26871000]

[50]   Simoni E, Caporaso R, Bergamini C, *et al.* Polyamine Conjugation as a Promising Strategy To Target Amyloid Aggregation in the Framework of Alzheimer's Disease. ACS Med Chem Lett 2016; 7(12): 1145-50.
[http://dx.doi.org/10.1021/acsmedchemlett.6b00339] [PMID: 27994754]

[51]   Neagu E, Paun G, Albu C, Radu GL. Assessment of acetylcholinesterase and tyrosinase inhibitory and antioxidant activity of *Alchemilla vulgaris* and *Filipendula ulmaria* extracts. J Taiwan Inst Chem Eng 2015; 52: 1-6.
[http://dx.doi.org/10.1016/j.jtice.2015.01.026]

[52]   Singleton VL, Orthofer R, Lamuela-Raventós RM. Analysis of total phenols and other oxidation substrates and antioxidants by means of folin-ciocalteu reagent. Methods Enzymol 1999; 299: 152-78.
[http://dx.doi.org/10.1016/S0076-6879(99)99017-1]

[53]   Ingkaninan K, Temkitthawon P, Chuenchom K, Yuyaem T, Thongnoi W. Screening for acetylcholinesterase inhibitory activity in plants used in Thai traditional rejuvenating and neurotonic remedies. J Ethnopharmacol 2003; 89(2-3): 261-4.
[http://dx.doi.org/10.1016/j.jep.2003.08.008] [PMID: 14611889]

[54]   Cristea V, Deliu C, Oltean B, *et al.* Soilless cultures for pharmaceutical use and biodiversity conservation. Acta Hortic 2009; (843): 157-64.
[http://dx.doi.org/10.17660/ActaHortic.2009.843.19]

[55]   Zengin G, Bulut G, Mollica A, Haznedaroglu MZ, Dogan A, Aktumsek A. Bioactivities of *Achillea phrygia* and *Bupleurum croceum* based on the composition of phenolic compounds: *in vitro* and *in silico* approaches. Food Chem Toxicol 2017; 107(Pt B): 597-608.
[http://dx.doi.org/10.1016/j.fct.2017.03.037] [PMID: 28343034]

[56]   Kuppusamy A, Arumugam M, George S. Combining in silico and *in vitro* approaches to evaluate the acetylcholinesterase inhibitory profile of some commercially available flavonoids in the management of Alzheimer's disease. Int J Biol Macromol 2017; 95: 199-203.
[http://dx.doi.org/10.1016/j.ijbiomac.2016.11.062] [PMID: 27871793]

[57]   Orhan IE, Jedrejek D, Senol FS, *et al.* Molecular modeling and *in vitro* approaches towards cholinesterase inhibitory effect of some natural xanthohumol, naringenin, and acyl phloroglucinol derivatives. Phytomedicine 2018; 42: 25-33.
[http://dx.doi.org/10.1016/j.phymed.2018.03.009] [PMID: 29655693]

[58]   Zengin G, Rodrigues MJ, Abdallah HH, *et al.* Combination of phenolic profiles, pharmacological properties and *in silico* studies to provide new insights on Silene salsuginea from Turkey. Comput Biol Chem 2018; 77: 178-86.
[http://dx.doi.org/10.1016/j.compbiolchem.2018.10.005] [PMID: 30336375]

[59]   Ferlemi AV, Katsikoudi A, Kontogianni VG, *et al.* Rosemary tea consumption results to anxiolytic- and anti-depressant-like behavior of adult male mice and inhibits all cerebral area and liver cholinesterase activity; phytochemical investigation and in silico studies. Chem Biol Interact 2015; 237: 47-57.
[http://dx.doi.org/10.1016/j.cbi.2015.04.013] [PMID: 25910439]

[60]   Singleton VL, Esau P. Phenolic substances in grapes and wine, and their significance. Adv Food Res Suppl 1969; 1: 1-261.
[PMID: 4902344]

[61]   Ben IO, Woode E, Abotsi WKM. Preliminary Phytochemical Screening and *in vitro* Antioxidant Prop-. J Med Biomed Sci 2013; 2: 6-15.

[62]   Lee J, Durst RW, Wrolstad RE. Determination of total monomeric anthocyanin pigment content of fruit juices, beverages, natural colorants, and wines by the pH differential method: collaborative study. J AOAC Int 2005; 88(5): 1269-78.

[http://dx.doi.org/10.1093/jaoac/88.5.1269] [PMID: 16385975]

[63]   Ellman GL, Courtney KD, Andres V Jr, Feather-Stone RM. A new and rapid colorimetric determination of acetylcholinesterase activity. Biochem Pharmacol 1961; 7: 88-95.
[http://dx.doi.org/10.1016/0006-2952(61)90145-9] [PMID: 13726518]

[64]   Ślusarczyk S, Senol Deniz FS, Woźniak D, *et al.* Selective *In vitro* and in silico cholinesterase inhibitory activity of isoflavones and stilbenes from Belamcandae chinensis rhizoma. Phytochem Lett 2019; 30: 261-72.
[http://dx.doi.org/10.1016/j.phytol.2019.02.006]

[65]   Nicolaou KC, Kang Q, Wu TR, Lim CS, Chen DYK. Total synthesis and biological evaluation of the resveratrol-derived polyphenol natural products hopeanol and hopeahainol A. J Am Chem Soc 2010; 132(21): 7540-8.
[http://dx.doi.org/10.1021/ja102623j] [PMID: 20462209]

[66]   Lee G-S, Lee H-S, Kim S-H, Suk D-H, Ryu D-S, Lee D-S. Anti-cancer activity of the ethylacetate fraction from Orostachys japonicus for modulation of the signaling pathway in HepG2 human hepatoma cells. Food Sci Biotechnol 2014; 23: 269-75.
[http://dx.doi.org/10.1007/s10068-014-0037-0]

[67]   Yoon NY, Min BS, Lee HK, Park JC, Choi JS. A potent anti-complementary acylated sterol glucoside from Orostachys japonicus. Arch Pharm Res 2005; 28(8): 892-6.
[http://dx.doi.org/10.1007/BF02973873] [PMID: 16178413]

[68]   Lee H, Kang C, Jung ES, Kim J-S, Kim E. Antimetastatic activity of polyphenol-rich extract of Ecklonia cava through the inhibition of the Akt pathway in A549 human lung cancer cells. Food Chem 2011; 127(3): 1229-36.
[http://dx.doi.org/10.1016/j.foodchem.2011.02.005] [PMID: 25214119]

[69]   Zhang H, Oh J, Jang T-S, Min BS, Na M. Glycolipids from the aerial parts of Orostachys japonicus with fatty acid synthase inhibitory and cytotoxic activities. Food Chem 2012; 131: 1097-103.
[http://dx.doi.org/10.1016/j.foodchem.2011.09.058]

[70]   Kim HJ, Lee JY, Kim SM, *et al.* A new epicatechin gallate and calpain inhibitory activity from Orostachys japonicus. Fitoterapia 2009; 80(1): 73-6.
[http://dx.doi.org/10.1016/j.fitote.2008.10.003] [PMID: 18977282]

[71]   NONAKA G, NISHIOKA I. Tannins and related compounds. LXXVII. Novel chalcan-flavan dimers, assamicains A, B and C, and a new flavan-3-ol and proanthocyanidins from the fresh leaves of *Camellia sinensis* L. var. assamica Kitamura. Chem Pharm Bull (Tokyo) 1989; 37: 77-85.
[http://dx.doi.org/10.1248/cpb.37.77]

[72]   Ivanov SA, Nomura K, Malfanov IL, Sklyar IV, Ptitsyn LR. Isolation of a novel catechin from Bergenia rhizomes that has pronounced lipase-inhibiting and antioxidative properties. Fitoterapia 2011; 82(2): 212-8.
[http://dx.doi.org/10.1016/j.fitote.2010.09.013] [PMID: 20923698]

[73]   Sun YN, Li W, Kim JH, *et al.* Chemical constituents from the root of Polygonum multiflorum and their soluble epoxide hydrolase inhibitory activity. Arch Pharm Res 2015; 38(6): 998-1004.
[http://dx.doi.org/10.1007/s12272-014-0520-4] [PMID: 25413971]

[74]   Kim JH, Lee S-H, Lee HW, *et al.* (-)-Epicatechin derivate from Orostachys japonicus as potential inhibitor of the human butyrylcholinesterase. Int J Biol Macromol 2016; 91: 1033-9.
[http://dx.doi.org/10.1016/j.ijbiomac.2016.06.069] [PMID: 27341781]

[75]   Omar SH, Scott CJ, Hamlin AS, Obied HK. Biophenols: Enzymes (β-secretase, Cholinesterases, histone deacetylase and tyrosinase) inhibitors from olive (*Olea europaea* L.). Fitoterapia 2018; 128: 118-29.
[http://dx.doi.org/10.1016/j.fitote.2018.05.011] [PMID: 29772299]

[76]   Cox CJ, Choudhry F, Peacey E, *et al.* Dietary (-)-epicatechin as a potent inhibitor of βγ-secretase

amyloid precursor protein processing. Neurobiol Aging 2015; 36(1): 178-87.
[http://dx.doi.org/10.1016/j.neurobiolaging.2014.07.032] [PMID: 25316600]

[77] Faghih Z, Fereidoonnezhad M, Tabaei SMH, Rezaei Z, Zolghadr AR. The binding of small carbazole derivative (P7C3) to protofibrils of the Alzheimer's disease and β-secretase: Molecular dynamics simulation studies. Chem Phys 2015; 459: 31-9.
[http://dx.doi.org/10.1016/j.chemphys.2015.07.026]

[78] MacMillan KS, Naidoo J, Liang J, *et al.* Development of proneurogenic, neuroprotective small molecules. J Am Chem Soc 2011; 133(5): 1428-37.
[http://dx.doi.org/10.1021/ja108211m] [PMID: 21210688]

[79] Shimmyo Y, Kihara T, Akaike A, Niidome T, Sugimoto H. Flavonols and flavones as BACE-1 inhibitors: structure-activity relationship in cell-free, cell-based and *in silico* studies reveal novel pharmacophore features. Biochim Biophys Acta 2008; 1780(5): 819-25.
[http://dx.doi.org/10.1016/j.bbagen.2008.01.017] [PMID: 18295609]

[80] Singh M, Arseneault M, Sanderson T, Murthy V, Ramassamy C. Challenges for research on polyphenols from foods in Alzheimer's disease: bioavailability, metabolism, and cellular and molecular mechanisms. J Agric Food Chem 2008; 56(13): 4855-73.
[http://dx.doi.org/10.1021/jf0735073] [PMID: 18557624]

[81] Tandon A, Singh SJ, Gupta M, *et al.* Notch pathway up-regulation *via* curcumin mitigates bisphenol-A (BPA) induced alterations in hippocampal oligodendrogenesis. J Hazard Mater 2020; 392122052
[http://dx.doi.org/10.1016/j.jhazmat.2020.122052] [PMID: 32151947]

[82] Hirata Y, Ito Y, Takashima M, *et al.* Novel oxindole-curcumin hybrid compound for antioxidative stress and neuroprotection. ACS Chem Neurosci 2020; 11(1): 76-85.
[http://dx.doi.org/10.1021/acschemneuro.9b00619] [PMID: 31799835]

[83] Catalogna G, Moraca F, D'Antona L, *et al.* Review about the multi-target profile of resveratrol and its implication in the SGK1 inhibition. Eur J Med Chem 2019; 183111675
[http://dx.doi.org/10.1016/j.ejmech.2019.111675] [PMID: 31539779]

[84] Bahri S, Jameleddine S, Shlyonsky V. Relevance of carnosic acid to the treatment of several health disorders: Molecular targets and mechanisms. Biomed Pharmacother 2016; 84: 569-82.
[http://dx.doi.org/10.1016/j.biopha.2016.09.067] [PMID: 27694001]

[85] Candiracci M, Piatti E, Dominguez-Barragán M, *et al.* Anti-inflammatory activity of a honey flavonoid extract on lipopolysaccharide-activated N13 microglial cells. J Agric Food Chem 2012; 60(50): 12304-11.
[http://dx.doi.org/10.1021/jf302468h] [PMID: 23176387]

[86] Pacifico S, Gallicchio M, Lorenz P, *et al.* Neuroprotective potential of Laurus nobilis antioxidant polyphenol-enriched leaf extracts. Chem Res Toxicol 2014; 27(4): 611-26.
[http://dx.doi.org/10.1021/tx5000415] [PMID: 24547959]

[87] Cabrera Pérez LC, Gutiérrez Sánchez M, Mendieta Wejebe JE, Hernández Rgodríguez M, Fragoso Vázquez MJ, Salazar JR, *et al.* Novel 5-aminosalicylic derivatives as anti-inflammatories and myeloperoxidase inhibitors evaluated *in silico*, *in vitro* and *ex vivo.* Arab J Chem 2019; 12: 5278-91.
[http://dx.doi.org/10.1016/j.arabjc.2016.12.026]

[88] Lecanu L, Yao W, Piechot A, Greeson J, Tzalis D, Papadopoulos V. Identification, design, synthesis, and pharmacological activity of (4-ethyl-piperazin-1-yl)-phenylmethanone derivatives with neuroprotective properties against β-amyloid-induced toxicity. Neuropharmacology 2005; 49(1): 86-96.
[http://dx.doi.org/10.1016/j.neuropharm.2005.01.028] [PMID: 15992583]

[89] Yao J, Zhang B, Ge C, Peng S, Fang J. Xanthohumol, a polyphenol chalcone present in hops, activating Nrf2 enzymes to confer protection against oxidative damage in PC12 cells. J Agric Food Chem 2015; 63(5): 1521-31.
[http://dx.doi.org/10.1021/jf505075n] [PMID: 25587858]

[90]   Carvalho D, Paulino M, Polticelli F, Arredondo F, Williams RJ, Abin-Carriquiry JA. Structural evidence of quercetin multi-target bioactivity: A reverse virtual screening strategy. Eur J Pharm Sci 2017; 106: 393-403.
[http://dx.doi.org/10.1016/j.ejps.2017.06.028] [PMID: 28636950]

[91]   Smith B, Medda F, Gokhale V, Dunckley T, Hulme C. Recent advances in the design, synthesis, and biological evaluation of selective DYRK1A inhibitors: a new avenue for a disease modifying treatment of Alzheimer's? ACS Chem Neurosci 2012; 3(11): 857-72.
[http://dx.doi.org/10.1021/cn300094k] [PMID: 23173067]

[92]   Chua SW, Cornejo A, van Eersel J, *et al.* The Polyphenol Altenusin Inhibits *In vitro* Fibrillization of Tau and Reduces Induced Tau Pathology in Primary Neurons. ACS Chem Neurosci 2017; 8(4): 743-51.
[http://dx.doi.org/10.1021/acschemneuro.6b00433] [PMID: 28067492]

[93]   Yao J, Gao X, Sun W, Yao T, Shi S, Ji L. Molecular hairpin: a possible model for inhibition of tau aggregation by tannic acid. Biochemistry 2013; 52(11): 1893-902.
[http://dx.doi.org/10.1021/bi400240c] [PMID: 23442089]

[94]   Guéroux M, Pinaud-Szlosek M, Fouquet E, De Freitas V, Laguerre M, Pianet I. How wine polyphenols can fight Alzheimer disease progression: Towards a molecular explanation. Tetrahedron 2015; 71: 3163-70.
[http://dx.doi.org/10.1016/j.tet.2014.06.091]

[95]   Chakraborty S, Basu S. Multi-functional activities of citrus flavonoid narirutin in Alzheimer's disease therapeutics: An integrated screening approach and *In vitro* validation. Int J Biol Macromol 2017; 103: 733-43.
[http://dx.doi.org/10.1016/j.ijbiomac.2017.05.110] [PMID: 28528948]

[96]   Pugazhendhi A, Beema Shafreen R, Pandima Devi K, Suganthy N. Assessment of antioxidant, anticholinesterase and antiamyloidogenic effect of Terminalia chebula, Terminalia arjuna and its bioactive constituent 7-Methyl gallic acid – An *in vitro* and in silico studies. J Mol Liq 2018; 257: 69-81.
[http://dx.doi.org/10.1016/j.molliq.2018.02.081]

[97]   Chakraborty S, Bandyopadhyay J, Chakraborty S, Basu S. Multi-target screening mines hesperidin as a multi-potent inhibitor: Implication in Alzheimer's disease therapeutics. Eur J Med Chem 2016; 121: 810-22.
[http://dx.doi.org/10.1016/j.ejmech.2016.03.057] [PMID: 27068363]

[98]   Khan S, Nazir M, Raiz N, Saleem M, Zengin G, Fazal G, *et al.* Phytochemical profiling, *in vitro* biological properties and in silico studies on Caragana ambigua stocks (Fabaceae): A comprehensive approach. Ind Crops Prod 2019; 131: 117-24.
[http://dx.doi.org/10.1016/j.indcrop.2019.01.044]

[99]   Chang W, Huang D, Lo YM, *et al.* Protective effect of caffeic acid against alzheimer's disease pathogenesis *via* modulating cerebral insulin signaling, β-Amyloid accumulation, and synaptic plasticity in hyperinsulinemic rats. J Agric Food Chem 2019; 67(27): 7684-93.
[http://dx.doi.org/10.1021/acs.jafc.9b02078] [PMID: 31203623]

[100]   Mazumder MK, Choudhury S. Tea polyphenols as multi-target therapeutics for Alzheimer's disease: An in silico study. Med Hypotheses 2019; 125: 94-9.
[http://dx.doi.org/10.1016/j.mehy.2019.02.035] [PMID: 30902161]

[101]   Rodrigues MJ, Slusarczyk S, Pecio Ł, *et al. In vitro* and *in silico* approaches to appraise *Polygonum maritimum* L. as a source of innovative products with anti-ageing potential. Ind Crops Prod 2018; 111: 391-9.
[http://dx.doi.org/10.1016/j.indcrop.2017.10.046]

[102] Mahomoodally MF, Zengin G, Zheleva-Dimitrova D, *et al.* Metabolomics profiling, bio-pharmaceutical properties of Hypericum lanuginosum extracts by *in vitro* and *in silico* approaches. Ind Crops Prod 2019; 133: 373-82.
[http://dx.doi.org/10.1016/j.indcrop.2019.03.033]

SUBJECT INDEX

A

Acid(s) 4, 7, 8, 69, 144, 178, 190, 206, 212, 219, 220, 221, 222, 223, 224, 226, 229, 231, 240, 242, 243, 246
  acylquinic 246
  ascorbic 213, 214
  benzoic 220, 222, 240
  butyric 4
  caffeic 219, 220, 222, 223, 224, 226, 242, 243
  carnosic 223, 224, 231
  chebulinic 240
  coumaric 240
  ethylenediamine tetraacetic 226
  hydroxybenzoic 221
  lactic 178
  methionine amino 190
  monohydroxybenzoic 212
  nordic hydroguaiaretic 214
  nucleic 144
  phenolic 206, 229, 246
  polyglutamic 69
  protocatechuic 212, 220, 221
  rosmarinic 218, 220, 222, 223, 224
  saturated fatty (SFA) 7, 8
  trans-cinnamic 220
Actions, neuroprotective 156, 231
Activity 62, 63, 90, 92, 186, 204, 215, 218, 219, 220, 223, 225, 229, 230, 232, 233, 238, 242, 243, 245
  anti-aggregating 218
  anti-amyloidogenic 215, 238
  anti-cholinesterase 220
  anti-inflammatory 233
  catalase 242, 243
  inflammatory 186
  oxidant 232
Adipokines 5, 6, 14, 24
  inflammatory 14
Age-related 141, 143, 149
  diseases development 143

  illnesses 149
  neuroinflammation 141
Akt activation 156
AKT 13, 14, 17, 153, 156
  pathway 13, 14, 153, 156
  signaling pathway 17
Alleviation of endothelial dysfunction 23
Alzheimer's 147
  dementia 147
  diagnosis 147
AMPA receptors 15, 61, 74, 81
AMPK signaling 18
Amyloid 8, 57, 78, 80, 82, 93, 150, 178, 186, 205, 227, 243
  peptides 178
  precursor protein (APP) 8, 57, 78, 80, 82, 93, 150, 186, 205, 227, 243
Analysis 37, 147, 212, 227, 229
  histopathological 147
  immunohistochemical 229
  stereological 37
  western blot 212, 229
Anthocyanin 227
Anti-amyloidogenic effect 212, 213, 241
Anti-depressant effects 223
Anti-inflammatory effects 9, 22, 187, 232
Antioxidant(s) 9, 10, 19, 140, 142, 145, 156, 186, 207, 230, 232, 233, 238, 239, 241, 242, 246, 247
  activity 230, 233, 238, 239, 241, 246, 247
  profile 156
  properties 207, 246
Anti-radical effects 232
Antisense oligonucleotides (ASOs) 56, 92, 183
Anti-tau 56, 94, 96
  active immunotherapy 96
  immunotherapy 94
  therapy 56
Anxiety 34, 152
  disorders 34
  symptoms 152

Dr. José Juan Antonio Ibarra Arias (Ed.)

www.ingramcontent.com/pod-product-compliance
Lightning Source LLC
Chambersburg PA
CBHW050818220326
41598CB00006B/252